Introduction to Coordination Chemistry

Inorganic Chemistry

A Wiley Series of Advanced Textbooks
ISSN: 1939-5175

Previously Published Books in this Series

Chirality in Transition Metal Chemistry
Hani Amouri & Michel Gruselle; ISBN: 978-0-470-06054-4

Bioinorganic Vanadium Chemistry
Dieter Rehder; ISBN: 978-0-470-06516-7

Inorganic Structural Chemistry, Second Edition
Ulrich Müller; ISBN: 978-0-470-01865-1

Lanthanide and Actinide Chemistry
Simon Cotton; ISBN: 978-0-470-01006-8

Mass Spectrometry of Inorganic and Organometallic Compounds:
Tools – Techniques – Tips
William Henderson & J. Scott McIndoe; ISBN: 978-0-470-85016-9

Main Group Chemistry, Second Edition
A. G. Massey; ISBN: 978-0-471-49039-5

Synthesis of Organometallic Compounds: A Practical Guide
Sanshiro Komiya; ISBN: 978-0-471-97195-5

Chemical Bonds: A Dialog
Jeremy Burdett; ISBN: 978-0-471-97130-6

Molecular Chemistry of the Transition Elements: An Introductory Course
François Mathey & Alain Sevin; ISBN: 978-0-471-95687-7

Stereochemistry of Coordination Chemistry
Alexander Von Zelewsky; ISBN: 978-0-471-95599-3

Bioinorganic Chemistry: Inorganic Elements in the Chemistry of Life – An Introduction
and Guide
Wolfgang Kaim; ISBN: 978-0-471-94369-3

For more information on this series see: www.wiley.com/go/inorganic

Introduction to Coordination Chemistry

Geoffrey A. Lawrance

University of Newcastle, Callaghan, NSW, Australia

WILEY

A John Wiley and Sons, Ltd., Publication

Library of Congress Cataloging-in-Publication Data

Lawrance, Geoffrey A.
    Introduction to coordination chemistry / Geoffrey A. Lawrance.
        p. cm.
    Includes bibliographical references and index.
    ISBN 978-0-470-51930-1 – ISBN 978-0-470-51931-8
    1. Coordination compounds.   I. Title.
QD474.L387 2010
541′.2242–dc22
                                        2009036555

A catalogue record for this book is available from the British Library.

ISBN: 978-0-470-51930-1 (HB)
          978-0-470-51931-8 (PB)

Typeset in 10/12pt Times by Aptara Inc., New Delhi, India.

Contents

Preface

This textbook is written with the assumption that readers will have completed an introductory tertiary-level course in general chemistry or its equivalent, and thus be familiar with basic chemical concepts including the foundations of chemical bonding. Consequently, no attempt to review these in any detail is included. Further, the intent here is to avoid mathematical and theoretical detail as much as practicable, and rather to take a more descriptive approach. This is done with the anticipation that those proceeding further in the study of the field will meet more stringent and detailed theoretical approaches in higher-level courses. This allows those who are not intending to specialize in the field or who simply wish to supplement their own separate area of expertise to gain a good understanding largely free of a heavy theoretical loading. While not seeking to diminish aspects that are both important and central to higher-level understanding, this is a pragmatic approach towards what is, after all, an introductory text. Without doubt, there are more than sufficient conceptual challenges herein for a student. Further, as much as is practicable in a chemistry book, you may note a more relaxed style which I hope may make the subject more approachable; not likely to be appreciated by the purists, perhaps, but then this is a text for students.

The text is presented as a suite of sequential chapters, and an attempt has been made to move beyond the pillars of the subject and provide coverage of synthesis, physical methods, and important bioinorganic and applied aspects from the perspective of their coordination chemistry in the last four chapters. While it is most appropriate and recommended that they be read in order, most chapters have sufficient internal integrity to allow each to be tackled in a more feral approach. Each chapter has a brief summary of key points at the end. Further, a limited set of references to other publications that can be used to extend your knowledge and expand your understanding is included at the end of each chapter. Topics that are important but not central to the thrust of the book (nomenclature and symmetry) are presented as appendices.

Supporting Materials

Self-assessment of your understanding of the material in each chapter has been provided for, through assembly of a set of questions (and answers). However, to limit the size of this textbook, these have been provided on the supporting web site at www.wiley.com/go/lawrance

This book was written during the depths of the worst recession the world has experienced since the 1930s. Mindful of the times, in which we have seen a decay of wealth, all figures in the text are printed in greyscale to keep the price for the user down. Figures and drawings herein employed mainly ChemDraw and Chem3DPro; where required, coordinates for structures come from the Cambridge Crystallographic Data Base, with some protein views in Chapter Eight drawn from the Protein Data Bank (http://www.rcsb.org/pdb). Provision has been made for access to colour versions of all figures, should you as the reader feel these will assist understanding. For colour versions of figures, go to www.wiley.com/go/lawrance. Open access to figures is provided.

Acknowledgements

For all those who have trodden the same path as myself from time to time over the years, I thank you for your companionship; unknowingly at the time, you have contributed to this work through your influence on my path and growth as a chemist. This book has been written against a background of informal discussions in recent years with a number of colleagues on various continents at various times, and comments on the outline from a panel of reviewers assembled by the publishers. However, the three who have contributed their time most in reading and commenting on draft chapters of this book are Robert Burns, Marcel Maeder and Paul Bernhardt; they deserve particular mention for their efforts that have enhanced structure and clarity. The publication team at Wiley have also done their usual fine job in production of the textbook. While this collective input has led to a better product, I remain of course fully responsible for both the highs and the lows in the published version.

Most of all, I could not possibly finish without thanking my wife Anne and family for their support over the years and forbearance during the writing of this book.

Geoffrey A. Lawrance
Newcastle, Australia – October, 2009

Preamble

Coordination chemistry in its 'modern' form has existed for over a century. To identify the foundations of a field is complicated by our distance in time from those events, and we can do little more than draw on a few key events; such is the case with coordination chemistry. Deliberate efforts to prepare and characterize what we now call coordination complexes began in the nineteenth century, and by 1857 Wolcott Gibbs and Frederick Genth had published their research on what they termed 'the ammonia–cobalt bases', drawing attention to 'a class of salts which for beauty of form and colour . . . are almost unequalled either among organic or inorganic compounds'. With some foresight, they suggested that 'the subject is by no means exhausted, but that on the contrary there is scarcely a single point which will not amply repay a more extended study'. In 1875, the Danish chemist Sophus Mads Jørgensen developed rules to interpret the structure of the curious group of stable and fairly robust compounds that had been discovered, such as the one of formula $CoCl_3 \cdot 6NH_3$. In doing so, he drew on immediately prior developments in organic chemistry, including an understanding of how carbon compounds can consist of chains of linked carbon centres. Jørgensen proposed that the cobalt invariably had three linkages to it to match the valency of the cobalt, but allowed each linkage to include chains of linked ammonia molecules and or chloride ions. In other words, he proposed a carbon-free analogue of carbon chemistry, which itself has a valency of four and formed, apparently invariably, four bonds. At the time this was a good idea, and placed metal-containing compounds under the same broad rules as carbon compounds, a commonality for chemical compounds that had great appeal. It was not, however, a great idea. For that the world had to wait for Alfred Werner, working in Switzerland in the early 1890s, who set this class of compounds on a new and quite distinctive course that we know now as coordination chemistry. Interestingly, Jorgensen spent around three decades championing, developing and defending his concepts, but Werner's ideas that effectively allowed more linkages to the metal centre, divorced from its valency, prevailed, and proved incisive enough to hold essentially true up to the present day. His influence lives on; in fact, his last research paper actually appeared in 2001, being a determination of the three-dimensional structure of a compound he crystallized in 1909! For his seminal contributions, Werner is properly regarded as the founder of coordination chemistry.

Coordination chemistry is the study of coordination compounds or, as they are often defined, coordination complexes. These entities are distinguished by the involvement, in terms of simple bonding concepts, of one or more coordinate (or dative) covalent bonds, which differ from the traditional covalent bond mainly in the way that we envisage they are formed. Although we are most likely to meet coordination complexes as compounds featuring a metal ion or set of metal ions at their core (and indeed this is where we will overwhelmingly meet examples herein), this is not strictly a requirement, as metalloids may also form such compounds. One of the simplest examples of formation of a coordination compound comes from a now venerable observation – when BF_3 gas is passed into a liquid trialkylamine, the two react exothermally to generate a solid which contains equimolar amounts of each precursor molecule. The solid formed has been shown to consist of molecules $F_3B–NR_3$, where what appears to be a routine covalent bond now links the boron and nitrogen centres. What is peculiar to this assembly, however, is that electron book-keeping suggests that the boron commences with an empty valence orbital whereas the nitrogen commences with one lone pair of electrons in an orbital not involved previously in bonding. Formally, then, the new bond must form by the two lone pair valence electrons on the nitrogen being inserted or donated into the empty orbital on the boron. Of course, the outcome is well known – a situation arises where there is an increase in shared electron density between the joined atom centres, or formation of a covalent bond. It is helpful to reflect on how this situation differs from conventional covalent bond formation; traditionally, we envisage covalent bonds as arising from two atomic centres each providing an electron to form a bond through sharing, whereas in the coordinate covalent bond one centre provides both electrons (the donor) to insert into an empty orbital on the other centre (the acceptor); essentially, you can't tell the difference once the coordinate bond has formed from that which would arise by the usual covalent bond formation. Another very simple example is the reaction between ammonia and a proton; the former can be considered to donate a lone pair of electrons into the empty orbital of the proton. In this case, the acid–base character of the acceptor–donor assembly is perhaps more clearly defined for us through the choice of partners. Conventional Brønsted acids and bases are not central to this field, however; more important is the Lewis definition of an acid and base, as an electron pair acceptor and electron pair donor respectively.

Today's coordination chemistry is founded on research in the late nineteenth and early twentieth century. As mentioned above, the work of French-born Alfred Werner, who spent most of his career in Switzerland at Zürich, lies at the core of the field, as it was he who recognized that there was no required link between metal oxidation state and number of ligands bound. This allowed him to define the highly stable complex formed between cobalt(III) (or Co^{3+}) and six ammonia molecules in terms of a central metal ion surrounded by six bound ammonia molecules, arranged symmetrically and as far apart as possible at the six corners of an octahedron. The key to the puzzle was not the primary valency of the metal ion, but the apparently constant number of donor atoms it supported (its 'coordination number'). This 'magic number' of six for cobalt(III) was confirmed through a wealth of experiments, which led to a Nobel Prize for Werner in 1913. Whereas his discoveries remain firm, modern research has allowed limited examples of cobalt(III) compounds with coordination numbers of five and even four to be prepared and characterized. As it turns out, Nature was well ahead of the game, since metalloenzymes with cobalt(III) at the active site discovered in recent decades have a low coordination number around the metal, which contributes to their high reactivity. Metals can show an array of preferred coordination numbers, which vary not only from metal to metal, but can change for a particular metal with formal oxidation state of a metal. Thus Cu(II) has a greater tendency towards five-coordination than Mn(II), which prefers six-coordination. Unlike six-coordinate Mn(II), Mn(VII) prefers four-coordination. Behaviour in the solid state may differ from that in solution, as a result of the availability of different potential donors resulting from the solvent itself usually being a possible ligand. Thus $FeCl_3$ in the solid state consists of Fe(III) centres surrounded octahedrally by six Cl^- ions, each shared between two metal centres; in aqueous acidic solution, '$FeCl_3$' is more likely to be met as separate $[Fe(OH_2)_6]^{3+}$ and Cl^- ions.

Inherently, whether a coordination compound involves metal or metalloid elements is immaterial to the basic concept. However, one factor that distinguishes the chemistry of the majority of metal complexes is an often incomplete d (for transition metals) or f (for lanthanoids and actinoids) shell of electrons. This leads to the spectroscopic and magnetic properties of members of these groups being particularly indicative of the compound under study, and has driven interest in and applications of these coordination complexes. The field is one of immense variety and, dare we say it, complexity. In some metal complexes it is even not easy to define the formal oxidation state of the central metal ion, since electron density may reside on some ligands to the point where it alters the physical behaviour.

What we can conclude is that metal coordination chemistry is a demanding field that will tax your skills as a scientist. Carbon chemistry is, by contrast, comparatively simple, in the sense that essentially all stable carbon compounds have four bonds around each carbon centre. Metals, as a group, can exhibit coordination numbers from two to fourteen, and formal oxidation states that range from negative values to as high as eight. Even for a particular metal, a range of oxidation states, coordination numbers and distinctive spectroscopic and chemical behaviour associated with each oxidation state may (and usually does) exist. Because coordination chemistry is the chemistry of the vast majority of the Periodic Table, the metals and metalloids, it is central to the proper study of chemistry. Moreover, since many coordination compounds incorporate organic molecules as ligands, and may influence their reactivity and behaviour, an understanding of organic chemistry is also necessary in this field. Further, since spectroscopic and magnetic properties are keys to a proper understanding of coordination compounds, knowledge of an array of physical and analytical methods is important. Of course coordination chemistry is demanding and frustrating – but it rewards the student by revealing a diversity that can be at once intriguing, attractive and rewarding. Welcome to the wild and wonderful world of coordination chemistry – let's explore it.

1 The Central Atom

1.1 Key Concepts in Coordination Chemistry

The simple yet distinctive concept of the *coordinate bond* (also sometimes called a *dative bond*) lies at the core of coordination chemistry. Molecular structure, in its simplest sense, is interpreted in terms of covalent bonds formed through shared pairs of electrons. The coordinate bond, however, arises not through the sharing of electrons, one from each of two partner atoms, as occurs in a standard covalent bond, but from the donation of a pair of electrons from an orbital on one atom (a lone pair) to occupy an empty orbital on what will become its partner atom.

First introduced by G.N. Lewis almost a century ago, the concept of a covalent bond formed when two atoms share an electron pair remains as a firm basis of chemistry, giving us a basic understanding of single, double and triple bonds, as well as of a lone pair of electrons on an atom. Evolving from these simple concepts came valence bond theory, an early quantum mechanical theory which expressed the concepts of Lewis in terms of wavefunctions. These concepts still find traditional roles in coordination chemistry. However, coordination chemistry is marked by a need to employ the additional concept of coordinate bond formation, where the bond pair of electrons originates on one of the two partner atoms alone. In coordinate bond formation, the bonding arrangement between electron-pair acceptor (designated as A) and electron-pair donor (designated as :D, where the pair of dots represent the lone pair of electrons) can be represented simply as Equation (1.1):

$$A + :D \rightarrow A : D \tag{1.1}$$

The product alternatively may be written as $A \leftarrow :D$ or $A \leftarrow D$, where the arrow denotes the direction of electron donation, or, where the nature of the bonding is understood, simply as $A-D$. This latter standard representation is entirely appropriate since the covalent bond, once formed, is indistinguishable from a standard covalent bond. The process should be considered reversible in the sense that, if the $A-D$ bond is broken, the lone pair of electrons originally donated by :D remains entirely with that entity.

In most coordination compounds it is possible to identify a central or core atom or ion that is bonded not simply to one other atom, ion or group through a coordinate bond, but to several of these entities at once. The central atom is an acceptor, with the surrounding species each bringing (at least) one lone pair of electrons to donate to an empty orbital on the central atom, and each of these electron-pair donors is called a *ligand* when attached. The central atom is a metal or metalloid, and the compound that results from bond formation is called a coordination compound, coordination complex or often simply a *complex*. We shall explore these concepts further below.

Introduction to Coordination Chemistry Geoffrey A. Lawrance
© 2010 John Wiley & Sons, Ltd

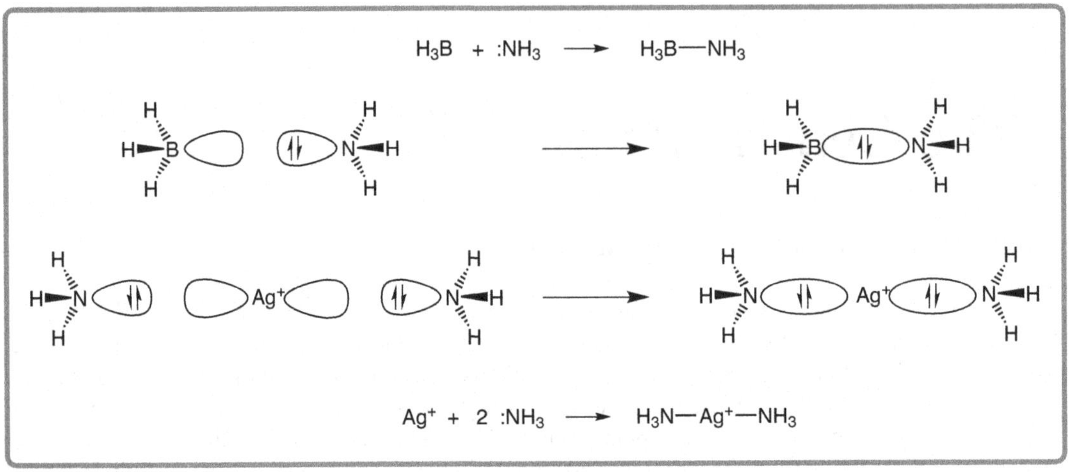

$$H_3B \;+\; :NH_3 \;\longrightarrow\; H_3B\!-\!NH_3$$

Figure content (schematic diagrams).

Figure 1.1
A schematic view of ammonia acting as a donor ligand to a metalloid acceptor and to a metal ion acceptor to form coordinate bonds.

$$Ag^+ \;+\; 2\;:NH_3 \;\longrightarrow\; H_3N\!-\!Ag^+\!-\!NH_3$$

The species providing the electron pair (the electron-pair donor) is thought of as being coordinated to the species receiving that lone pair of electrons (the electron-pair acceptor). The coordinating entity, the *ligand*, can be as small as a monatomic ion (e.g. F^-) or as large as a polymer – the key characteristic is the presence of one or more lone pairs of electrons on an electronegative donor atom. Donor atoms often met are heteroatoms like N, O, S and P as well as halide ions, but this is by no means the full range. Moreover, the vast majority of existing organic molecules can act as ligands, or else can be converted into molecules capable of acting as ligands. A classical and successful ligand is ammonia, NH_3, which has one lone pair (Figure 1.1). Isoelectronic with ammonia is the carbanion $^-CH_3$, which can also be considered a ligand under the simple definition applied; even hydrogen as its hydride, H^-, has a pair of electrons and can act as a ligand. It is not the type of donor atom that is the key, but rather its capacity to supply an electron pair.

The acceptor with which a coordinate covalent bond is formed is conventionally either a metal or metalloid. With a metalloid, covalent bond formation is invariably associated with an increase in the number of groups or atoms attached to the central atom, and simple electron counting based on the donor–acceptor concept can account for the number of coordinate covalent bonds formed. With a metal ion, the simple model is less applicable, since the number of new bonds able to be generated through complexation doesn't necessarily match the number of apparent vacancies in the valence shell of the metal; a more sophisticated model needs to be applied, and will be developed herein. What is apparent with metal ions in particular is the strong drive towards complexation – 'naked' ions are extremely rare, and even in the gaseous state complexation will occur. It is a case of the whole being better than the sum of its parts, or, put more appropriately, coordinate bond formation is energetically favourable.

A more elaborate example than those shown above is the anionic compound SiF_6^{2-} (Figure 1.2), which adopts a classical octahedral shape that we will meet also in many metal complexes. Silicon lies below carbon in the Periodic Table, and there are some limited similarities in their chemistry. However, the simple valence bond theory and octet rule that

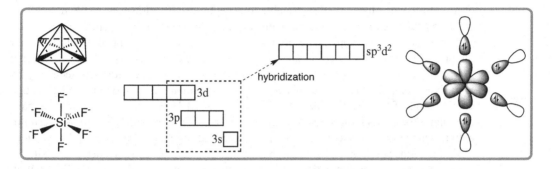

Figure 1.2
The octahedral $[SiF_6]^{2-}$ molecular ion, and a simple valence bond approach to explaining its formation. Overlap of a p orbital containing two electrons on each of the six fluoride anions with one of six empty hybrid orbitals on the Si(IV) cation, arranged in an octahedral array, generates the octahedral shape with six equivalent covalent σ bonds.

works so well for carbon cannot deal with a silicon compound with six bonds, particularly one where all six bonds are equivalent. One way of viewing this molecular species is as being composed of a Si^{4+} or Si(IV) centre with six F^- anions bound to it through each fluoride anion using an electron pair ($:F^-$) to donate to an empty orbital on the central Si(IV) ion, which has lost all of its original four valence electrons in forming the Si^{4+} ion. Using traditional valence bond theory concepts, a process of hybridization is necessary to accommodate the outcome (Figure 1.2). The generation of the shape arises through asserting that the silicon arranges a combination of one 3s, three 3p and two of five available 3d valence orbitals into six equivalent sp^3d^2 hybrid orbitals that are directed as far apart as possible and towards the six corners of an octahedron. Each empty hybrid orbital then accommodates an electron pair from a fluoride ion, each leading in effect to a coordinate covalent bond that is a σ bond because electron density in the bond lies along the line joining the two atomic centres. The shape depends on the type and number of orbitals that are involved in the hybridization process. Above, a combination resulting in an octahedral shape (sp^3d^2 hybrids) is developed; however, different combinations of orbitals yield different shapes, perhaps the most familiar being the combination of one s and three p orbitals to yield tetrahedral sp^3; others examples are linear (sp hybrids) and trigonal planar (sp^2 hybrids) shapes.

A central atom or ion with vacant or empty orbitals and ionic or neutral atoms or molecules joining it, with each bringing lone pairs of electrons, is the classic requirement for formation of what we have termed coordinate bonds, leading to a coordination compound. The very basic valence bonding model described above can be extended to metal ions, as we will see, but with some adjustments due to the presence of electrons in the d orbitals; more sophisticated models are required. Of developed approaches, molecular orbital theory is the most sophisticated, and is focused on the overlap of atomic orbitals of comparable energy on different atoms to form molecular orbitals to which electrons are allocated. While providing accurate descriptions of molecules and their properties, it is relatively complicated and time-consuming, and somewhat difficult to comprehend for large complexes; consequently, simpler models still tend to be used.

In the simple theory based on Lewis' concepts exemplified above, the key aspects are an empty orbital on one atom and a filled orbital (with a pair of electrons present, the lone pair) on the other. Many of the ligand species providing the lone pair are considered bases in the classical Brønsted–Lowry concept of acids and bases (which has as its focus the transfer of a proton), since these species are able to accept a proton. However, in the description we have developed here, no proton is involved, but the concept of accepting an electron-deficient species does apply. The broader and more general concept of an electron-pair donor as a base and an electron-pair acceptor as the acid evolved, and these are called a *Lewis base* (electron-pair donor) and a *Lewis acid* (electron-pair acceptor). Consequently, an H_3B-NH_3 compound is traditionally considered a coordination compound, arising through coordination of the electron deficient (or Lewis acid) H_3B and the electron lone-pair-containing (or Lewis base) compound $:NH_3$ (Figure 1.1). It is harder, in part as a result of entrenched views of covalent bonding in carbon-based compounds, to accept $[H_3C-NH_3]^+$ in similar terms purely as a H_3C^+ and $:NH_3$ assembly. This need to consider and debate the nature of the assembly limits the value of the model for non-metals and metalloids. With metal ions, however, you tend to know where you stand – almost invariably, you may start by considering them as forming coordination compounds; perhaps it is not surprising that coordination chemistry is focused mainly on compounds of metals and their ions.

Coordination has a range of consequences for the new assembly. It leads to structural change, seen in terms of change in the number of bonds and/or bond angles and distances. This is inevitably tied to a change in the physical properties of the assembly, which differ from those of its separate components. With metal atoms or ions at the centre of a coordination complex, even changing one of a set of ligands will be reflected in readily observable change in physical properties, such as colour. With growing sophistication in both synthesis and our understanding of physical methods, properties can often be 'tuned' through varying ligands to produce a particular result, such as a desired reduction potential.

It should also be noted that a coordination compound adopts one of a limited number of basic shapes, with the shape determined by the nature of the central atom and its attached ligands. Moreover, the physical properties of the coordination compound depend on and reflect the nature of the central atom, ligand set and molecular shape. Whereas only one central atom occurs in many coordination compounds (a compound we may thus define as a monomer), it should also be noted that there exists a large and growing range of compounds where there are two or more 'central atoms', either of the same or different types. These 'central atoms' are linked together through direct atom-to-atom bonding, or else are linked by ligands that as a result are joined to at least two 'central atoms' at the same time. This latter arrangement, where one or even several ligands are said to 'bridge' between central atoms, is the more common of these two options. The resulting species can usually be thought of as a set of monomer units linked together, leading to what is formally a polymer or, more correctly when only a small number of units are linked, an oligomer. We shall concentrate largely on simple monomeric species herein, but will introduce examples of larger linked compounds where appropriate.

Although, as we have seen, the metalloid elements can form molecular species that we call coordination compounds, the decision on what constitutes a coordination compound is perhaps more subtle with these than is the case with metals. Consequently, in this tale of complexes and ligands, it is with metals and particularly their cations as the central atom that we will almost exclusively meet examples.

1.2 A Who's Who of Metal Ions

The Periodic Table of elements is dominated by metals. Moreover, it is a growing majority, as new elements made through the efforts of nuclear scientists are invariably metallic. If the Periodic Table was a parliament, the non-metals would be doomed to be forever the minority opposition, with the metalloids a minor third party who cannot decide which side to join. The position of elements in the Periodic Table depends on their electronic configuration (Figure 1.3), and their chemistry is related to their position. Nevertheless, there are common features that allow overarching concepts to be developed and applied. For example a metal from any of the s, p, d or f blocks behaves in a common way – it usually forms cations, and it overwhelmingly exists as molecular coordination complexes through combination with other ions or molecules. Yet the diversity of behaviour underlying this commonality is both startling and fascinating, and at the core of this journey.

The difficulty inherent in isolating and identifying metallic elements meant that, for most of human history, very few were known. Up until around the mid-eighteenth century, only gold, silver, copper and iron of the d-block elements were known and used as isolated metals. However, in an extraordinary period from around 1740 to 1900, all but two of the naturally existing elements from the d block were firmly identified and characterized, and it was the synthesis and identification of technetium in 1939, the sole 'missing' element in the core of this block because it has no stable isotopes, that completed the series. In almost exactly 200 years, what was to become a large block of the Periodic Table was cemented in place; this block has now been expanded considerably with the development of higher atomic number synthetic elements. Along with this burst of activity in the identification of elements came, in the late nineteenth century, the foundations of modern coordination chemistry, building on this new-found capacity to isolate and identify metallic elements.

Almost all metals have a commercial value, because they have found commercial applications. It is only the more exotic synthetic elements made as a result of nuclear reactions that have, as yet, no real commercial valuation. The isolation of the element can form the starting point for applications, but the chemistry of metals is overwhelmingly the chemistry of metals in their ionic forms. This is evident even in Nature, where metals are rarely found in their elemental state. There are a few exceptions, of which gold is the standout example, and it was this accessibility in the metallic state that largely governed the adoption and use in antiquity of these exceptions. Dominantly, but not exclusively, the metal is found in a positive oxidation state, that is as a cation. These metal cations form, literally, the core of coordination chemistry; they lie at the core of a surrounding set of molecules or atoms, usually neutral or anionic, closely bound as ligands to the central metal ion. Nature employs metal ions in a variety of ways, including making use of their capacity to bind to organic molecules and their ability to exist, at least for many metals, in a range of oxidation states.

The origins of a metal in terms of it Periodic Table position has a clear impact on its chemistry, such as the reactions it will undergo and the type of coordination complexes that are readily formed. These aspects are reviewed in Chapter 6.2, after important background concepts have been introduced. At this stage, it is sufficient to recognize that, although each metallic element is unique, there is some general chemical behaviour, that relates to the block of the Periodic Table to which it belongs, that places both limitations on and some structure into chemical reactions in coordination chemistry.

1.2.1 Commoners and 'Uncommoners'

Because we meet them daily in various forms, we tend to think of metals as common. However, 'common' is a relative term – iron may be more common than gold in terms of availability in the Earth's crust, but gold is itself more common than rhenium. Even for the fairly well-known elements of the first row of the d block of the Periodic Table, abundance in

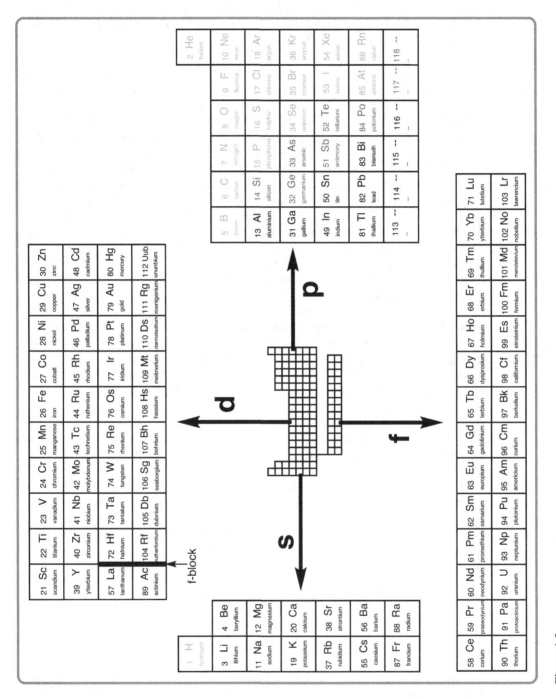

Figure 1.3
The location of metals in the Periodic Table of the elements.

the Earth's crust varies significantly, from iron (41 000 ppm) to cobalt (20 ppm); moreover, what we think of as 'common' metals, like copper (50 ppm abundance) and zinc (75 ppm abundance), are really hardly that. Availability of an element is not driven by how much is present on average in the Earth's crust, of course, but by other factors such as its existence in sufficiently high concentrations in accessible ore bodies and its commercial value and applicability (see Chapter 9). Iron, more abundant than the sum of all other d-block elements, is mined from exceedingly rich ore deposits and is of major commercial significance. Rhenium, the rarest transition metal naturally available, is a minor by-product of some ore bodies where other valuable metals are the primary target, and in any case has limited commercial application. Nevertheless, our technology has advanced sufficiently that there is not one metal available naturally on Earth that is not isolated in some amount or form, and for which some commercial applications do not now exist. Even synthetic elements are available and applicable. Complexes of an isotope of technetium, the only d-block element with no stable isotopes that consequently does not exist in the Earth's crust and must be made in a nuclear reactor, are important in medical γ-ray imaging; in fact, sufficient technetium is produced so that it may be considered as accessible as its rare, naturally available, partner element rhenium. As another example, an isotope of the synthetic f-block actinoid element americium forms the core of the ionization mechanism operating in the sensor of household smoke detectors.

These observations have one obvious impact on coordination chemistry – every metallic element in the Periodic Table is accessible and in principle able to be studied, and each offers a suite of unique properties and behaviour. As a consequence, they are in one sense all now 'common'; what distinguishes them are their relative cost and the amounts available. In the end, it has been such commercially-driven considerations that have led to a concentration on the coordination chemistry of the more available and applicable lighter elements of the transition metals, from vanadium to zinc. Of course Nature, again, has made similar choices much earlier, as most metalloenzymes employ light transition elements at their active sites.

1.2.2 Redefining Commoners

Apart from availability (Section 1.2.1), there is another more chemical approach to commonality that we should dwell on, an aspect that we have touched upon already. This is a definition in terms of oxidation states. With the most common of all metals in the Earth's crust, the main group element aluminium, only one oxidation state is important – Al(III). However, for the most common transition metal (iron), both Fe(II) and Fe(III) are common, whereas other higher oxidation states such as Fe(IV) are known but very uncommon. With the rare element rhenium, the reverse trend holds true, as the high oxidation state Re(VI) is common but Re(III) and Re(II) are rare. What is apparent from these observations is that each metal can display one or more 'usual' oxidation states and a range of others met much more rarely, whereas some are simply not accessible.

What allows us to see the uncommon oxidation states is their particular environment in terms of groups or atoms bound to the metal ion, and in general there is a close relationship between the groups that coordinate to a metal and the oxidation states it can sustain, which we will explore later. The definition of 'common' in terms of metal complexes in a particular oxidation state is an ever-changing aspect of coordination chemistry, since it depends in part on the amount of chemistry that has been performed and reported; over time, a metal in a particular oxidation state may change from 'unknown' to 'very rare' to 'uncommon' as more chemists beaver away at extending the chemistry of an element. At

Sc scandium	Ti titanium	V vanadium	Cr chromium	Mn manganese	Fe iron	Co cobalt	Ni nickel	Cu copper	Zn zinc
		0 d^5	0 d^6	0 d^7	0 d^8	0 d^9	0 d^{10}		
		1 d^4	1 d^5	1 d^6	1 d^7	1 d^8	1 d^9	1 d^{10}	
	2 d^2	2 d^3	2 d^4	2 d^5	2 d^6	2 d^7	2 d^8	2 d^9	2 d^{10}
3 d^0	3 d^1	3 d^2	3 d^3	3 d^4	3 d^5	3 d^6	3 d^7	3 d^8	
	4 d^0	4 d^1	4 d^2	4 d^3	4 d^4	4 d^5	4 d^6		
		5 d^0	5 d^1	5 d^2	5 d^3	5 d^4			
			6 d^0	6 d^1	6 d^2				
				7 d^0					

Figure 1.4
Oxidation states met amongst complexes of transition metal elements; *d*-electron counts for the particular oxidation states of a metal appear below each oxidation state. [Oxidation states that are relatively common with a range of known complexes are in black, others in grey.]

this time, a valid representation of the status of elements of the first row of the d block with regard to their oxidation states is shown in Figure 1.4. Clearly, oxidation states two and three are the most common. Notably, hydrated transition metal ions of charge greater than 3+ (that is, oxidation state over three) are not stable in water, so higher oxidation state species invariably involve other ligands apart from water. Differences in the definition of what amounts to a common oxidation state leads to some variation, but the general trends remain constant.

What is immediately apparent from Figure 1.4 is that most metals offer a wealth of oxidation states, with the limit set by simply running out of d electrons (i.e. reaching the d^0 arrangement) or else reaching such a high reduction potential that stability of the ion is severely compromised (that is it cannot really exist, because it involves itself immediately in oxidation–reduction reactions that return the metal to a lower and more common stable oxidation state). Notably, it gets harder to 'use up' all d electrons on moving from left to right across the Periodic Table, associated with both the rising number of d electrons and lesser screening from the charge on the nucleus. Still, you are hardly spoilt for choice as a coordination chemist!

The standard reduction potential (E^0) provides a measure of the stability of a metal in a particular oxidation state. The E^0 value is the voltage generated in a half-cell coupled with the standard hydrogen electrode (SHE), which itself has a defined half-cell potential of 0.0 V. Put simply, the more positive is E^0 the more difficult is it for metal oxidation to a hydrated metal ion to occur. Alternatively, we could express it by saying that the less positive is E^0, the more stable is the metal in the higher oxidation state of its couple and consequently the less easily is it reduced to the lower oxidation state. Metal activity can be related to reactivity with a protic solvent (like water) or hydrogen ions, and correlates with electronegativity. Very electropositive metals (reduction potentials of cations $<-1.6\,V$) have low electronegativities; these include the s block and all lanthanide metals. Electropositive metals display cation reduction potentials up to $\sim 0\,V$, and include the first row of the d-block and some p-block elements. Electronegative metals have positive cation reduction potentials; these include most of the second and third rows of the d block. Reactivities in redox processes differ for these different classes; electronegative metals are not corroded by oxygen, for example, unlike electropositive metals.

Yet another way of defining commonality with metal ions relates to how many ligand donor groups may be attached to the central metal. This was touched on in the Preamble, and we'll use and expand on the same example again. Cobalt(III) was shown decades ago to have what was then thought to be invariably six donor groups or atoms bound to the central metal ion, or a *coordination number* of six. While this is still the overwhelmingly common coordination number for cobalt in this oxidation state, there are now stable examples for Co(III) of coordination numbers of five and even four. In its other common oxidation state, as Co(II), there are two 'common' coordination numbers, four and six; it is hardly a surprise, then, that more and more examples of the intermediate coordination number five have appeared over time. Five-coordination has grown to be almost as common for another metal ion, Cu(II), as four or six, illustrating that our definitions of common and uncommon do vary historically. That's a problem with chemistry generally – it never stands still. The number of research papers published with a chemical theme each year continues to grow at such a rate that it is impossible to read a single year's complete offerings in a decade, let alone that year.

1.3 Metals in Molecules

Metals in the elemental form typically exhibit bright, shiny surfaces – what we tend to expect of a 'metallic' surface. In the atmosphere, rich with oxygen and usually containing water vapour, these surfaces may be prone to attack, depending on the E^0 value; this leads to the bright surface changing character as it becomes oxidized. Although a highly polished steel surface is attractive and valued, the same surface covered in an oxide layer (better known in this particular case as rust) is hardly a popular fashion statement. Yet it is the formation of rust which is perfectly natural, with the shiny metal surface the unnatural form that needs to be carefully and regularly maintained to retain its initial condition. What we are witnessing with rust formation is a chemical process governed by thermodynamics (attainment of an equilibrium defined by the stability of the reaction products compared to the reactants) and kinetics (the rate at which change, or the chemical reaction, proceeds to equilibrium under the conditions prevailing). While the outcome may not be aesthetically appealing (unless one wants to make a virtue of rusted steel as a 'distressed' surface with character), chemistry is not given to making allowances for the sake of style or commerce – it is a demanding task to 'turn off' the natural chemistry of a system. Some metals, such as titanium, are less wilful than iron; they undergo surface oxidation, but form a tight monolayer of oxide that is difficult to penetrate and thus is resistant to further attack.

Of course, were the metal ions that exist in the oxidized surface to undergo attack in a different way, through complexation by natural ligands and subsequent dissolution, fresh metal surface would be exposed and available for attack. Such a process, occurring over long periods, would suggest that free active metals would increasingly end up dissolved as their ions in the ocean, and this is clearly not so – most metal ion concentrations in the ocean are very low (apart from alkali metal ions, being <0.001 ppm). In reality, because reactive elemental-state metals are made mainly through human action, the contribution to the biosphere even by reversion to ionic forms will be small. Most metals are locked up as ions in rocks – particularly as highly insoluble oxides, sulfides, sulfates or carbonates that will dissolve only with human interception, through reaction with strong acids or ligands. Even if they enter the biosphere as soluble complex ions, they are prone to chemistry that leads to re-precipitation. The classic example is dissolved iron(II), which readily undergoes aerial oxidation to Fe(III) and precipitation as a hydroxide, followed by dehydration to an oxide, all occurring below neutral pH.

Thus in the laboratory we tend to meet almost all metals in a pure form as synthetic cationic salts of common anions. These tend to be halides or sulfates, and it is these metal salts, hydrated or anhydrous, that form the entry point to almost all of metal coordination chemistry. In nature, it is no accident that metal ions that are relatively common tend to find roles, mediated of course by their chemical and electrochemical properties. Thus iron is heavily used not only because it is common, but also because it forms strong complexes with available biomolecules and has an Fe(II)/(III) redox couple that is accessible by biological oxidants and reductants and thus useful to drive some biochemical processes.

1.3.1   Metals in the Natural World

Most metals in the Earth's crust are located in highly inorganic environments – as components of rocks or soils on land or under water. Where metals are aggregated in local high concentrations through geological processes, these may be sufficient in amount and concentration to represent an ore deposit, which is really an economic rather than a scientific definition. In addition, metals are present in water bodies as dissolved cations; their concentrations can be in a very few cases substantial, as is the case with sodium ion in seawater. However, even if present in very low concentration, as for gold in seawater, the size of the oceans means that there is a substantial amount of gold (and other metals) dispersed in the aquatic environment. The other location of metals is within living organisms, where, of the transition metals, iron, zinc and copper predominate. On rare occasions the concentration of another metal may be relatively high; this is the case in some plants that tolerate and concentrate particular metal ions, such as nickel in *Hybanthus floribundus*, native to Western Australia, which can be hyper-accumulated up to ~50 mg per gram dry weight. Levels of metal ions in animals and in particular plants vary with species and environment. However, generally metals are present in nature in only trace amounts (Table 1.1). High levels of most metal ions are toxic to living species; for example ryegrass displays a toxicity order Cu > Ni > Mn > Pb > Cd > Zn > Al > Hg > Cr > Fe, with each species displaying a unique trend.

Metals were eventually recognized as having a presence in a range of biomolecules. Where metal cations appear in living things, their presence is rarely if ever simply fortuitous. Rather, they play a particular role, from simply providing an ionic environment through to being at the key active site for reactions in a large enzyme. Notably, it is the lighter alkali, alkaline earth and transition elements that dominate the metals present in living organisms. Of transition metals, although iron, copper and zinc are most dominant, almost all of the first row transition elements play some part in the functioning of organisms. Nevertheless, even heavier elements such as molybdenum and tungsten are found to have some roles.

Table 1.1 Typical concentrations (ppm) of selected metals ions in nature.

Metal	Earth's crust	Oceans	Plants (ryegrass)	Animals (human blood)
Na	23 000	10 500	1 000	2 000
K	21 000	1 620	28 000	1 600
Mg	23 000	1 200	2 500	40
Ca	41 000	390	12 500	60
Al	82 000	0.000 5	50	0.3
Sc	16	0.000 000 6	>0.01	0.008
Ti	5 600	0.000 48	2.0	0.055
V	160	0.001	0.07	<0.000 2
Cr	100	0.000 18	0.8	0.008
Mn	950	0.000 11	130	0.005
Fe	41 000	0.000 1	240	450
Co	20	0.000 001	0.6	0.01
Ni	80	0.000 1	6.5	0.03
Cu	50	0.000 08	9.0	1.0
Zn	75	0.000 05	31	7.0
Mo	1.5	0.01	1.1	0.001
Cd	0.11	0.000 001 1	0.07	0.0052
Pb	14	0.000 02	2.0	0.21
Sn	2.2	0.000 002 3	<0.01	0.38
Ce	68	0.000 002	<0.01	<0.001

Keys to metal ion roles are their high charge (and surface charge density), capacity to bind organic entities through strong coordinate bonds, and ability in many cases to vary oxidation states. We shall return to look at metals in biological environments in more detail in Chapter 8.

1.3.2 Metals in Contrived Environments

What defines chemistry over the past century has been our growing capacity to design and construct molecules. The number of new molecules that have been synthesized now number in the millions, and that number continues to grow at an astounding pace, along with continuing growth in synthetic sophistication; we have reached the era of the 'designer' molecule. Many of the new organic molecules prepared can bind to metal ions, or else can be readily converted to other molecules that can do so. This, along with the diversity caused by the capacity of a central metal ion to bind to a mixture of molecules at one time, means that the number of potential metal complexes that are not natural species is essentially infinite. Chemistry has altered irreversibly the composition of the world, if not the universe.

Discovering when the first synthetic metal complex was deliberately made and identified is not as easy as one might expect, because so much time has passed since that event. One popular candidate is *Prussian blue*, a cyanide complex of iron, developed as a commercial artist's colour in the early eighteenth century. A more reliable candidate is what we now know as hexaamminecobalt(III) chloride, discovered serendipitously by Tassaert in 1798, which set under way a quest to interpret its unique properties, such as how separately stable species NH_3 and $CoCl_3$ could produce another stable species $CoCl_3 \cdot 6NH_3$, and to discover similar species. As new compounds evolved, it was at first sufficient to identify them simply through their maker's name. Thus came into being species such as *Magnus's green salt* ($PtCl_2 \cdot 2NH_3$) and *Erdmann's salt* ($Co(NO_2)_3 \cdot KNO_2 \cdot 2NH_3$). This first attempt at nomenclature was doomed by profligacy, but as many compounds isolated were coloured, another way of identification arose based on colour; thus Tasseart's original yellow compound $CoCl_3 \cdot 6NH_3$ became luteocobaltic chloride, and the purple analogue $CoCl_3 \cdot 5NH_3$ was named purpureocobaltic chloride. This nomenclature also dealt with isomers, with two forms of $CoCl_3 \cdot 4NH_3$ identified and recognized – green praseocobaltic chloride and violet violeocobaltic chloride. Suffice to say that this nomenclature soon ran out of steam (or at least colours) also, and modern nomenclature is based on sounder structural bases, demonstrated in Appendix 1.

While some may quail at the outcomes of all this profligate molecule building, what remains a constant are the basic rules of chemistry. A synthetic metal complex obeys the same basic chemical 'rules' as a natural one. 'New' properties result from the character of new assemblies, not from a shift in the rules. As a classic example of how this works, consider the case of Vitamin B_{12}, distinguished by being one of a limited number of biomolecules centred on cobalt, and one of a rare few natural organometallic (metal–carbon bonded) compounds. This was discovered to exist with good stability in three oxidation states, Co(III), Co(II) and Co(I). Moreover, it was found to involve a C—Co(III) bond. At the time of these discoveries, examples of low molecular weight synthetic cobalt(III) complexes also stable in both Co(II) and Co(I) oxidation states were few if any in number, nor had the Co(III)–carbon bond been well defined. Such observations lent some support to a view that metals in biological entities were 'special'. Of course, time has removed the discrepancy, with synthetic Co complexes stable in all of the (III), (II) and (I) oxidation states well established, and examples of the Co(III)–carbon bond reported even with very simple ligands in other sites around the metal ion. The 'special' nature of metals in biology is essentially a consequence of their usually very large and specifically arranged macromolecular environments. While it is demanding to reproduce such natural environments in detail in the laboratory, it is possible to mimic them at a sufficient level to reproduce aspects of their chemistry.

Of course, the synthetic coordination chemist can go well beyond nature, by making use of facilities that don't exist in Earth's natural world. This can include even re-making elements that have disappeared from Earth. Technetium is radioactive in all its isotopic forms, and consequently has been entirely transmuted to other elements over time. However, it can be made readily enough in a nuclear reactor, and is now widely available. All of its chemistry, consequently, is synthetic or contrived. The element boron has given rise to a rich chemistry based on boron hydrides, most of which are too reactive to have any geological existence. Some boron hydrides as well as mixed carbon–boron compounds (carboranes) can bind to metal ions. Nitrogen forms a vast array of carbon-based compounds (amines) that are excellent at binding to metal ions; Nature also makes wide use of these for binding metal ions, but the construction of novel amines has reached levels that far exceed the limitations of Nature. After all, most natural chemistry has evolved at room temperature and pressure in near-neutral aqueous environments – limitations that do not apply in a chemical laboratory. What the vast array of synthetic molecules for binding metal ions provides is a capacity to control molecular shape and physical properties in metal-containing compounds not envisaged possible a century ago. These have given rise to applications and technologies that seem to be limited only by our imagination.

1.3.3   Natural or Made-to-Measure Complexes

Metal complexes are natural – expose a metal ion to molecules capable of binding to that ion and complexation almost invariably occurs. Dissolve a metal salt in water, and both cation and anion are hydrated through interaction with water. In particular, the metal ion acts as a

Lewis acid and water as a Lewis base, and a structure of defined coordination number with several M^{n+} ←:OH_2 bonds results; an experimentally-determined M—O—H angle of ~130° is consistent with involvement of a lone pair on the approximately tetrahedral oxygen. The coordinate bond is at the core of all natural and synthetic complexes.

While metals are usually present in minute amounts in living organisms, techniques for isolation and concentration have been developed that allow biological complexes to be recovered. An array of metalloproteins now offered commercially by chemical companies is evidence of this capacity. However, relying on natural sources for some compounds is both limiting and expensive. Many drugs and commercial compounds of natural origins are now prepared reliably and cheaply synthetically. Drugs originally from natural sources are made synthetically because the amount required to satisfy global demand makes isolation from natural sources impractical. This can also apply both to molecules able to attach to metal ions, and to their actual metal complexes. Simple across-the-counter compounds of metals find regular medical use; zinc supplements, for example, are actually usually supplied as a simple synthetic zinc(II) amino acid complex. Aspects of biological coordination chemistry are covered in Chapter 8.

Isolation of metal ions from ores by hydrometallurgical (water-based) processing often relies on complexation as part of the process. For example gold recovery from ore currently employs oxygen as oxidant and cyanide ion as ligand, leading selectively to a soluble gold(I) cyanide complex. Copper(II) ion dissolved from ore is recovered from an aqueous mixture by solvent extraction as a metal complex into kerosene, followed by decomposition and back extraction into aqueous acid, from which it is readily isolated by reduction to the metal. Pyrometallurgical (high temperature) processes for isolation of metals, on the other hand, usually rely on reduction reactions of oxide ores at high temperature. Electrochemical processes are also in regular industrial use; aluminium and sodium are recovered via electrochemical processes from molten salts. An overview of applied coordination chemistry is covered in Chapter 9.

1.4 The Road Ahead

Having identified the important role of metals as the central atom in coordination chemistry, it is appropriate at this time to recognize that the metal has partners and to reflect on the nature of the partnership. The partners are of course the ligands. A coordination complex can be thought of as the product of a molecular marriage – each partner, metal and ligand, brings something to the relationship, and the result of the union involves compromises that, when made, mean the union is distinctly different from the prior independent parts. While this analogy may be taking anthropomorphism to the extreme (unless one wants to carry it below even the atomic level), it is nevertheless not a bad analogy and not so unreasonable an outlook to think of a complex as a 'living' combination. After all, as we shall touch on later, it is not a totally inert combination. Metal complexes undergo ligand exchange (dare we talk of divorce and remarriage?) and can change their shape depending in part on their oxidation state and in part on their partner's preferences (shades of molecular-level compromise here?). With the right partner, the union will be strong, something that we can actually measure experimentally. It's no doubt stretching the analogy to talk of a perfect match, but the concept of fit and misfit between metals and ligands has been developed. What all of this playing with common human traits is about, is alerting you to core aspects of coordination chemistry – partnership and compromise.

In the rest of this book we will be examining in more detail ligands, metal–ligand assembly and the consequences. These include molecular shape, stability, properties and how we can measure and interpret these. Further, we will look at metal complexes in place – in nature and in commerce, and speculate on the future. Overall, the intent is to give as broad and deep an overview as is both reasonable and proper in an introductory text. Pray continue.

Concept Keys

A coordination complex consists of a central atom, usually a metal ion, bound to a set of ligands by coordinate bonds.

A coordinate covalent bond is distinguished by the ligand donor atom donating both electrons (of a lone pair) to an empty orbital on the central atom to form the bond.

A ligand is a *Lewis base* (electron-pair donor), the central atom a *Lewis acid* (electron-pair acceptor).

A 'common' metal may be defined simply by its geo-availability, but from a coordination chemistry perspective it is more appropriate to define 'common' in terms of aspects such as preferred oxidation state, number of coordinated donors or even preferred donor types.

Metal ions may exist and form complexes in a number of oxidation states; this is particularly prevalent in the d block.

First row d-block metal ions are found dominantly in the M(II) or M(III) oxidation states. Heavier members of the d block tend to prefer higher oxidation states.

Further Reading

Atkins, P. and Jones, L. (2000) *Chemistry: Molecules, Matter and Change*, 4th edn, Freeman, New York, USA. An introductory general chemistry textbook appropriate for reviewing basic concepts.

Beckett, M. and Platt, A. (2006) *The Periodic Table at a Glance*, Wiley-Blackwell, Oxford, UK. This short undergraduate-focussed book gives a fine, well-illustrated introductory coverage of periodicity in inorganic chemistry.

Gillespie, R.J. and Popelier, P.L.A. (2002) *Chemical Bonding and Molecular Geometry*, Oxford University Press. A coverage from the fundamental level upward of various models of molecular bonding.

Housecroft, C.E. and Sharpe, A.G. (2008) *Inorganic Chemistry*, 3rd edn, Pearson Education. Of the large and sometimes daunting general advanced textbooks on inorganic chemistry, this is a finely-written and well-illustrated current example, useful as a resource book.

2 Ligands

2.1 Membership: Being a Ligand

A ligand is an entity that binds strongly to a central species. This broad general definition allows extension of the concept beyond chemistry – you will meet it in molecular biology and biochemistry, for example where the 'complex' formed by a 'ligand' with a biomolecule involves weaker noncovalent interactions like ionic and hydrogen bonding. In coordination chemistry, a ligand is a molecule or ion carrying suitable donor groups capable of binding (or coordinating covalently) to a central atom. This central atom that is the focus of ligand coordination is most commonly a metal, although a central metalloid atom can take on the same role. The term ligand first appeared early in the twentieth century, and achieved popular use by mid-century. Strictly speaking, a molecule or ion doesn't become a ligand until it is bound, and prior to than is formally called a proligand; for simplicity, and in line with common usage, we shall put aside this distinction. For an atom or molecule, being a ligand is a lot like a plant being green – surprisingly common. That a metal atom or ion is almost invariably found with a tied set of companion atoms or molecules has been known for a long time, but it was only from around the beginning of the twentieth century that a clear concept of what a ligand is and how it binds to a central atom began to develop.

2.1.1 What Makes a Ligand?

The range of molecules that can bind to metal ions as ligands is diverse, and includes inorganic atoms, ions and molecules as well as organic molecules and ions. With metal–metal bonds also well known, one could argue in a simplistic sense that metals themselves can act as ligands, but this is not a direction we shall take. The number of molecules known to undergo, or are potentially capable of, ligation is extremely large – apart from inorganic systems, most organic molecules can either act as ligands directly or else are able to be converted into other molecules that can do so. This is because membership has but one basic requirement – in the simple valence bond model, the key to an atom or molecule acting as a ligand is the presence of at least one lone pair of electrons, as indicated in the cartoon in Figure 2.1. The atom that carries the lone pair is termed the *donor atom*, and is the atom bonded to the metal; where it is part of a well-recognized functional group (like an amine, $R-NH_2$, or carboxylate, $R-COO^-$), we speak of a *donor group*, but must recognize that it is typically one particular donor atom of the group that is bound to the metal. In its initial development, the general view of a ligand was that its role, beyond electron pair donation, is somewhat passive – a spectator rather than a player, if you wish. We now know that ligands influence the central metal ion significantly and can display reactivity and undergo chemistry of their own while still bound to the metal ion, which

Introduction to Coordination Chemistry   Geoffrey A. Lawrance
© 2010 John Wiley & Sons, Ltd

Metal ions are hardly ever found naked. They are always clothed with ligands.

Figure 2.1
An anthropomorphic view of being a metal coordination complex.

we shall return to in Chapter 6; this does not alter the fundamentals of their behaviour as ligands, however.

Ligands (often represented by the general symbol L) may present a single donor atom with a lone pair for binding to a metal ion and thus occupy only one coordination site, $M^{n+} \leftarrow :L$; this is then called a *monodentate* ligand. Classical examples of monodentate ligands include ammonia ($:NH_3$), water ($:OH_2$), and chloride ion ($:Cl^-$), although the latter two in fact have more than one lone pair available on the donor atom. Many ligands offer more than one donor group, each with a lone pair capable of binding to the same metal – a potential *polydentate* ligand. As a general rule, heteroatoms (particularly O, N, S and P) in organic molecules carry one or more lone pairs of electrons, the key requirement for being a donor atom in a ligand; thus, identifying the presence of these atoms is a good start to identifying whether a molecule may act as a ligand, and to seeing how many donor groups it may offer to a metal ion for binding. Of course, location, local environment and relative orientation of potential donors in a larger molecule play a role in how many nominally 'available' donor groups may bind collectively to a single metal ion, but identification of their presence is always the first step.

2.1.2 Making Attachments – Coordination

Only in the gas phase is a ligand likely to meet a 'naked' metal or its ion, and this is a highly contrived situation. In the liquid or solid state, where we overwhelmingly meet coordination complexes, a potential ligand will normally be confronted by a metal already carrying a set of ligands. In many cases in solution these other ligands are solvent molecules themselves – but no less legitimate as ligands simply because they can serve two roles, as ligand or solvent. Binding of a new ligand in what is termed the *inner coordination sphere* of the metal ion usually requires that it replace an existing ligand, a process termed *substitution*. What drives this process we shall deal with mainly in Chapter 5.

The strong drive towards complexation of nominally 'naked' metal ions is readily observed. For example if a simple hydrated salt like copper(II) sulfate ($CuSO_4 \cdot 5H_2O$) is dried in a vacuum oven to the point where no attached water molecules are present, a colourless anhydrous salt $CuSO_4$ is obtained. If this is dissolved in water, a pale blue solution is immediately formed as the metal ion is hydrated, a process which simply involves a set of water molecules rapidly binding to the metal ion as ligands. An energy change is associated with this process – the heat of hydration. If the solvent is removed by evaporation and the residual solid gently dried, a blue solid is recovered. This is the hydrated, or complexed, salt that has the formulation $CuSO_4 \cdot 5H_2O$. That the water molecules are tightly bound to the copper ion can be shown by simply measuring weight change as temperature is slowly raised. What is observed is that all water is not removed simply by heating to 100 °C, but is eventually removed fully only following heating to over 200 °C for an extended period, with recovery after that stage of the anhydrous species. Application of an array of advanced experimental methods allows us to observe the species in solution also; not only can we observe the presence of separate cations and anions, but the size, shape and environment of the ions can be elucidated. This confirms that the copper ion exists with a well-defined sheath of water molecules, the inner coordination sphere, which are in effect simple ligands, each water molecule attached to the central metal through a coordinate covalent bond via an oxygen lone pair. When this entity is ionic, as is the case for copper(II), this complex is surrounded by a partially ordered *outer* (or *secondary*) *coordination sphere* where water molecules are hydrogen-bonded to the inner-sphere ligated water molecules; a third and subsequent sheath surrounds the second layer, the process continuing until the layers become indistinguishable from the bulk water. The various layers moving outwards from the centre undergo successively decreasing compression as a result, in the simplest view, of the progressively diminishing electrostatic influence of the metal ion.

It is also important to think about the lifetime of a particular complex ion. For an aquated metal ion in pure water, there is but one ligand type available. However, it is not correct to assume that, once formed, a complex ion inevitably remains with the same set of water molecules for ever. In solution, it is possible (indeed usual) for water molecules in the outer coordination sphere to change places with water molecules in the inner coordination sphere. Obviously, this is a difficult process to observe, since there has been no real change in the metal environment when one water molecule replaces another – a little like taking a cold can out of a refrigerator and replacing it with another warm one of the same kind, so that no one can tell unless they pick up the warm can. At the molecular level, one can adopt the cold/warm can concept to probe what is called ligand exchange, by adding water with a different oxygen isotope present and following its uptake into the coordination sphere. The facility of this water exchange process varies significantly with the type and oxidation state of the metal ion. Moreover, the rate of exchange varies not only with metal but with ligand – to the point where longevity of a particular complex can indeed be extreme, or the coordination sphere is for all intents and purposes fixed. We shall return to the concept of ligand exchange again in Chapter 5.

2.1.3   Putting the Bite on Metals – Chelation

The classic simple ligand is ammonia, since it offers but one lone pair of electrons, and thus cannot form more than one coordinate covalent bond (Figure 2.2). A water molecule has two lone pairs of electrons on the oxygen, yet also usually forms one coordinate covalent bond. If one looks at the arrangement of lone pairs, this is hardly surprising; once one coordinate bond is formed, the remaining lone pair points in the wrong direction to allow it to become attached to the same metal ion – only through attachment to a different metal could this lone pair achieve coordination (a situation for the ligand called *bridging*). We shall return to examine whether this can actually happen for a water molecule later.

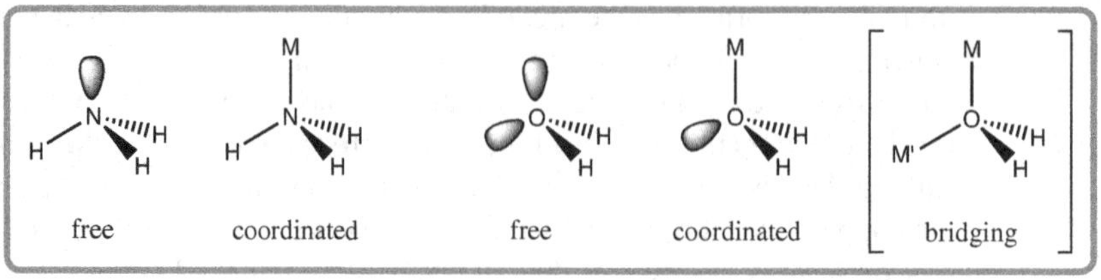

Figure 2.2
Free and coordinated ammonia and water molecules. The second lone pair of water is oriented in a direction prohibiting its interaction with the same metal centre as the first. However, it does have the potential, in principle, to use this lone pair to bind to a second metal centre in a *bridging* coordination mode. (Other groups bound to the metals are left off to simplify the views.)

Let's try to make it a bit easier for two lone pairs to interact with a single metal ion by putting them onto different atoms, and examine the result. We'll start with two ammonia residues linked by a single carbon atom – not a particularly chemically stable entity, but one that will suffice for illustrative purposes. Either the lone pair on the first N atom or the lone pair on the second N atom could form a single bond to a metal ion initially. While the second amine group is free to rotate about the resulting fixed M—N—C assembly, if the second lone pair is oriented in the same plane, it is now pointing more in the direction of the metal that was the case with the second lone pair on the water molecule. If the existing covalent bonds are somewhat deformed, coordination of both lone pairs to the same metal may be achieved (Figure 2.3).

Another and more stable example is the carboxylate group (R—COO⁻), which can coordinate in at least three ways – to one metal through one oxygen, bridging to two metals with each bound to one oxygen, or bound to one metal via both oxygen atoms (Figure 2.4). Note that the ring of atoms which includes the metal and donor atoms formed in both Figures 2.3 and 2.4 is identical in size, but differs in the type of donor atoms.

Where the one ligand employs two different donors to attach to the same metal, we have a situation called *chelation* – a *chelate ring* has been formed. The name derives from the concept of a lobster using both claws to get a better grip on its prey, put forward by Morgan and Drew in a research paper in 1920; not a bad analogy, given that chelates usually form much stronger complexes than an equivalent pair of simple monodentate ligands. A chelate ring is defined formally as the cyclic system that includes the two donor atoms, the metal ion, and the part of the ligand framework joining the two coordinated donors. The size of the chelate ring is then obtained by simply counting up the number of atoms linked covalently

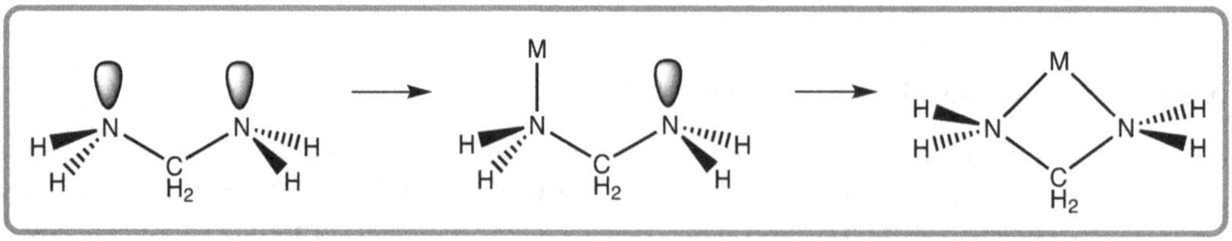

Figure 2.3
Diaminomethane, a molecule with two amine groups. Once the first is coordinated, the second lone pair from the other amine can be oriented in a direction more appropriate for bonding than is the case for two lone pairs on a single atom, with limited bond angle distortion permitting both to coordinate, illustrated at right, in a *chelated* coordination mode.

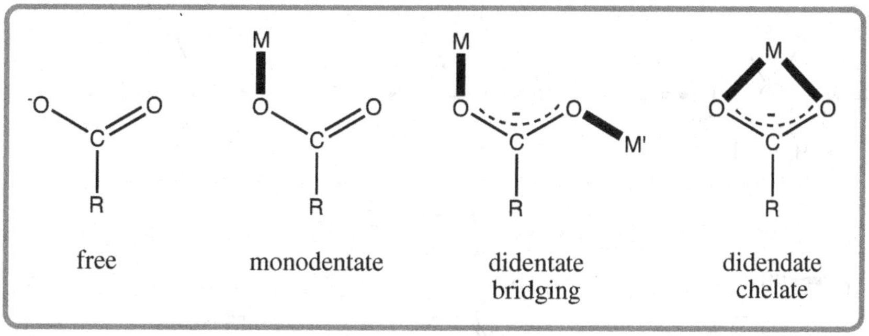

Figure 2.4
Metal ion binding options for a carboxylate group, featuring various monodentate and didentate coordination modes.

in the continuous ring, starting and ending at the metal, and including it. For example the chelated carboxylate in Figure 2.4 involves the sequence M→O→C→O→, with the last O returning us to the M, so four atoms are involved in the continuous ring, meaning it is a four-membered chelate ring.

For the diaminomethane of Figure 2.3, like the carboxylate discussed above, the chelate ring is a four-membered ring, as it involves four atoms (including the metal) linked together in a ring by four covalent bonds, two of which are coordinate bonds. Just in the way that ring structures of a certain range of sizes in organic compounds are inherently stable, chelation leads to enhanced stability in metal complexes for chelate rings of certain sizes.

If, instead of diaminomethane, the much more chemically stable diaminoethane (formally named ethane-1,2-diamine, but also called ethylenediamine or often simply 'en') is employed, chelation leads to a five-membered chelate ring. For this to happen, first one nitrogen must form a bond to the metal, then the remaining lone pair must be rotated to an appropriate orientation and the nitrogen approach the metal so as to lead to effective binding and hence chelation. The anchoring of the first nitrogen to the metal means the second one cannot be too far away in any orientation, facilitating its eventual coordination (Figure 2.5).

Looking along the C—C bond of diaminoethane, the two amines must adopt a *cis* disposition for chelation; in the *trans* disposition (shown at centre left in Figure 2.6), only bridging to two separate metals can result. With a flexible ligand like this, rotation about the C—C bond readily permits change from one conformation to another in the free ligand

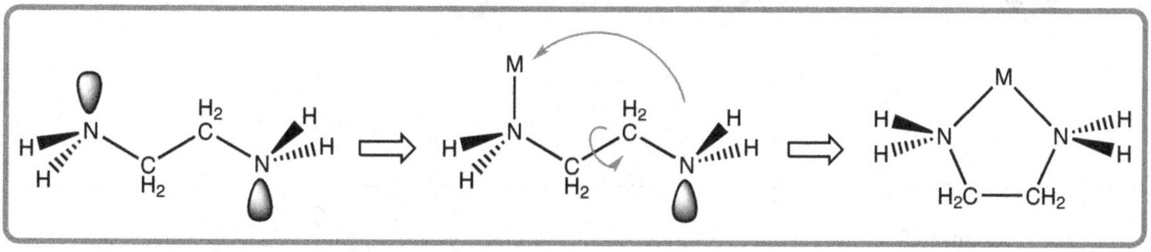

Figure 2.5
The stepwise process for chelation of diaminoethane. This features initial monodentate formation, rearrangement and orientation of the second lone pair, and its subsequent binding to form the chelate ring.

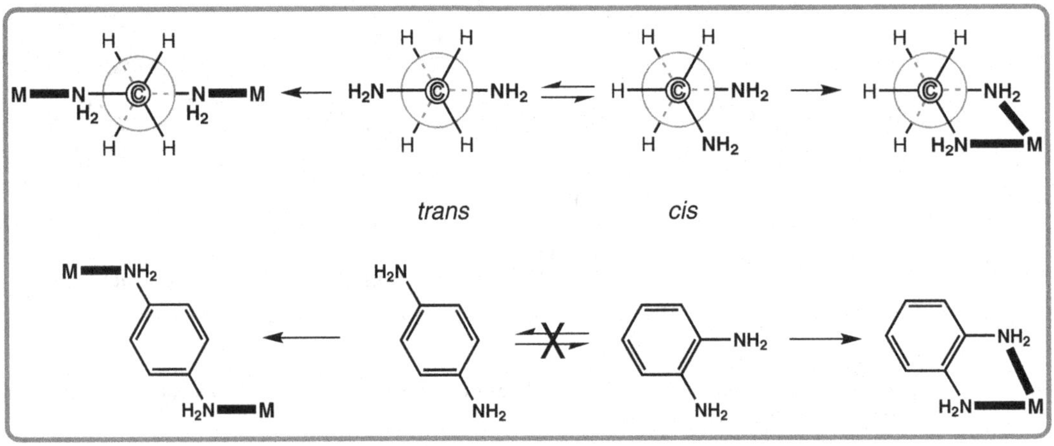

Figure 2.6
Freedom to rotate about the C—C bond in diaminoethane permits *cis* or *trans* isomers, capable of chelation and bridging respectively (top). For rigid diaminobenzene (bottom), rearrangement is not possible, and the two isomers shown have exclusive, different coordinating functions as didentate ligands.

(Figure 2.6); this will not be possible with rigid ligands like diaminobenzene, where the 1,4- (*para* or *trans*) isomer and the 1,2- (*ortho* or *cis*) isomer are distinctly different molecules, the former able only to bridge whereas the latter may chelate (although both are called didentate ligands (*di* = two) since they each bind both of their two nitrogen donors).

The chelate ring formed with 1,2-diaminobenzene is flat, because of the dominating influence of the flat, rigid aromatic ring. However, the ring with diaminoethane is not flat, since each N and C centre in the ring is seeking to retain its normal tetrahedral shape. Looking into the ring with the N—M—N plane perpendicular to the plane of the paper, the shape of the ring is clearer; one C is up above this plane, the other down – the ring is said to be *puckered* (Figure 2.7). If planarity of the carbon joined to the donor atom is enforced, such as is the case for the planar sp^2-hybridized carbon in a carboxylate, planarity

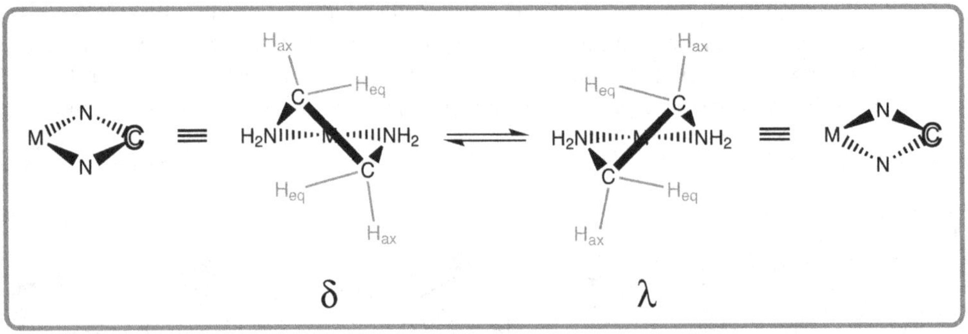

Figure 2.7
Chelate ring conformations in chelated diaminoethane (ethane-1,2-diamine, en), designated as δ and λ. Views looking into the N—M—N plane (centre; H atoms bound to C atoms disposed roughly in the plane, H_{eq}, and perpendicular to the plane, H_{ax}, also included) and along the C—C bond (sides; H atoms removed for clarity) are shown. Ready interconversion between the two conformations (which are mirror images) is possible, as only a small energy barrier exists between them.

in the chelate ring arises. The glycine anion ($H_2N-CH_2-COO^-$), with one tetrahedral and one trigonal planar carbon, forms a five-membered chelate ring with less puckering than diaminoethane, whereas the oxalate dianion ($^-OOC-COO^-$), with two trigonal planar carbons, is completely flat in its chelated form.

For the puckered diaminoethane, there are some further observations to make. The chelate ring is more rigid than the freely-rotating unbound ligand, so that the protons on each carbon are nonequivalent, as one points essentially vertically (axial, H_{ax}), the other sideways approximately parallel to the N—M—N plane (equatorial, H_{eq}). Nevertheless it is sufficiently flexible that it can invert – one carbon moving upwards while the other moves downward to yield the other form. These two forms are examples of two different *conformations*; one is called λ, the other δ, by convention; they are mirror images of each other. Any chelate ring that is not flat may have such conformers.

A vast array of didentate chelates exist, so that this one type alone can be daunting because of the variety. However, there are a number of essentially classical and popular examples, many of which tend to form flat chelate rings rather than puckered ones as a result of the shape of the donor group or enforced planarity of the whole assembly due to conjugation. A selection of common ligands appears in Figure 2.8, along with 'trivial' or abbreviated names often used to identify these molecules as ligands. One aspect of the set of examples is that the chain of atoms linking the donor atoms can vary – they do not all lead to the same chelate rings size; however, it is notable that a four-atom chain leading to five-membered chelate rings are most common. This aspect is addressed in the next section.

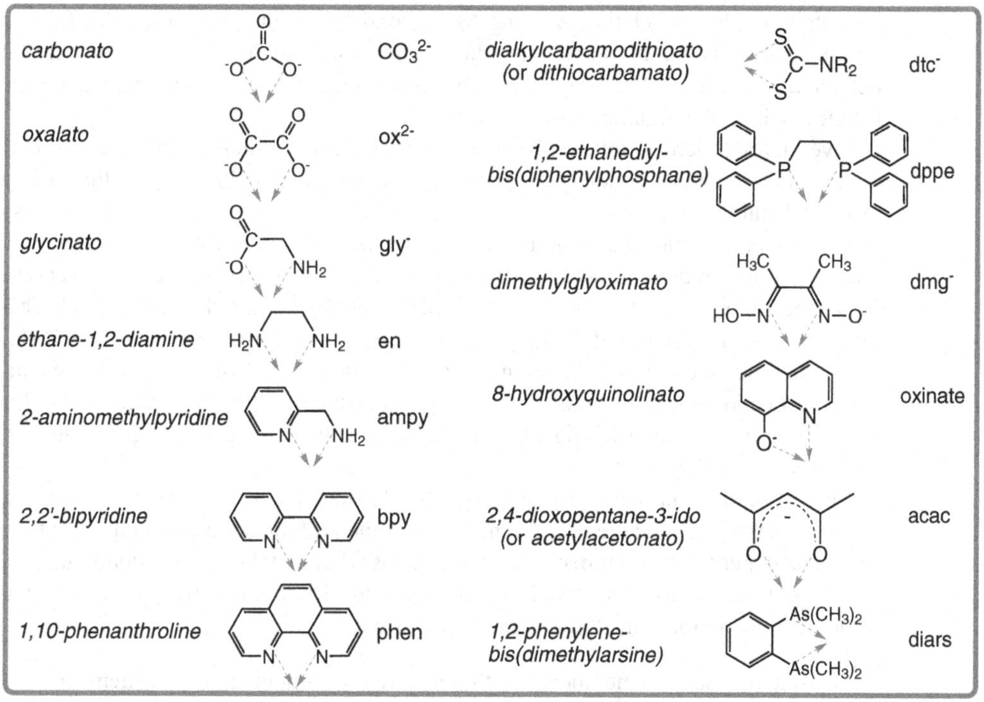

Figure 2.8
Some common didentate chelating ligands. Common abbreviations used for the ligands are given to the right of each line drawing.

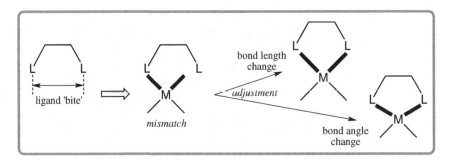

Figure 2.9
Chelate ring formation may not be ideal in terms of the 'fit' of the ligand to the metal. Potential mismatch resolution involves in large part (but not exclusively) adjustment in the metal–donor distances and the angles around the metal.

2.1.4 Do I Look Big on That? – Chelate Ring Size

As the size of the chain of atoms linking a pair of donor atoms grows, the size of the chelate ring that the molecule forms grows. This affects the 'bite' of the chelate, which is the preferred separation of the donors in the chelate, with concomitant effects on the stability and strength of the assembly. Altering bond distances and angles in the organic framework of a ligand involve a greater expenditure of energy than is the case with distances and angles around the metal, and so adjustments are often greater around the metal centre. Where a mismatch occurs in chelation, varying M–L bond length or L–M–L angles (or usually both in concert) is the dominant way adjustment is made to deal with the nonideal 'fit' of metal and potential chelate (Figure 2.9), although some limited adjustments in angles and distances within the organic ligand do occur.

Nevertheless, there is inherently no real upper limit on chelate ring size, except that as the chains between donors get very long, the size of the ring is such that it imparts no special stability on the complex, and thus no benefit is obtained – nor is it as easy for chelation to occur, since the second donor may be located well away from the anchored first donor and thus not in a preferred position for binding. Examples of three- to seven-chelate rings appear in Figure 2.10; note how the experimentally measured L–M–L angle changes with ring size. For simple didentate ligands of the type represented in Figure 2.10, you may sometimes meet a classification in terms of the number of atoms in the chain that separates the donor groups; thus one forming a four-membered ring is called a 1,1-ligand, the five-membered ring a 1,2-ligand, and the six-membered ring a 1,3-ligand and so on; it is not heavily used, however, and we shall not employ it here.

What we find is that there is a preferred chelate ring size; as the ring size rises, there is a rise in stability of the assembled complex, and then a fall as the ring continues to grow. This trend depends on a number of factors, such as what metal ion, what donor groups, and what ligand framework is involved. Nevertheless, for the common lighter metals (first row of the periodic d block) the trend is fairly consistent:

3-membered < 4-membered < 5-membered > 6-membered > 7-membered.

Overall, the five-membered chelate ring is preferred. We can actually measure this trend experimentally, such as for the series of O-donors in Figure 2.10 below. This is seen experimentally in terms of the stability of metal complexes (you can consider this as

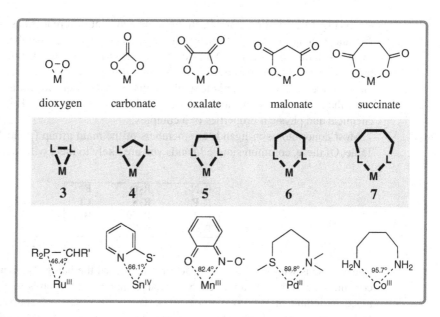

Figure 2.10
Examples of from three- to seven-membered chelate rings. Note how ring size impacts on the L—M—L intraligand angle, shown in the lower set.

a measure of the desire to form a complex), as exemplified for oxalate, malonate and succinate with a range of divalent transition metal ions in Figure 2.11.

In all cases above, the trend 5-membered > 6-membered > 7-membered ring is preserved. The size of the metal ion and its preferred bond lengths also play a role – for example while the trend is consistent for the lighter metals, larger metal ions may prefer a different ring size. There is another trend exhibited here that we will return to later also – a consistent variation with metal, irrespective of ligand.

2.1.5 Different Tribes – Donor Group Variation

What should already be obvious is that there can be different types of donors, since we have by now introduced examples of molecules where N, O, S, P and even C atoms bind to the

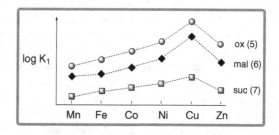

Figure 2.11
Variation with chelate ring size of the stability of complexes for various metal(II) ions for binding of the O,O-chelates oxalate (ox, 5-membered ring), malonate (mal, 6) and succinate (suc, 7).

metal (Figure 2.10, for example). In the same way that different tribes are all members of the same human species, molecules with different donors are all members of the same basic species – in this case ligands. You expect different tribes to have distinctive differentiating features, and the same applies to ligands with different donors. Because the donor atom is directly attached to the central metal, the metal feels the influence of the donor atom much more than any other atoms and groups linked to that donor atom. This is reflected in the chemical and physical properties of a complex.

Most donor atoms in ligands are members of the main group (p block) of the Periodic Table. Of these, common simple ligands you are likely to meet will be the following:

R_3N	R_2O	F^-
R_3P	R_2S	Cl^-
R_3As	R_2Se	Br^-
R_3Sb	R_2Te	I^-

Those ligands with N, O, P and S donors, as well as the halogen anions, are particularly common. These donor atoms cover the large majority of ligands you are ever likely to meet, including in natural biomolecules. There are others, of course; even carbon, as H_3C^-, for example, is an effective donor, and there is an area of coordination chemistry (organometallic chemistry) devoted to compounds that include M—C bonds, addressed later in Section 2.5.

Very often, ligands contain more than one potential donor group. Where these are not identical, we have a *mixed-donor* ligand. These are extremely common, indeed the dominant class; classic examples are the amino acids [H_2N—CH(R)—COOH], which present both N (amine) and O (carboxylate) donors. Where there are choices of donor groups available to a metal ion, it is hardly surprising that some preference may exist – we shall return to a discussion of this aspect in Chapter 3.5.

2.1.6   Ligands with More Bite – Denticity

So far, we have restricted our discussion to ligands that attach to a metal at one or two coordination sites. While these are very common, they by no means represent the limit. It is possible to have ligands that can provide sufficient donor groups in the one molecule to satisfy the coordination sphere of a metal ion fully. Indeed, they may have sufficient to bind several metal ions. *Denticity* is the way we define the number of coordinated donors of a ligand; denticity simply refers to the number of donor groups of any one ligand molecule that are bound to the metal ion, which need not be all of the donors offered by a ligand. Where coordination is by just one donor group, we are dealing with *monodentate* ligands. When two donors of a ligand are bound, it is a *didentate* ligand; three bound yields a *tridentate*, and four bound is called a *tetradentate* ligand. The terminology is outlined in more depth in Appendix One. Denticity represents the way we regularly discuss ligand binding in spoken form; it provides brevity at least, as we can then say 'a didentate ligand' rather than 'a ligand bound through two donor groups'.

It is important to recognize that denticity and the number of potential donor groups of a ligand are not always the same thing. We can see this with a very simple example, the amino acid glycine, which can bind to a metal via only the carboxylate group, only the amine group, or as a chelate through both at once. The first two modes are examples of

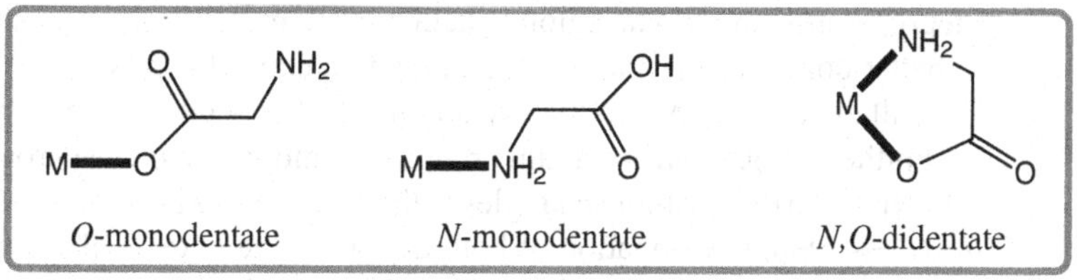

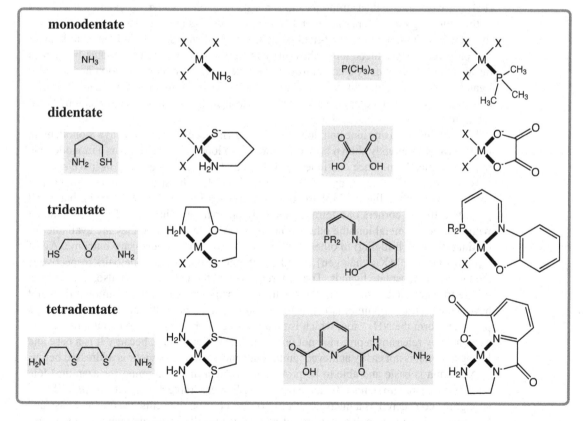

Figure 2.12
Possible coordination modes for the amino acid ligand glycine.

the molecule acting as a monodentate ligand, the last, of the molecule acting as a didentate chelate ligand (Figure 2.12).

The groups that bind to a metal are influenced also by the shape of the ligand and their location on the ligand framework. Some further examples of ligands of different denticity are given in Figure 2.13. To retain our focus on the concept at this stage, the examples involve coordination to a simple flat square shape that we shall meet again later; of course, other shapes are able to accommodate ligands as well. Note that some ligands undergo

Figure 2.13
Examples of ligands of different denticity. Several undergo deprotonation prior to coordination, as indicated by the binding of an anionic form. X-Groups inserted in other available metal coordination sites identify those sites not used by the target ligand.

deprotonation to form an anion in achieving complexation, consistent with the anion being a better donor for the cationic metal than a neutral ligand on purely electrostatic grounds. Not all donors groups can readily do this, as it depends on the acidity of the group.

Further, some ligands contain groups with more than one heteroatom in principle capable of acting as a donor atom; examples in the figure above are carboxylate ($-COO^-$) and amido ($-N^--CO-$). Coordination of both donors to the one metal ion may be forbidden by the shape of the ligand molecule, or else occurs under only special synthetic conditions. We shall explore polydentate ligands in more detail in Section 2.3.

2.2 Monodentate Ligands – The Simple Type

2.2.1 Basic Binders

The simplest class of ligands offers a single lone pair for binding. The ammonia molecule (NH_3) is the classical example, which we have introduced above. Other members of the same column of the Periodic Table offer the same property, such as phosphine, PH_3. However, their ability to act as ligands is related to their own inherent basicity, stability and chemistry (PH_3 is a weaker base than ammonia and is not stable in air or water, for example), and also to their meeting the right partner metal. Whereas N-donors are excellent choices for light, first-row transition metal ions, P-donors are not as well suited, and bind better to heavier and/or lower-charged metal ions. These are concepts that we will return to later, but it does throw up one of chemistry's conundrums – if water is 55 M in concentration and a useful ligand in its own right, why is it that even low concentrations of some other molecules dissolved in water can compete very effectively as ligands for a dissolved metal ion? We'll return to this issue in Chapter 3.5.

Having more than one lone pair does not exclude a molecule from acting as a monodentate ligand. We have already shown how water, with two lone pairs, can bind as a monodentate ligand to a metal ion. Once one lone pair is bound, the other points away in space and can only be employed, if at all, by a different metal ion. It is most usual to meet water as a simple monodentate ligand. In the extreme, a halogen ion has four lone pairs, but as all point to different corners of a tetrahedron, only one pair at a time can effectively interact with a particular metal ion, the others pointing away from it. Again, they are available for binding (through bridging) to other metal ions, but this is not a necessary outcome. All of R_3N, R_2O and X^- (X = halogen) ligands, with one, two and four lone pairs respectively, are met as monodentate ligands. The difference is that the latter two can also, in principle, bridge to other metal ions using their extra lone pairs, an option that ammonia does not have. However, even ammonia can achieve more lone pairs by the device of removing a proton, to form the NH_2^- ion, which is then capable of bridging to two metal ions. Acting as an acid by releasing a proton is not our usual view of ammonia, because it is a base and losing a proton is not easily achieved; ammonia's more usual behaviour is addressed below.

Ammonia is basic and able to be protonated to form the ammonium (or azanium) ion, with its base strength defined by an experimentally measurable parameter, the pK_a. If base strength is considered as a measure of affinity for a proton, it seems reasonable to assume that this is also a reasonable measure of affinity for small, positively charged metal ions. This works reasonably well as a gross measure. For example, NH_3 is much more basic than PH_3, which can thus account for ammonia being a better ligand for first row transition metal ions than phosphine. Further, the oxygen analogue OH_3^+ (oxonium or oxidanium)

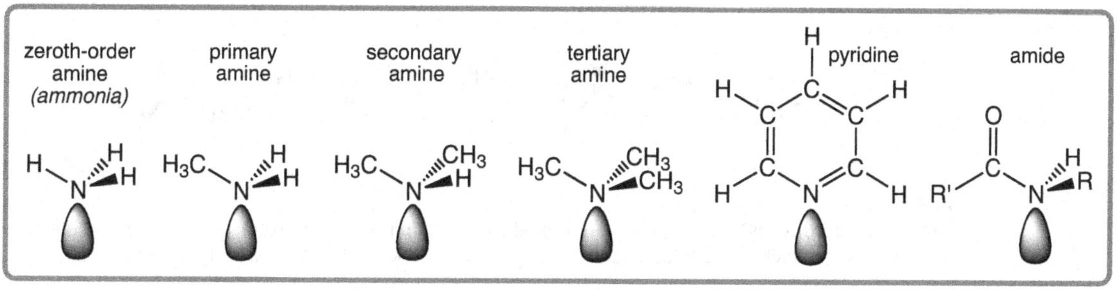

Figure 2.14
Examples of the N-donor family of monodentate ligands.

has one lone pair remaining but it is an extremely poor base and is not able to be protonated again to form OH_4^{2+} (oxidanediium); it is also totally ineffective as a ligand to metal ions. We are now beginning to define some ligands as 'better' than others on the basis of certain chemical and physical properties – this is a move towards understanding selectivity and preference, or why molecules dissolved in a solvent can compete with that solvent as ligands for a metal ion.

2.2.2 Amines Ain't Ammines – Ligand Families

One should not be fooled by assuming that amines are ammines – in other words, as an example, although ammonia (NH_3) and methanamine (CH_3NH_2) carry a common N donor atom and act as monodentate ligands they differ in terms of the groups that are attached to the N-donor. One has three H atoms bound, the other two H atoms and one methyl group (Figure 2.14). So what does this difference do? First, the molecular sizes differ. Second, electronic effects differ as a result of attaching different species to the N atom; an N—H bond and an N—C bond involve differing polarity effects. One result is that the acidity of the amine changes; the pK_a of a zeroth-order amine (one without any N—R group, i.e. ammonia) is different from that of a primary amine (one with substitution of one H by an R-group, such as methanamine). While both ammonia and methanamine can and do act as efficient ligands, their preferences are not exactly identical, as a result of inherent variations in their properties.

This can be carried further by considering a secondary nitrogen compound such as $(CH_3)_2NH$ and a tertiary nitrogen compound such as $(CH_3)_3N$. Both can in principle bind to metals, but there are consequences of these higher level substitutions, and we will come back to them later. Further, there are other types of nitrogen donors; consider pyridine (C_6H_5N), which has a nitrogen in a benzene-like ring and involves C—N=C character, or an amide group (R—HN—CO—R'), which has an accessible deprotonated form (R—$^-$N—CO—R') that is a more effective ligand. Each type of N-donor has characteristics as a ligand that do not equate with other N-donors – variety is indeed the spice of life in coordination chemistry.

2.2.3 Meeting More Metals – Bridging Ligands

We have identified ammonia (NH_3) as the classical example of a monodentate ligand, with only a single lone pair. However, as already mentioned, if ammonia is stripped of a proton to form the anion NH_2^-, it now offers two lone pairs. Although accessible in aqueous solution only at high pH, this ion is a strong base and is known for its capacity to bind

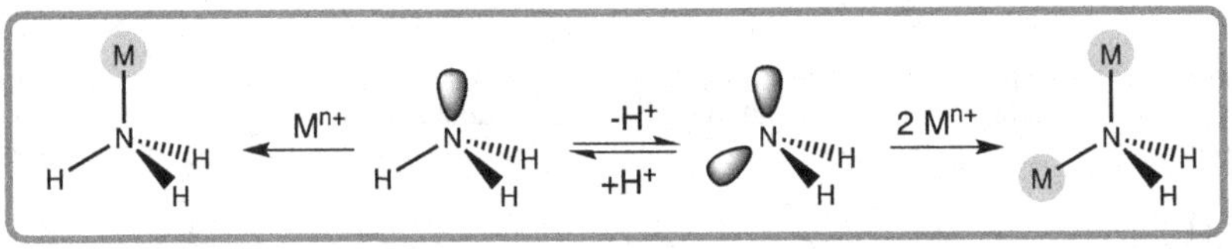

Figure 2.15
Ammonia coordinates as a monodentate ligand to one metal. When a proton is removed it exhibits the capacity to attach the resultant additional lone pair to a second metal ion in a bridging mode.

efficiently to two metal ions in a bridging mode. This is an example of how the creation of more lone pairs associated with forming a new anion can expand access to more metal ions, through these ligands making additional attachments to other metal ions at the same time – the process of *bridging*. This is illustrated for ammonia and its anion in Figure 2.15.

The water molecule (OH_2) is able to lose protons sequentially to form hydroxide (OH^-) and oxide (O^{2-}). As the protons are stripped off, the number of lone pairs increases, with the hydroxide offering three and oxide offering four. Just having more than one lone pair is no guarantee of more bonds forming. Binding as a monodentate ligand to a single metal occurs for all of OH_2, OH^- and O^{2-}, in all cases involving an oxygen donor atom. However, these ligands can, in principle, expand their linkages by making additional *bridging* attachments to other metal ions. We see this with the anions, but rarely with water itself. It is useful to examine why neutral H_2O doesn't bridge between two metal ions. When a coordinate covalent bond is formed to a positively charged metal ion, there is a strong tendency for electron density to 'shift' towards the metal ion; we can visualize this as a diminution in the size of the remaining lone pair, making it less accessible and attractive to a second metal ion. If a proton is removed from the coordinated water, creating a second free lone pair, there is a significant increase in electron density on the oxygen and so attractiveness for a second metal cation to bind and form a second coordinate covalent bond is increased. The negative charge on the HO^- ligand acts to balance the positive charges on the two metal ions forced to be located closely together (Figure 2.16). Bridging groups are represented by the Greek letter μ (mu) as a prefix in formulae (see Appendix 1).

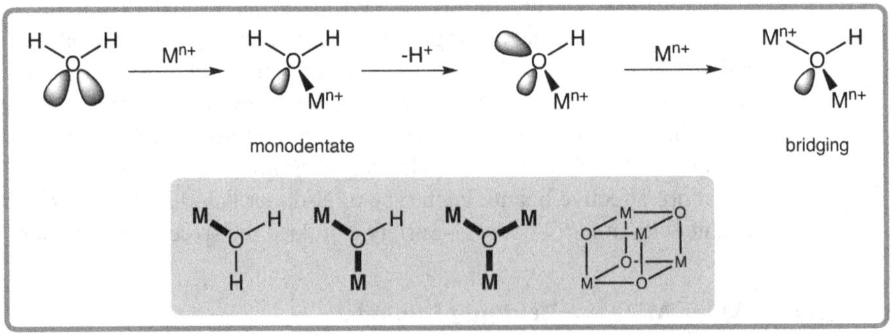

Figure 2.16
Deprotonation of water enhances the prospect of bridging between two metal ions by increasing electron density on the O atom. Successive deprotonation can permit multiple bridging to metal ions, resulting in small metal-oxide clusters.

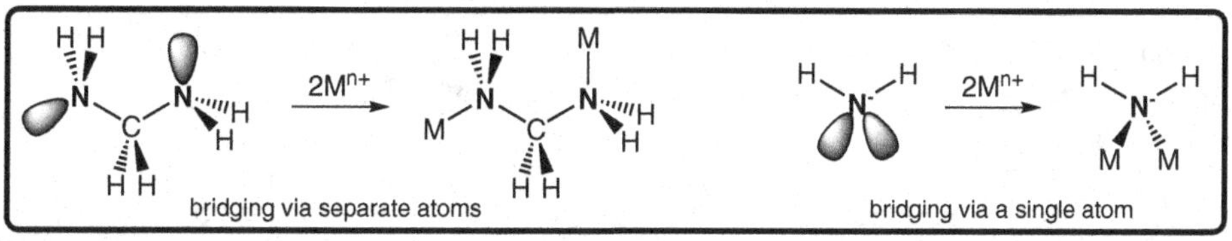

Figure 2.17
Modes through which bridging between two different metals can arise, illustrated for simple N-donor ligands.

An extension of this process is illustrated in the lower part of Figure 2.16 (ignoring nonbonding or 'spectator' lone pairs) starting from water bound in its usual form as a monodentate ligand to one metal. As each proton is successively removed, the capacity to attach the resultant lone pair to a second and then even, with the next deprotonation, to a third metal ion arises. The O donor is acting as a monodentate ligand to more than one metal at once. This is well known behaviour. What isn't shown (for simplicity) are the other bonds to the metals that can lead to large clusters in some cases, such as the box-like framework shown at bottom right in Figure 2.16, nor those additional ligands not involved in bridging.

Thus ligands can accommodate two metal ions in two distinct ways: by using two different donor atoms on the one ligand to bind to two separate metal ions; or by using two different lone pairs on the same donor group to bind to two separate metal ions (Figure 2.17).

The former leads to greater distance between the like-charged metal centres, and is thus likely to be a less demanding process. The latter brings the metal ions much closer together but can only occur, of course, with donor groups that have the potential to provide at least two lone pairs, and that are usually anionic so as to help two like-charged metal centres approach each other closely. Systems where several metal ions are bound in close proximity employ the same basic concepts discussed here.

2.3   Greed is Good – Polydentate Ligands

2.3.1   The Simple Chelate

The concept of a chelate ring has been introduced earlier in this chapter. The simplest chelates consist of just two potential donors attached to a (usually, but not necessarily) carbon chain in positions that permit both donors to attach to the same metal ion through bonding at two adjacent sites around that central metal. There is no compulsion for such a ligand to chelate, but the fact that they usually do is an indication that such arrangements are especially favourable, or the complex assembly is more stable as a result of chelation. Consider an ethane-1,2-diamine molecule binding to a metal ion. The free ligand is flexible, and rotation about C—C and C—N bonds is facile. This allows it to move around to adopt shapes that suit it best for coordination and chelation. When a molecule of the free ligand and an aquated metal ion react, the most stable product is a chelated species; this *overall* reaction is represented by the solid arrow in Figure 2.18. This representation is not an indication of how the process occurs, however, and it is very unlikely that the two amines bind to the metal ion at once in a concerted process because the organization of the molecules required to permit this to occur is too great and the probability of it happening too low. Rather, the process of chelation occurs through a stepwise process. The sequence of steps

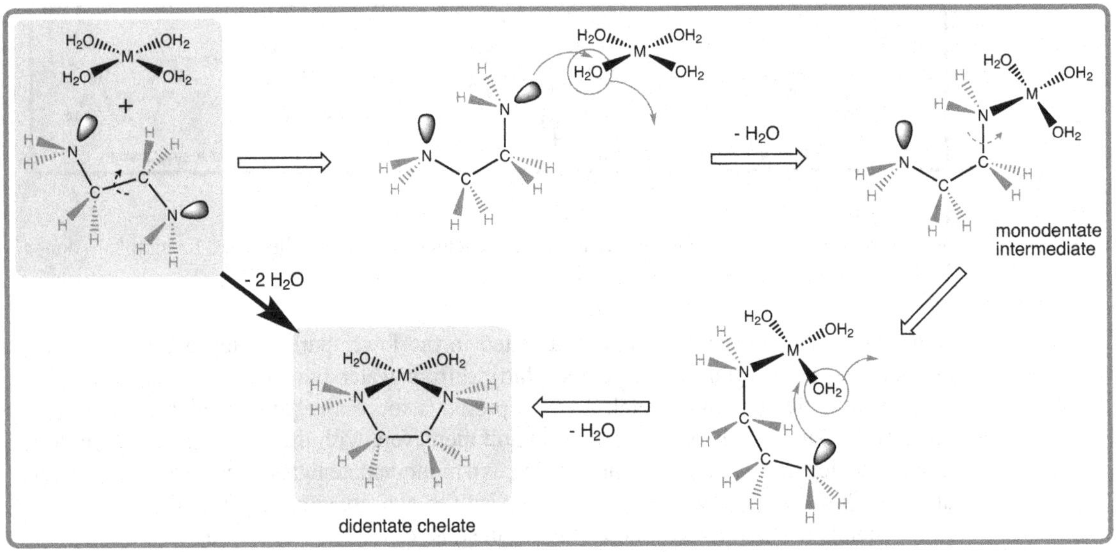

Figure 2.18
The process of chelation. A single concerted step (solid arrow) from reactants to products is *not* favoured, but rather a stepwise process (hollow arrows) is involved.

represented by the pathway of hollow arrows in Figure 2.18 is a plausible *mechanism* for the reaction. Along the way, an *intermediate* compound presumed to form is the monodentate ethane-1,2-diamine complex (albeit short-lived, because it reacts on rapidly). This is a viable entity, as this mode of coordination has been seen in some isolable complexes where only one accessible binding site exists around a metal ion due to other ligands already being firmly and essentially irreversibly bound in other sites.

The step-by-step process of chelation described above is believed to be the usual way all polydentate ligands go about coordinating to metal ions in general – 'knitting' themselves onto the metal ion 'stitch by stitch' (or, more correctly, undergoing *stepwise coordination*). It also provides a way whereby compounds can achieve only partial coordination of a potential donor set, when there are insufficient coordination sites for the full suite of potential donor groups available on a ligand.

Although most of the chelate ligands you will encounter have a carbon backbone and heteroatoms like nitrogen, oxygen or sulfur as donors, this is not essential. This can best be illustrated by comparing the two pairs of chelated systems in Figure 2.19, which have identical ring sizes and shapes, yet one of each pair contains no carbon atoms at all.

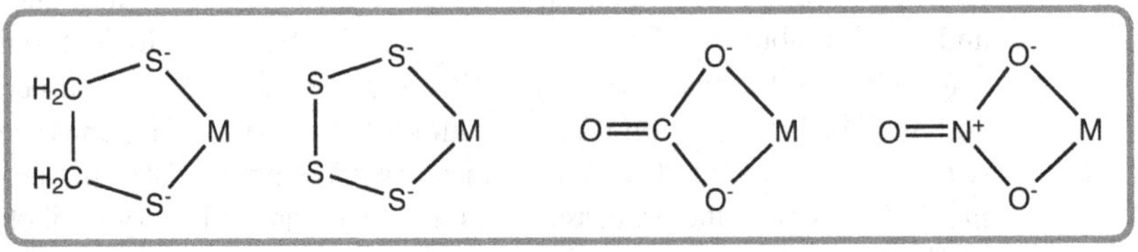

Figure 2.19
Simple chelate rings with common donor atoms but different chain atoms.

There are a range of ligand systems without carbon chains known, but they need no special consideration. They can be treated at an elementary level just like the more common carbon-backboned ligands. Herein, it is this latter and most common type upon which we shall concentrate.

2.3.2 More Teeth, Stronger Bite – Polydentates

As a general observation, the more donor atoms by which a molecule is bound to a metal ion, the stronger will be the assembly. There is basic common sense in this, of course, since more donors bound means more coordinate bonds linking the ligand to the metal ion. To separate metal and ligand, all of these bonds would need to be broken, which one would obviously anticipate costing more energy than breaking just a single bond to separate a metal ion from a monodentate ligand. A case of more is better, or greed is good. There is a conventional nomenclature to identify the number of bonds formed between a ligand and a central metal ion, shown in Table 2.1. Here, *denticity* defines the number of bound donor groups; this can equal the number of potential donor groups available (as in examples in the table), or be a lesser number if not all groups available are coordinated. See Appendix 1 for some other notes on polydenticity.

Table 2.1 identifies molecules that can bind at up to six coordination sites. This is by no means the limit, as inherently a molecule (such as a peptide polymer) may offer a vast number of potential donors. It is not uncommon for polydentate ligands to use only some and not all of their donors in forming complexes; this may be driven by the relative location of the potential donors on the ligand (and the overall shape and folding of the ligand), which affects their capacity to all bind at once, and the number of donors, as more may be offered than can be accommodated around the metal ion. As an example, a ligand like $^-OOC-CH_2-NH-CH_2-COO^-$ has one amine and two carboxylate groups, all of which are capable of coordination. However, it may bind through one, two or all three of these donors, acting, in turn, as a mono-, di- and tri-dentate ligand. Where more are available than required, the set that eventually bind are usually those that produce the thermodynamically most stable assembly. Some molecules may easily accommodate more than one metal ion; a polymer with an array of donors would be an obvious example, with different parts of the chain binding to different metal ions – we shall look at this prospect soon.

Table 2.1 Naming of different numbers of donor groups bound to a single metal ion, with examples of typical simple linear ligands of each class.

No. bound donors	Ligand denticity	Examples of ligands (donor atoms highlighted in bold)
One	Monodentate	$\mathbf{N}H_3$ $\mathbf{O}H_2$ $\mathbf{F}^-$
Two	Didentate	$H_2\mathbf{N}-CH_2-CH_2-\mathbf{N}H_2$ $H_2\mathbf{N}-CH_2-CO-\mathbf{O}^-$
Three	Tridentate	$H_2\mathbf{N}-CH_2-CH_2-\mathbf{N}H-CH_2-CH_2-\mathbf{N}H_2$
		$^-\mathbf{O}-CO-CH_2-\mathbf{N}H-CH_2-CO-\mathbf{O}^-$
		$H_2\mathbf{N}-CH_2-CH_2-\mathbf{S}-CO-CH_2-\mathbf{P}(CH_3)_2$
Four	Tetradentate	$H_2\mathbf{N}-CH_2-CH_2-\mathbf{N}H-CH_2-CH_2-\mathbf{N}H-CH_2-CH_2-\mathbf{N}H_2$
		$^-\mathbf{O}-CO-CH_2-\mathbf{N}^--CO-CH_2-\mathbf{N}^--CO-CH_2-\mathbf{N}H_2$
Five	Pentadentate	$H_2\mathbf{N}-CH_2-CH_2-\mathbf{N}H-CH_2-CH_2-\mathbf{N}H-CH_2-CH_2-\mathbf{N}H-CH_2-CH_2-\mathbf{N}H_2$
		$^-\mathbf{O}-CO-CH_2-\mathbf{N}H-CH_2-CH_2-\mathbf{S}-CH_2-CH_2-\mathbf{N}H-CH_2-CH_2-\mathbf{O}^-$
Six	Hexadentate	$^-\mathbf{O}-CO-CH_2-\mathbf{N}H-CH_2-CH_2-\mathbf{N}H-CH_2-CH_2-\mathbf{N}H-CH_2-CH_2-\mathbf{N}H-CH_2-CO-\mathbf{O}^-$

2.3.3   Many-Armed Monsters – Introducing Ligand Shape

High denticity ligands need not be simply composed of linear chains like the examples in Table 2.1. The shape of the ligand has an important role to play in the manner of coordination and the strength of the complex, but there is no particular shape that is forbidden to a ligand. Molecules may be linear, branched, podal, cyclic, polycyclic, helical, as well as aliphatic or aromatic or mixtures of components of these – the over-arching requirement is simply a set of donor groups with lone pairs of electrons. In some cases, they may be dinucleating or polynucleating, that is capable of binding to more than one metal at the same time. Some fairly simple examples and their family name appear in Figure 2.20. We shall deal more with shape later, since ligand shape or topology can have an important influence on coordination, but it may be useful to introduce some concepts now.

Let's consider two different types of hexadentate ligands: a linear polyamine and a branched polyaminoacid. To coordinate a metal ion and use all six donors, the linear molecule must wrap around the metal ion like a ribbon, whereas the branched molecule wraps up the metal ion more like a hand with fingers (Figure 2.21). On a macroscopic scale, the latter is like holding a ball in the palm of your hand with the fingers wrapped tightly around it – a nice, secure way of holding it. On the molecular scale, branching can likewise lead to a more 'secure' complex, which makes the branched polyaminoacid (called EDTA) represented in its anionic form in Figure 2.21 such an effective, strong ligand for binding most metal ions.

The shape of the ligand places demands on how it can coordinate to a metal ion. For a many-branched ligand, where the branches or 'fingers' are terminated with donor groups capable of binding to a metal ion, the ligand can achieve a firm grip by employing these 'fingers'. Looking at the way such ligands wrap around a metal ion, you may also get an idea of how they can be removed – by detaching each 'finger' in turn, then taking the metal ion out of the 'palm' of the ligand. Initial coordination is in effect the reverse of this process – step by step binding to eventually achieve the final complex. Of course, because the metal ion is not 'naked' in the first place but carries a suite of usually simple ligands when it encounters the polydentate, each step of attachment of a donor of the polydentate involves departure of a simple ligand; it is a substitution process.

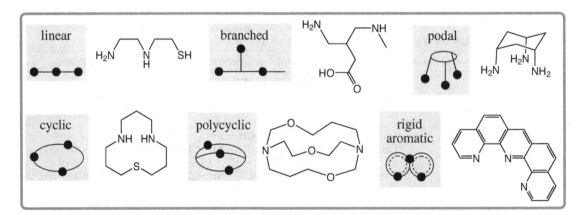

Figure 2.20
Examples of polydentate ligands of selected different basic shapes.

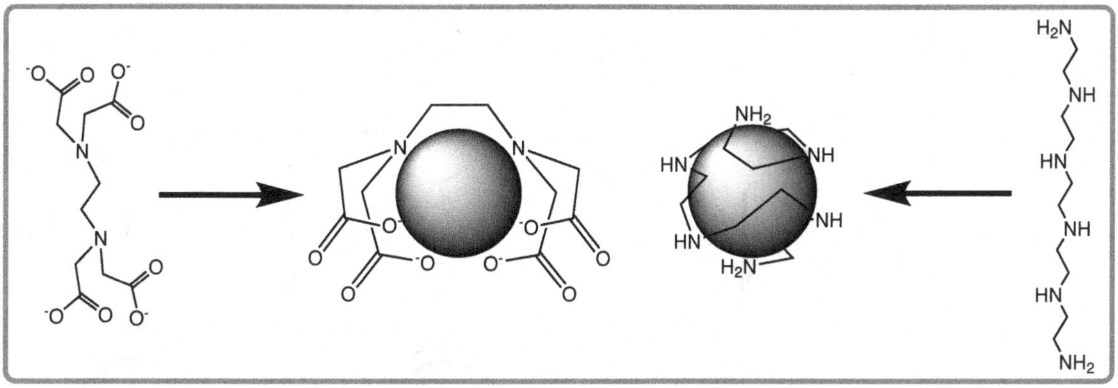

Figure 2.21
The process of complexation of polydentate ligands involves an aspect of 'wrapping' around the metal ions, usually with gross rearrangement of the ligand shape.

2.4 Polynucleating Species – Molecular Bigamists

2.4.1 When One is Not Enough

Small ligands like the didentate ethane-1,2-diamine may not satisfy the coordination sphere of a metal ion, so that, provided sufficient ligand is available, the metal may choose to attach more than one ligand, leading not only to ML species but ML_2 and ML_3 or even more attached ligands. One consequence of working with big ligands that carry many potential donor groups is that they may decide to do the exact opposite – use their suite of donor groups to attach to more than one metal ion at once. Thus, instead of a simple ML species, they may form M_2L or even M_3L or higher.

The whole situation is about getting satisfaction, or meeting the desires of both ligand and metal. In this marriage of metal and ligand, there needs to be some form of give and take to reach an agreeable outcome. The metal is seeking to maintain its desired number of metal–donor bonds, and the ligand is seeking to employ as many of its donor groups as practicable. They are after the same general outcome, but come to assembly with different demands and expectations. Normally, they reach a satisfactory match, which may be just a simple 1:1 assembly; however, it often really is a case of one is not enough. Satisfaction may require either the use of more ligands than metals or the exact opposite.

Let's look more closely at a situation where more than one metal ion could be accommodated, with a relatively simple set of ligands. The polyaminoacid anion $EDTA^{4-}$ on the left in Figure 2.22 we have met already above, and it has a flexible aliphatic linker between the two $N(CH_2COO^-)_2$ heads. Thus (as described above), it can wrap up a single metal ion effectively, forming a 1:1 M:L complex species. However, the molecule on the right has the flexible linker replaced by a rigid, flat aromatic ring. Now the two nitrogen atoms cannot bend around to coordinate to two adjacent sites of the metal and thus cannot form the initial central chelate ring. Consequently, this ligand has more success binding a single metal ion in each of its two $N(CH_2COO^-)_2$ head units; this leads to fewer donor groups coordinated to each metal, but this is the sole way that all six donor groups can be employed with a minimum number of metal ions, leading to a M_2L complex species. Molecular bigamy has led to a successful compromise.

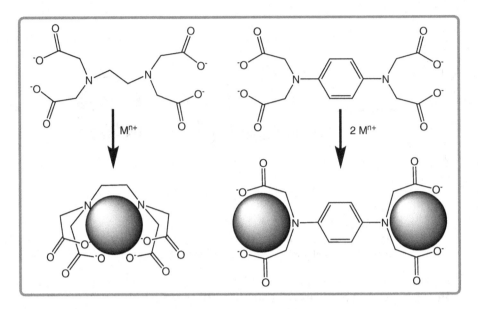

Figure 2.22
How ligand flexibility influences the process of complexation of polydentate ligands to metal ions.
The rigid aromatic link in the right-hand example prohibits 'wrapping' of the two arms around a
single metal, and coordination to two separate metal ions may be preferred.

Complexation of the ligand on the right in Figure 2.22 provides the ligand with a choice;
it may form a 1:1 M:L complex and leave some donor groups uncoordinated, or form a
2:1 M:L complex and employ all donor groups. In reality, both forms may exist in solution
at once, depending on the relative concentrations of ligand and metal, and we will explore
those types of situation separately in Chapter 5.

2.4.2 Vive la Difference – Mixed-metal Complexation

As mentioned already, the way a ligand goes about binding several metal ions is inevitably
a stepwise process, as the probability of all metal ions attaching to the ligand at once is
so low that it can be ignored. This suggests that one may be able to not only see some
intermediate species, as they could be inherently stable, but also that one should be able to
intercept these species and insert a different metal ion into an empty cavity to produce, not
a M_2L species, but a $MM'L$ species (Figure 2.23). Because $MM'L$ carries different metal
ions, it will differ in chemical and physical properties from a M_2L or M'_2L species with
the same metal ions in each site.

We can carry that concept a little further, and examine a slightly different ligand set
(Figure 2.24). Here, each of the two ligand head units is different. This means that when
a first metal is coordinated, it has two inequivalent ML options – it can bind to the all-N
set, or else the mixed-O,N set. When the second and different metal enters and binds to
the remaining set, this then leads to two different $MM'L$ species, provided there is no high
selectivity of a metal for a particular head unit.

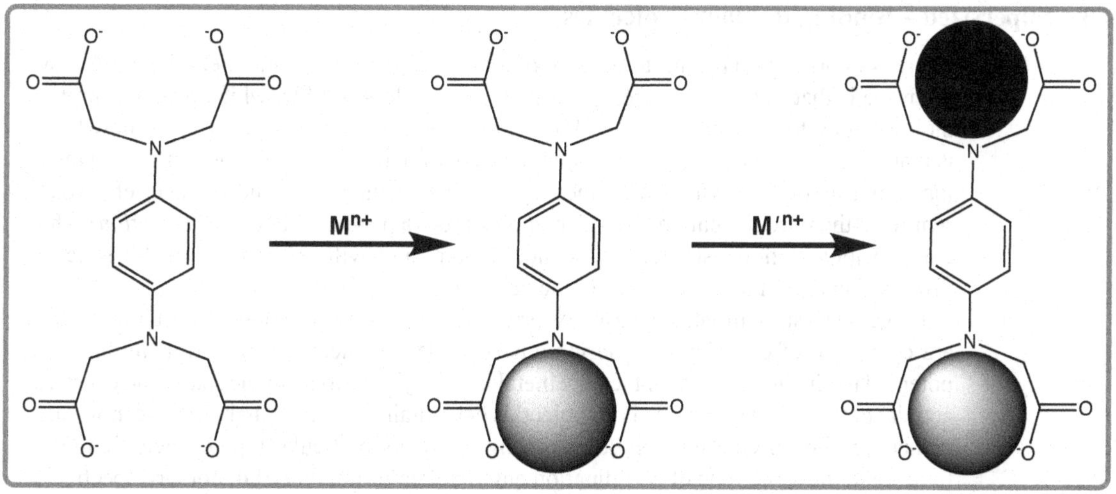

Figure 2.23
How the process of sequential complexation of metal ions may be applied to permit coordination to two different metal ions.

Thus, with this unsymmetrical ligand, there are two complexes possible with two different metal ions inserted in different compartments. This simple example is sufficient to illustrate the often vast range of options that becomes available when complicated ligands bind to metal ions. However, not all of these options will be equally probable, as the thermodynamic stability of each of the various options will differ, concepts that we will also deal with further in Chapter 5.

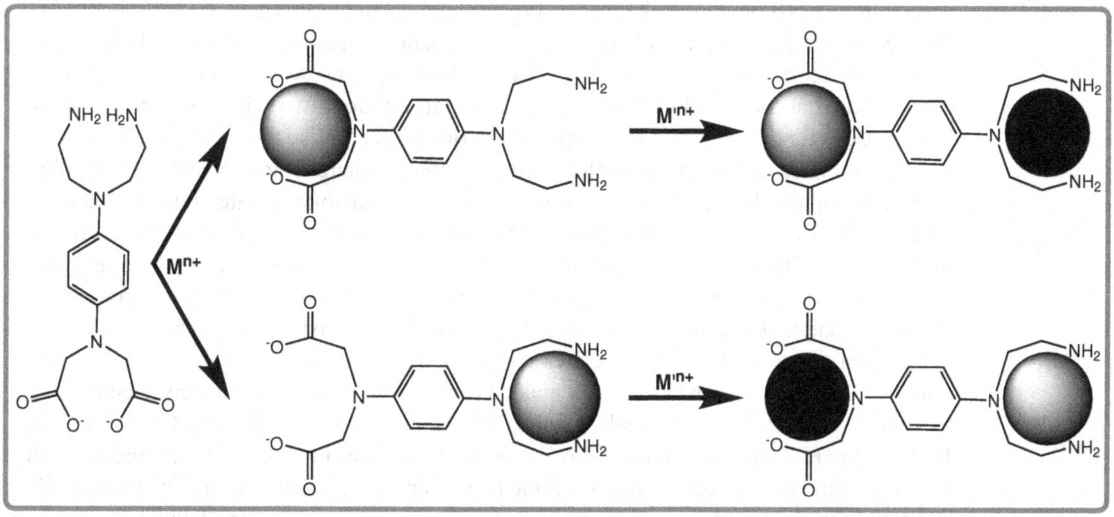

Figure 2.24
How the process of choice and selection during complexation of metal ions to an unsymmetrical ligand may lead to two differing MM′L forms.

2.4.3   Supersized – Binding to Macromolecules

There is in principle no limit to the size of a molecule that may bind metal ions. What we often see is that as molecules get larger they become less soluble, which places a limit on their capacity to coordinate to metal ions in solution. However, what we also now know is that even solids carrying groups capable of coordinating to metal ions can adsorb metal ions from a solution with which they are in contact onto the solid surface, effectively removing them from solution through complexation (a process called *chemisorption*). Thus complexation is not restricted to the liquid state, and will occur in the solid state; as discussed earlier, it also can occur in the gas phase.

Large, supersized molecules (usually termed *polymers*) are common. Many biomolecules are composed of complex polymeric units (peptides or nucleotides), and many have the potential to bind metal ions; in fact, the metal ions may contribute to the shape and chemical activity of the polymer. So size is no object. What remains true even in these huge molecules is that the rules of coordination do not change, and any particular type of metal ion tends to demand the same type of coordination environment or number of donors irrespective of the size of the ligand. This means that, if a metal ion prefers to form six bonds to six donor groups, it will usually achieve this whether only one donor group is offered by a molecule (by binding six separate molecules) or whether one hundred donor groups are offered by a molecule (by binding selectively only six of the one hundred donors available).

2.5   A Separate Race – Organometallic Species

We have mentioned earlier the prospect of the carbanion H_3C^- being an effective ligand, since it offers an electron pair donor in the same way that ammonia does. To emphasize this aspect, even simple compounds like $[M(CH_3)(NH_3)_5]^{n+}$ have been prepared in recent decades. A vast area of chemistry has grown featuring metal–carbon bonds called, not surprisingly, organometallic chemistry. Typical ligands met in this area are carbon monoxide, alkenes and arenes; it is also conventional practice to include hydride and phosphine ligands. There has been a tendency to regard this field as somehow separate from traditional compounds, but the division is somewhat artificial, although we will see there are some good reasons for this schism. One simple distinction is that, unlike the usually ionic Werner-style compounds, many organometallic compounds are neutral, low melting and boiling point compounds that dissolve in organic solvents, and display greater ligand reactivity. Further, they usually feature metals in low oxidation states, and in polymetallic systems are more likely to involve direct metal–metal bonds. They can also display ligand types and structural characteristics that were not anticipated from classical Werner-style coordination chemistry, and that require more sophisticated models for satisfactory interpretation.

One distinction that we meet relates to the mode of bonding in organometallic compounds. We have seen how H_3C^- can make use of its lone pair in the conventional manner to coordinate, forming a σ covalent bond. This approach is less obvious for a molecule like ethene ($H_2C=CH_2$), which has been shown to usually involve side-on bonding with the two carbon centres equidistant from the metal centre. This is not an arrangement easily dealt with by the simple covalent bonding model, which can accommodate bonding only by regarding the molecule as formally $H_2C^- - {}^+CH_2$ with the carbanion carrying the lone pair alone able to bond covalently. There are clearly problems with this approach, since the bonding arrangement in the resulting complex involves a single σ bond, leaving the

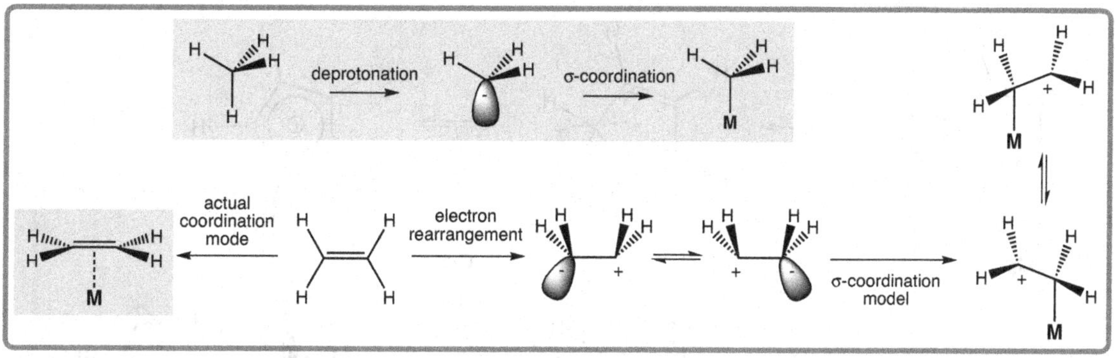

Figure 2.25
Coordination of H_3C^- and $H_2C{=}CH_2$ to metal centres. Whereas the former can be interpreted using a traditional σ covalent bonding model, this is far less successful for the latter.

carbocation displaced away from the metal centre; even allowing an equilibrium to operate between the two carbon sites, the model is not very compelling. What is found in reality is that not only is each carbon equidistant from the metal centre but also the carbon-carbon bond length when coordinated remains much like a standard double bond, with the whole ethylene molecule remaining flat and showing no tetrahedral distortion (Figure 2.25). It appears that some form of π bonding may be involved, and a different bonding model involving a molecular orbital approach is required.

The 'side-on' bonding situation is exaggerated further when coordinating one of the classical ligands of organometallic chemistry, the cyclopentadienyl ion, $C_5H_5^-$ (Cp^-) (Figure 2.26). This organic anion is formed by deprotonation of cyclopentadiene; in a 'frozen' format, it could be considered to have one carbon with a lone pair, which could bind to a metal ion through this single carbon in a simple covalent σ-bonded mode. However, this is not the only or even usual form of coordination. Rather, it commonly binds to a metal ion 'side on', with all five carbon atoms equidistant from the metal ion, in what can be considered a π-bonding mode. In this mode of coordination, it effectively occupies a face of an octahedron, in a somewhat similar manner to how a cyclic, saturated triamine ligand can do. In Figure 2.26 (where H atoms are usually not shown for simplicity), comparison is made between a metal ion bound to two cyclic triamine ligands, one each of a triamine and a cyclopentadienyl anion, and two cyclopentadienyl anions. You may see how the Cp^- anion effectively replaces three traditional σ-donor groups on a 'face', leading eventually to a $[M^{n+}(Cp^-)_2]$ compound called, for obvious reasons, a 'sandwich' complex. The Cp^- and other unsaturated carbon-bonding ligands can exhibit variable coordination through the use of different numbers of available carbons (in the same way that a traditional polydentate ligand may not employ all of its potential donor groups in coordination), and the mode of coordination is defined by a η^x nomenclature, where the superscript x represents the number of carbons in close bonding contact with the metal atom. For example the mode with a single M—C bond shown below would be a η^1 bonding form, whereas the usual facially-bound form would be a η^5 bonding mode.

Apart from unsaturated alkenes and cycloalkenes acting as ligands, the simplest ligand of organometallic chemistry is carbon monoxide, which usually acts as a monodentate σ-bonded ligand through the carbon atom, forming a linear M—C≡O bonding arrangement, although our covalent bonding model sits uneasily with this representation, and a molecular

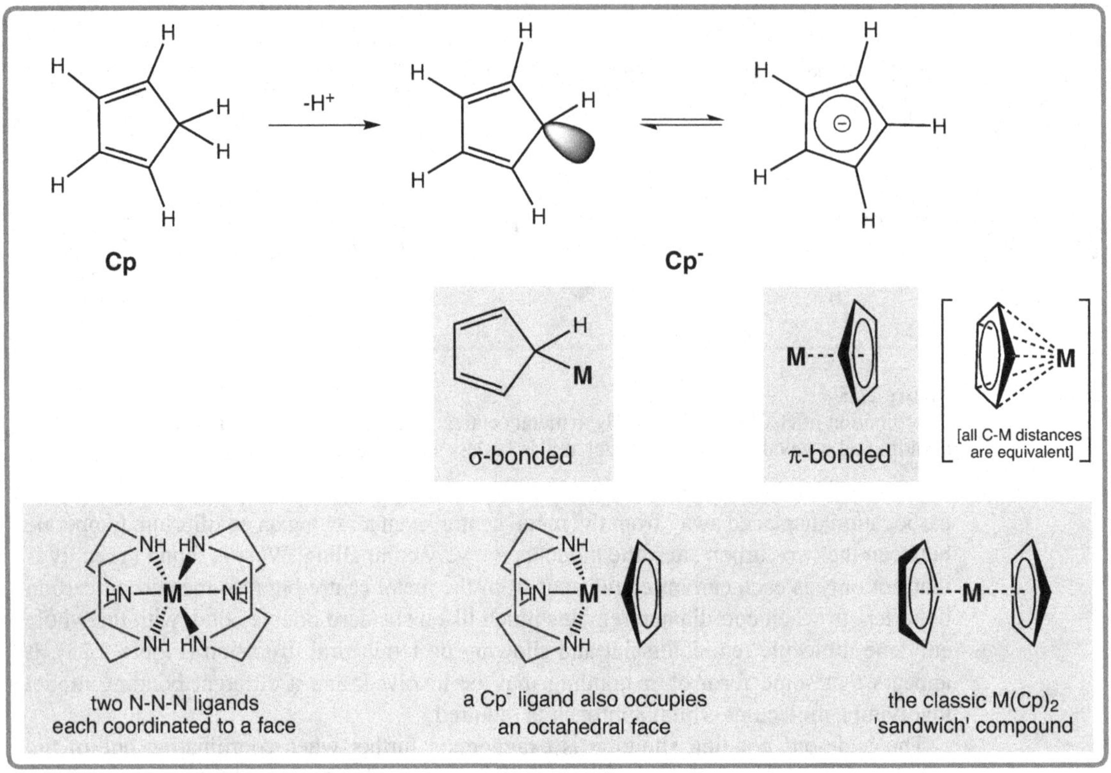

Figure 2.26
Formation and coordination of $H_5C_5^-$ as a σ-bonded ligand and as a π-bonded ligand. In the latter mode, the ligand occupies the face of an octahedron in the same way that a conventional cyclic triamine does.

orbital model is more useful. In this field we also see an example of a rare cationic ligand, NO^+. With organometallic chemistry such an extensive and demanding field in its own right, we shall deliberately limit our discussion of these somewhat unusual ligands and their compounds in this introductory textbook. However, there are aspects of their mode of coordination and spectroscopic properties that have driven extension of bonding models, and this will be addressed in Chapter 3.

Concept Keys

Metal ions are almost invariably met with a set of coordinated ligands, each of which provides one or more lone pairs of electrons to make one or more covalent bonds.

Denticity defines the number of donor groups of a ligand that are coordinated. Apart from one (monodentate), those providing two (didentate) or more (polydentate) donors may involve chelate ring formation. For each chelate ring, two linked donor groups bind the central atom, forming a cyclic unit that includes the central atom.

Usually, the higher the denticity, the more stable is the complex formed.

> Chelate ring size and stability will vary with the size of the chain linking the pair of donor groups. In terms of stability, five-membered rings are usually most stable for the lighter transition metal ions.
>
> Ligands with at least two donor groups may be involved in bridging between two metal ions as an option to chelation of one metal ion.
>
> The type of donor atom influences the stability and properties of complexes.
>
> The shape of a ligand influences the mode of coordination and stability of formed complexes.
>
> Organometallic complexes are those in which a metal–carbon bond is involved. The organic ligand may be σ-bonded (as occurs for $^-CH_3$ and CO) or π-bonded (as occurs for $H_2C{=}CH_2$ and $C_5H_5{}^-$).

Further Reading

Constable, E.C. (1996) *Metals and Ligand Reactivity: An Introduction to the Organic Chemistry of Metal Complexes*, Wiley-VCH Verlag GmbH, Weinheim, Germany. Although at a more advanced level, the early chapters give a useful introduction to metal–ligand systems and some key aspects of their chemistry.

Crabtree, R.H. (2009) *The Organometallic Chemistry of the Transition Metals*, 5th edn, John Wiley & Sons, Inc., Hoboken, USA. A popular, advanced comprehensive coverage of this adjunct field to coordination chemistry, with a fine clear introduction to the principles of the field that students may find revealing.

Hill, A.F. (2002) *Organotransition Metal Chemistry*, Royal Society of Chemistry, Cambridge, UK. A valuable undergraduate-level introduction to the broad field of organometallic compounds, with practical and applied aspects also covered; beyond the focus of the present text, but useful for those who wish to extend their knowledge.

King, R.B. (ed.) (2005) *Encyclopaedia of Inorganic Chemistry*, John Wiley & Sons, Ltd, Chichester, UK. A ten-volume set that provides short, readable articles mostly at an approachable level on specific topics in alphabetical order; these stretch from ligands to coordination, organometallic, and bioinorganic chemistry, and beyond. A good resource set for this and other chapters.

3 Complexes

3.1 The Central Metal Ion

The first two chapters introduced aspects of coordination chemistry from the perspective of a central atom (usually a metal ion) and of a ligand. This chapter will expand on their interrelationship. A key observation of coordination chemistry is that the properties and reactivity of a bound ligand differ substantially from those of the free ligand. Concomitant with this outcome is a change in the properties of the central metal ion. This gross modification of the character of the molecular components upon binding of ligand to metal ion is at the heart of the importance of the field. Thus understanding metal–ligand assemblies is vital.

We have considered variable oxidation states in Chapter 1, and should now be aware that many metal ions can exist in more than one oxidation state. Let's recall how this can arise by looking at manganese, which has an electronic configuration $[Ar]4s^2 3d^5$. With the 4s and 3d electronic levels being similar in energy, electrons in these levels can be considered together as the valence electrons. Thus there are seven electrons 'available', and successive removal of electrons will take us from Mn(I), on removal of the first electron, through to Mn(VII), on removal of the seventh electron. Each successive removal costs more energy; successive ionization enthalpies for manganese rise from $0.724\,MJ\,mol^{-1}$ for the first through 1.515, 3.255, 4.95, 6.99, 9.20 to $11.514\,MJ\,mol^{-1}$ for the seventh. By this stage, the electron configuration has reached that of the inert gas [Ar], and breaking into this inert core costs too much energy; thus the highest accessible oxidation state for manganese is Mn(VII), or Mn^{7+} as the free ion. In general, as more and more valence electrons are removed and the positive charge rises, the energy cost of electron removal becomes higher and higher. Therefore, it is no surprise to find that, whereas elements in the d block up to Mn are known in oxidation states up to that with an [Ar]-only core, this is not achieved for elements after Mn. The highly-charged metal ions are powerful oxidants, seeking to reduce their oxidation state by re-acquisition of electrons. All eventually reach an oxidation state where they are such powerful oxidants that they oxidize practically any molecule they meet, including solvents. This means they can have no stable existence, and can only be found in highly contrived environments, such as trapped in a crystalline lattice following *in situ* generation in that environment. As a consequence, very high oxidation states are not often met, at least with the lighter metal ions. It is only with heavier elements of the Periodic Table, such as the second and third rows of the d-block metals, that higher oxidation states are more sustained, an outcome that may be related to the greater number of filled-shell electrons shielding the more exposed outer valence shell electrons from the nuclear charge.

From experimental evidence, the first row transition metals clearly prefer oxidation states +II and +III. Higher and lower formal oxidation states, even formally negative ones (such as −I, −II), are known, although oxidation states less than +I are almost exclusively

Introduction to Coordination Chemistry   Geoffrey A. Lawrance
© 2010 John Wiley & Sons, Ltd

found with organometallic compounds. Oxidation states in coordination compounds are sometimes hard to define. This arises where it may be difficult to decide whether an electron resides exclusively on the metal or exclusively on a ligand, or even is shared. Most reported oxidation states are what can be termed spectroscopic oxidation states – that is, defined by an array of spectroscopic properties as most consistent with a particular metal d-electron set. The main point about oxidation states is that, for every metal, its chemistry in different oxidation states is distinctly different. For example, Co(III) is slow in its ligand substitution reactions, whereas Co(II) is rapid; each has different preferences for ligand type and stereochemistry.

Because we are usually dealing with ionic salts that are water-soluble when we begin working with transition metal ions, the complexes formed in pure water where water molecules are also the ligand set are a common starting point for initiating other coordination chemistry. Most first row transition metals in the two commonest oxidation states form a stable complex ion with water as a ligand, usually of formulation $[M(OH_2)_6]^{n+}$ ($n = 2$ or 3) These ions are usually coloured, and display a wide variation in the potential for both the M(III)/(II) and M(II)/(0) redox couples, as shown in Table 3.1. This suite of observations will form the background to discussion later, where coordinated water ligands are replaced by other ligands, with associated change in physical properties such as colour and redox potential. Theories that we develop must be able to accommodate these changes.

3.2   Metal–Ligand Marriage

3.2.1   The Coordinate Bond

'Bond – Coordinate Bond'. Maybe it doesn't exactly roll off the tongue, but it's hard to avoid this adaptation of the personal introduction used by perhaps our best-known and most enduring screen spy to introduce this section – it serves its purpose to remind us of the endurance and strength of bonds between metals and ligands, which at a basic level we can consider as a covalent bond. Moreover, it isn't just any bond, but a specially-constructed coordinate bond – hence the name of this field, coordination chemistry. Unfortunately, the simple covalent bonding concept does not provide valid interpretations for all of the physical properties of coordination complexes, and more sophisticated theories are required. We shall examine a number of bonding models for coordination complexes in this chapter.

3.2.2   The Foundation of Coordination Chemistry

Before we travel too far, however, it may be valuable to look back to the beginnings of coordination chemistry. While examples of coordination compounds were slowly developed during the nineteenth century, it wasn't until the twentieth century that the nature of these materials was understood. They were at a very early stage named 'complex compounds', a reflection of their unexplained structures, and we still call them 'complexes' today. Around the beginning of the twentieth century, the wealth of instrumental methods we tend to take for granted today simply didn't exist. Chemists employed chemical tests, including elemental analyses, to probe formulation and structure, augmented by limited physical measurements such as solubility and conductivity in solution. Analyses defined the components, but not their structure. As a consequence, it became usual to represent

Table 3.1 Colours of the M^{2+}_{aq} and M^{3+}_{aq} ions of the first row of the d block, along with their standard reduction potentials.

Parameter	Sc	Ti	V	Cr	Mn	Fe	Co	Ni	Cu	Zn
$[M(OH_2)_6]^{3+}$ colour	Colourless	Violet	Blue	Violet	Red-brown	Very pale purple	Blue[a]	—[a]	—[a]	—[a]
$[M(OH_2)_6]^{2+}$ colour	—[a]	—[a]	Violet	Blue	Very pale pink	Pale green	Pink	Green	Blue-green	Colourless
$E^{\circ}(M^{III/II})$, V in aq. acid	—	−0.37	−0.25	−0.41	+1.59	+0.77	+1.84	>+2	>+2	≫+2
$E^{\circ}(M^{II/0})$, V in aq. acid	—	<−2	−1.19	−0.91	−1.18	−0.44	−0.28	−0.24	+0.15[b]	—

[a]Unknown, unstable or highly reactive.
[b]$Cu^{II/I}$ couple in this case.

them in a simple way as, for example, $CoCl_3 \cdot 6NH_3$ for Tassaert's pioneering complex that we now know as the ionic octahedral cobalt(III) compound $[Co(NH_3)_6]Cl_3$.

Simple tests of halide-containing compounds, involving gravimetric analysis of the amount of silver halide precipitated upon addition of silver ion to a solution and comparison with the known total amount of halide ion present from a more robust microanalysis method, identified the presence in some cases of both reactive and unreactive halide. For example, $CoCl_3 \cdot 6NH_3$ and $IrCl_3 \cdot 3NH_3$ precipitated three and zero chlorides respectively on addition of silver ion, meaning only one type of halide was present in each, but of different reactivity. Other species such as $CoCl_3 \cdot 4NH_3$ precipitated only one of the three halides readily, so that there were clearly two types of chloride present in the one compound. It was surmised that one type is held firmly in the complex and so is unavailable, whereas the other is readily accessed and behaves more like chloride ion in ionic sodium chloride. We now know these two classes as coordinated chloride ion (held via a M—Cl coordinate bond) and ionic chloride (present as simply the counter-ion to a positively-charged complex cationic species).

The availability of equipment to measure molar conductivity of solutions was turned to good use. It is interesting to note that coordination chemists still make use of physical methods heavily in their quest to assign structures – it's just that the extent and sophistication of instrumentation has grown enormously in a century. What conductivity could tell the early coordination chemist was some further information about the apparently ionic species inferred to exist through the silver ion precipitation reactions. This is best illustrated for a series of platinum(IV) complexes with various amounts of chloride ion and ammonia present (Figure 3.1). From comparison of measured molar conductivity with conductivities of known compounds, the number of ions present in each of the complexes could be determined. We now understand these results in terms of modern formulation of the complexes as octahedral platinum(IV) compounds with coordinated ammonia, where coordinated chloride ions make up any shortfall in the fixed coordination number of six. This leaves in most cases some free ionic chloride ions to balance the charge on the complex cation.

This type of what we now consider simple experiments provided a foundation for coordination chemistry as we now know it. What became compelling following the discovery of these intriguing compounds was to develop a theory that would account for these

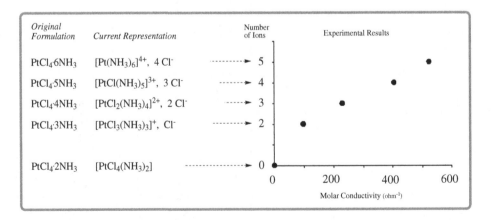

Figure 3.1
Identifying ionic composition from molar conductivity experiments for a series of platinum(IV) complexes of ammonia and chloride ion.

observations. One of the key aspects relates to shape in these molecules – and it is a wondrously variable area to behold.

3.2.3 Complex Shape – Not Just Any Which Way

It is now over two hundred years since the realization that chemical compounds have a distinctive three-dimensional shape was expressed by Wollaston in a paper published in 1808. As early as 1860, Pasteur suggested a tetrahedral grouping around carbon, and both Kekulé (in 1867) and Paterno in (1869) developed three-dimensional models involving tetrahedral carbon. From 1864, Crum Brown was developing graphical representations of inorganic compounds. By the second half of the nineteenth century, the tetrahedral carbon had been well defined experimentally by van't Hoff and LeBel (in 1874) through optical resolution of compounds. Following this approach, the tetrahedral nitrogen was established (by LeBel in 1891), then other tetrahedral centres, and subsequently other stereochemistries. The use of single crystal X-ray diffraction commenced with the work of William and Lawrence Bragg in the early twentieth century, and from the 1920s was established as the supreme method for determination of the absolute three-dimensional shape of molecules in the solid state.

Shapes of transition metal compounds received limited attention until late in the nineteenth century. From 1875 for several decades, Sophus Mads Jørgensen championed graphical chain formulae for metal complexes, building on an original proposal of C.W. Blomstrand of 1871. For cobalt(III), three direct bonds to the cobalt was assumed by defining a relationship between metal oxidation state and number of bonds, with other groups linked in chains reminiscent of the chains in organic compounds that had been developed a little earlier. Three representations are shown in Figure 3.2, along with modern formulations; for these chain formulae, a key aspect was that $Co-Cl$ bonds were considered unable to ionize in solution, whereas NH_3-Cl bonds could do so. This fitted with the ionization behaviour of the first three compounds presented – a case of a model interpreting (albeit very limited) experimental evidence satisfactorily. However, the nonelectrolyte $CoCl_3 \cdot (NH_3)_3$ could not be fitted at all with this model with its three-bond limit to cobalt(III), with at best a 1:1 electrolyte proposed. Alfred Werner rejected this model, and in a series of studies between 1891 and 1893 introduced two important concepts to coordination chemistry: stereochemistry and 'affinities'. His formulation for $CoCl_3 \cdot (NH_3)_3$ satisfies the nonelectrolytic nature of the compound (see Figure 3.2), and takes the 'modern' shape, also now proven absolutely by structural studies.

Werner's theory is generally seen as the birth of modern coordination chemistry. He proposed two valencies for metal ions: *primary* (ionizable; corresponds to oxidation state) and *secondary* (nonionizable; corresponds to coordination number), with the latter (usually four or six) able to be satisfied by neutral or anionic molecules, and with every element usually satisfying both its primary and secondary valence. Further, he recognized and proposed that these 'secondary valencies' would be distributed around the metal in fixed positions in space in such a way as to lead to the least repulsion between the groups or atoms attached. Subsequently, he devoted himself to finding experimental proof for his theory, using the limited methods available at the time, which included conductivity and resolution of complexes into optical isomers. His suite of studies of six-coordinate compounds, which he argued should exist in many cases as geometric and/or optical isomers, convincingly supported his predictions. By isolating optical isomers of a complex without any carbon atoms present, he also put to rest the view that optical activity was a property of carbon compounds

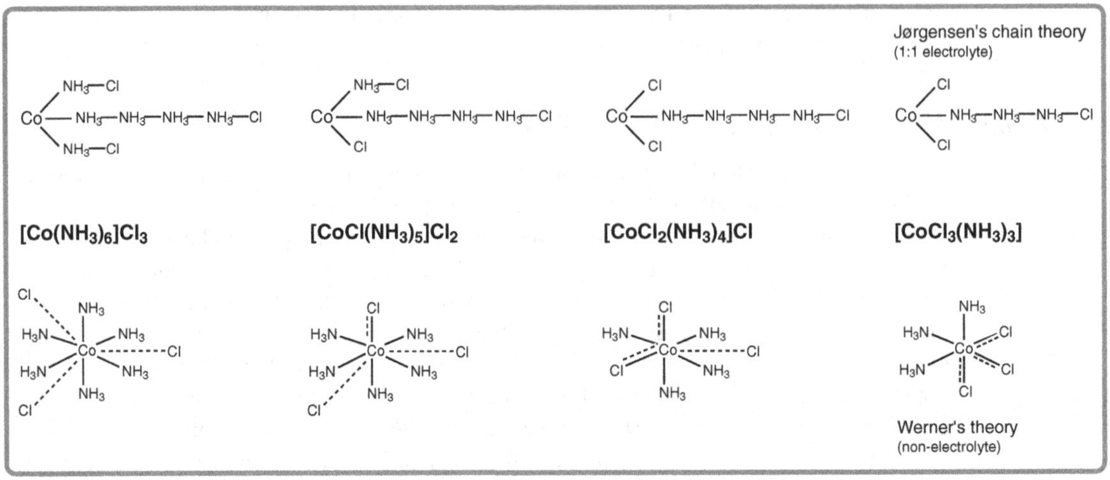

Figure 3.2
Jørgensen's chain theory for some cobalt(III) complexes (top), compared with the Werner representations (bottom). Werner's representation for $[CoCl_3(NH_3)_3]$ fits the nonelectrolyte behaviour seen experimentally, unlike the Jørgensen model.

alone. Werner also proposed in 1912 the concept of a second (outer) coordination sphere through directed but weaker interactions of groups in the first (inner) coordination sphere with species arranged in an outer shell around the main coordination sphere. These effects, involving what we would now call specific hydrogen bonding interactions, were later shown by Pfeiffer (in 1931) to exist through his work on optically active substances interacting with an optically inactive central metal complex, where an effect was induced in the metal complex despite its not being bonded coordinatively to the optically active species.

Werner's early representations of coordination complexes go a long way towards the representations we use today. This is illustrated in Figure 3.2 for what were initially formulated as $CoCl_3 \cdot 6NH_3$ and $CoCl_3 \cdot 4NH_3$. The former is formulated as $[Co(NH_3)_6]Cl_3$, and was assigned a primary valence (oxidation state) of three and a secondary valence (coordination number) of six. The three chloride ions that neutralize the charge on the cobalt(III) ion fully satisfy the primary valence, and the six ammonia molecules use the secondary valence fully, through being coordinated to the metal centre. Because the ammonias fully satisfy the coordination number, the chlorides do not participate in the same way and are disposed further away, as represented in Figure 3.2. This exposes them more to reaction, such as combining readily with silver ion to precipitate AgCl. For $CoCl_3 \cdot 4NH_3$, Werner applied his postulate that both primary and secondary valence seek to be satisfied. Thus, with only four ammonia molecules present, he included two of the three chloride ions as additional components of the coordination sphere to satisfy the coordination number of six. These two then serve a dual function, and satisfy both valencies; Werner represented them by a combined solid and dashed line, as exemplified in Figure 3.2. Unlike the remaining chloride, the two dual-purpose chlorides were considered firmly held and not available for reaction with silver ion to precipitate AgCl. The complex $CoCl_3 \cdot 3NH_3$, also represented in Figure 3.2, is not only a nonelectrolyte in Werner's theory, but unable to release chloride ion to form AgCl. Overall, his model interpreted all experimental observations of the time correctly, including being able to explain how $CoCl_3 \cdot 4NH_3$ can exist in two forms of what

Werner showed were geometric isomers, species with the same formula but different spatial dispositions of groups around the cobalt. Only for the octahedral shape are two isomers predicted for $CoCl_3 \cdot 4NH_3$, matching the experimental observations. Using either a flat hexagonal shape or else a trigonal prismatic shape, three isomers are predicted, supporting (but not proving) the octahedral shape for six-coordination. The proof came from Werner's success in resolving some six-coordinate complexes into optical isomers – something that could occur if they exist in the octahedral shape but not some other proposed shapes.

We now know conclusively that Werner was correct, or at least nearly so. Interestingly, although his octahedral shape is supremely dominant in six-coordination, we now know that there are trigonal prismatic six-coordinate shapes also – albeit with ligand systems to which Werner had no access. What sophisticated modern physical methods tell us conclusively is that coordination complexes don't just happen upon a shape. Each different coordination number known, and these vary from 2 to 14, supports a limited number of basic shapes. The shape, or stereochemistry, depends both on the metal and its oxidation state and on the form of the ligand – each contributes in different ways.

In 1940, Sidgwick and Powell proposed a simple approach to inorganic stereochemistry based on the concept that the broad features of molecular shape could be related to the distribution achieved by all bonding and nonbonding electron pairs (bond pairs and lone pairs) around a central atom as a result of minimized repulsion. A basic but very useful model for shape evolving out of this is the valence shell electron pair repulsion (VSEPR) model. This, an extension of Lewis' ideas, is based on a simple concept – it considers the electron pairs as point charges placed on a spherical surface with the central atom in the centre of the sphere. These charges of identical type will distribute themselves on the surface so as to minimize repulsive interactions. In other words, this is a simple *electrostatic* model. For any set of like charges, there is usually either only one lowest energy arrangement or a very limited set of arrangements of the same or at least very similar energy. This very simple model has proven to be of great value as a predictor of polyatomic molecular shape for p-block compounds, since in these any lone pairs present play an important structural and directional role in defining molecular shape. The concept remains valid decades after its inception. Although developed for main group (p-block) chemistry, the concepts can be adapted to some extent for metal complexes, and was first applied by Nyholm and Gillespie in the late 1950s to coordination complexes, although the strong role of lone pairs found in p-block chemistry is absent in the d block.

For d-block metal ions in particular, it was known that a series of structurally common $[M(OH_2)_6]^{2+}$ complexes form for metal ions with different numbers of d electrons, although from a VSEPR viewpoint one would expect structures to vary since electronic configurations differ – the nonbonding electrons are clearly not playing a major role. Kepert amended the VSEPR model for use with transition metals by ignoring nonbonding electrons, and considering only the set of ligand donor groups, treating them as point charges on a spherical surface as in the VSEPR model and considering repulsions between them. For two to six point charges, the outcomes are represented in Figure 3.3; only for five-coordination are two geometries of essential equal energy predicted. These dispositions of point charge on the surface then can be taken to represent the location of the donor atoms in the attached ligands, joined by coordinate bonds to the metal that is placed at the body centre of the sphere. Once bonds are inserted, this then represents the basic shape of the complex (Figure 3.3). The solid lines shown in the right half of Figure 3.3 represent the actual coordinate covalent bonds; the dashed lines in the left half of the figure only define the shape of the framework and do not relate to bonding at all.

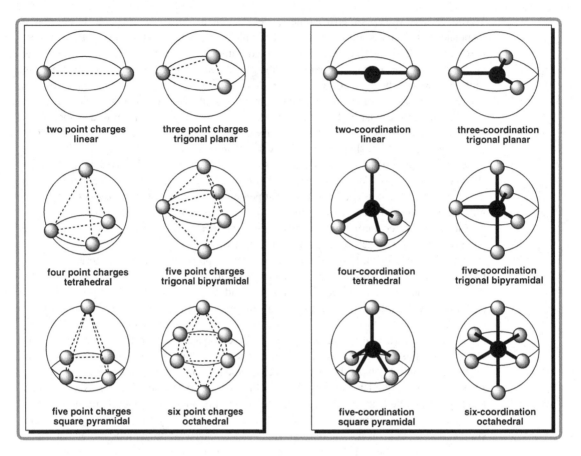

Figure 3.3
Distributions predicted for from two to six point charges on a spherical surface (left), and shapes of complexes evolving from this point-charge model, when a central atom is placed at the core of the sphere and considered to bond to each donor atom located where a point charge occurs.

As with any simple model, it is not universally correct, and for each coordination number, there may be alternate shapes. For example, the predicted shape for four-coordination is *tetrahedral*, but we know that one well-known shape found for four-coordination in metal complexes is *square planar*, a flat geometry where all four ligands lie in the same plane as the metal, disposed in a square arrangement with the metal ion at the centre. Obviously, there are other influences on the shape or stereochemistry of complexes. We shall look at real outcomes for each coordination number in turn in Chapter 4, but at this stage it is sufficient to recognize that there are several basic shapes, most of those being defined perfectly well by the simple model depicted in Figure 3.3 above.

It is also appropriate at this stage to recognize that the way we represent or draw molecules for display and discussion is important for communication of concepts. Technology has provided a number of ways in which molecules can be illustrated (Figure 3.4). For everyday use and discussion between chemists, it is most likely that the simple basic drawing will be used, as it can be hand or computer sketched rapidly. Views with perspective, or else ball and stick models, tend to be met in formal presentations, as will be the case in this book.

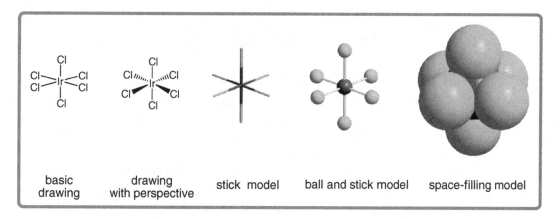

| basic drawing | drawing with perspective | stick model | ball and stick model | space-filling model |

Figure 3.4
Various ways in which metal complexes can be represented, illustrated for the simple octahedral complex ion $[IrCl_6]^{2-}$.

Space-filling models, while giving a view closer to the actual situation for the molecular assembly, are difficult to visualize, even for very simple molecules, because atoms at the front tend to obscure those behind. As a consequence, the ball and stick models are met more often in formal presentations.

3.3 Holding On – The Nature of Bonding in Metal Complexes

Having established the basic concepts of coordination complexes, it is now time to attempt to understand how these complexes hold together, or bond. To pursue this aspect, we need to develop models for bonding that not only provide a satisfactory basis for dealing with the array of shapes that exist, but also can provide interpretation of the spectroscopic and other physical properties of this class of compounds. It is useful to introduce the core concepts and models that we use to interpret observations immediately, as they pervade discussion throughout the field.

The metals of the d block characteristically exist in stable oxidation states for which the nd subshell has only partial occupancy by electrons. This differs from the situation with main group elements, for which it is common or at least more usual to have stable oxidation states associated with subshells that are either completely empty or filled. Importantly, the chemical and physical properties characteristic of the transition elements are determined by the partly filled nd subshells (likewise in the f block, but to a lesser extent, by partly-filled nf subshells). In the simple atomic model of the first-row d-block elements, however, the set of closely-spaced levels involving 4s, 4p and 3d orbitals can be considered as the valence orbitals. This provides the luxury of nine orbitals (one s, three p and five d), giving rise to what is called the nine-orbital (or 18-electron) rule that attempts to explain metal–donor coordination numbers of up to nine. The valence bond model approach to describing complexes is limited, but worth a short review. Valence bond theory, which describes bonding in terms of hybrid orbitals and electron pairs, evolved from the 1927 Heitler–London model for the covalent bond, evolving from the original 1902 concept of covalency proposed by Lewis, and augmented at a later date to include the hybridization concept by Pauling; it is obviously an early model.

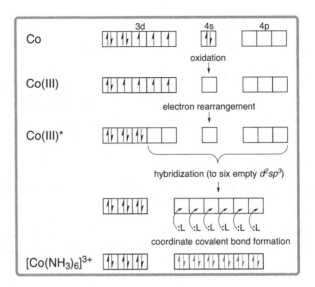

Figure 3.5
A simple valence bond description of bonding for the $[Co(NH_3)_6]^{3+}$ complex ion.

Consider cobalt as an example, forming the cobalt(III) cation and subsequently an octahedral $[Co(NH_3)_6]^{3+}$ complex ion where all Co—N bonds are identical. We commence with our basic model of the coordinate covalent bond, which requires the ligand donor group to supply a lone pair of electrons to an empty orbital on the metal. We can just about deal with this for the cobalt(III) cation, through a somewhat complicated process of ion formation, electron rearrangement, orbital hybridization and filling of empty hybrid orbitals, as depicted in Figure 3.5.

The model relies on energy expended in hybridization and electron relocation being recovered through the set of bond formation processes that occur. It is an adequate model, but rather limited. It can't accommodate, for example, six-coordination in complexes with more d electrons, even though there are many examples known experimentally, without recourse to employing the empty but higher energy 4d levels. The two systems, 3d—4s—4p and 4s—4p—4d, were introduced in the early 1950s by Nobel laureate Henry Taube and are termed *inner shell* and *outer shell* respectively to distinguish them and reflect reactivity differences for species fitted to the different models, but the need to move to the latter higher energy levels in the model is somewhat unsatisfactory. The concept is illustrated in Figure 3.6 for Co(III) in two different d-electron arrangements, as met in $[Co(NH_3)_6]^{3+}$ and $[CoF_6]^{3-}$. The model allows an understanding of magnetic properties relating to the number of unpaired electrons and aspects of reactivity, but deals very modestly with stereochemistry and spectroscopy. When a model meets with difficulty in explaining experimental observations, it becomes time to set it aside and look for alternative, perhaps more sophisticated, models. After all, the only value of a model is that it explains observations – otherwise, it's about as useful as a nose on a pumpkin.

The higher stability of inner shell complexes that employ just the nine 3d—4s—4p orbital set suggests that there may be some special stability associated with the systems that employ nine orbitals that can accommodate no more than 18 electrons. There arose the *18-electron rule*, which suggests that coordination complexes whose total number of valence

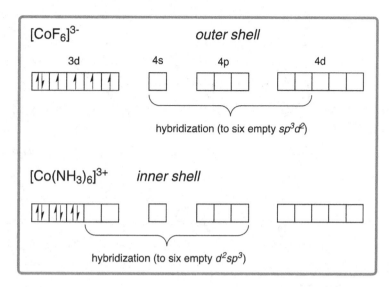

Figure 3.6
The extended valence bond description of *inner shell* and *outer shell* bonding for octahedral cobalt(III) complexes, required as a result of different *d*-electron arrangements on the metal. The six empty hybridized orbitals can in each case accommodate six bonding lone pairs from six ligand donor atoms.

electrons approaches or equals 18 (but does not exceed it) are more stable assemblies, with those actually achieving 18-electron sets being most stable. This rule works best for organometallic compounds, which are more covalent in character, whereas Werner-type compounds, with more ionic character, tend to break the rule more often, and consequently it is much less used for these. It may be helpful to illustrate it in operation, nevertheless. We shall do this with the simple organometallic compound, octahedral [Mn(CO)$_5$I]. There are two methods of electron counting: the closed shell (or oxidation state) method and neutral ligand (or formal charge method). Both exist because accounting for electrons is not always straightforward. Employing the former method as an illustration, [Mn(CO)$_5$I] may be considered as composed of Mn$^+$, five CO and one I$^-$ ligands. Each ligand contributes two electrons from a lone pair to the complex, and the metal contributes its valence electrons. This amounts to 6 electrons (for Mn$^+$), 10 electrons (for 5 × (CO)) and 2 electrons (for 1 × I$^-$), a total of 18 electrons. This suggests that [Mn(CO)$_5$I] should be stable, which is the case. This 'rule' more correctly acts as a guideline for stability, but we shall not dwell on it here, given our focus on Werner-type rather than organometallic-type complexes.

Here, it is perhaps valuable to remind ourselves of the nature of the d orbitals that we are employing. The five d orbitals (d_{xy}, d_{yz}, d_{xz}, $d_{x^2-y^2}$ and d_{z^2}) occupy different spatial orientations and involve two different basic shapes. These orbitals are shown in Figure 3.7. For elements of the f block, there are seven f orbitals (f_{xyz}, $f_{z(x^2-y^2)}$, $f_{y(z^2-x^2)}$, $f_{x(z^2-y^2)}$, f_{x^3}, f_{y^3} and f_{z^3}) which offer three different basic shapes and also, like the d orbitals, occupy quite different spatial positions. While we will not be discussing f orbitals in any detail, it is useful to be reminded that they do many of the things that d orbitals do, except that the 4f orbitals of the lanthanoids, for example, are less 'exposed' due to being screened by larger 5s, 5p and 6s shells, which leads to weaker interactions with their surrounding ligands and very similar chemistry across the series. The *n*d orbitals are more exposed, and hence more

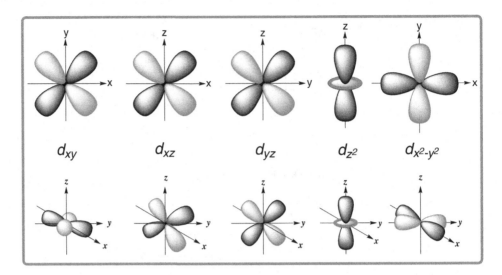

Figure 3.7
The five d orbitals of d-block elements, each represented in two different views.

involved in their element's chemistry. It is the set of five d orbitals that will be of particular interest in this book. To start to understand how they may differ, notice that two of the five d orbitals have lobes lying *along* axes of the defined coordinate system, whereas for the other three the lobes lie *between* axes. It is these two classes of spatial arrangement that are the key to much of the discussion that will follow in coming sections – suffice to say that this spatial difference will be significant.

Another way of viewing complex formation is to invoke a purely ionic model. In this, we recognize that the metal at the centre of coordination complexes is usually a cation, and that ligands are either anionic or else have regions of high electron density on the donors that make them attractive to the cation. Thus, we can conceive of a situation where a set of donors is arranged around a metal ion centre in an array defined mostly by their electrostatic attraction to the metal ion and electrostatic repulsion towards other ligands. While very limited in what it says about the metal–donor interaction, this ionic model does offer a useful and surprisingly successful way of predicting and understanding geometric shape in complexes, as we have seen in Figure 3.3, and we shall return to this in detail in Chapter 4.

It is at first sight an eccentricity of metal coordination compounds that we can invoke either a covalent bonding model or an ionic bonding model. However, if we think of ionic and covalent bonds as the two extremes of a bonding continuum, the apparent contradiction becomes more acceptable. It's almost a state of mind; we would rarely consider CH_4 as anything but covalently bonded, but at the same time recognize that representation as CH_3^- and H^+ at least provides a way of understanding some nucleophilic and electrophilic organic reactions. We are likely to be reasonably comfortable in considering a neutral compound such as $[Mo(CO)_6]$, composed formally of uncharged Mo(0) bonded to the carbon atoms of neutral CO molecules, as covalently bonded. When presented with a highly ionic assembly, such an $[Mn(OH)_6]^{2-}$ (stable only in very strongly basic solution), we can at least accept there is some wisdom in thinking of this as a Mn^{4+} ion with six hydroxide anions attached by strong ionic bonds. What is common to both approaches to bonding in coordination chemistry is that the bonding models tend to be holistic in nature rather than focused on individual atom-to-atom bonding; this is probably a better description of systems where a set of entities are assembled to form a much larger whole. Two theories have developed to explain the properties of metal complexes. For the former example, the molecular orbital theory represents an appropriate treatment of holistic covalent bonding, whereas for the latter example, the crystal field theory (CFT) is a useful holistic ionic bonding model. Both models have obvious limitations – not least of all an assumption of 'purity' in the bonding character that is unlikely to be so. But then, these are models, after all; complexes don't need either theory to go about their business. Our models are developed not only to allow some level of understanding of physical and chemical properties of complexes, but also to provide a means of prediction of properties where changes or new species are involved. A model is only useful if it performs the tasks we ask of it.

3.3.1 An Ionic Bonding Model – Introducing Crystal Field Theory

Crystal field theory for transition metal complexes simplifies the complex species to one featuring point charges involved in purely ionic bonds. The theory focuses on the usually partly filled valence level d sub-shell, with a key recognition being that the fivefold energy degeneracy of the d subset present in the 'bare' metal will be lost in a ligand environment, leading to nonequivalence of energies amidst the nd orbitals. This then demands consideration of how the electrons occupy these levels, since variation in this occupancy can be associated with variation in properties. Reducing the inherently spherical metal ion to a point positive charge appears less of an approximation than reducing ligand donor groups each to a point negative charge that represents the electron pair. Ignoring ligand bulk and shape aspects would seem a big leap, but as it turns out the model does yield useful outcomes in terms of interpreting physical properties such as colour and magnetic properties in metal complexes. Bonding is considered to arise dominantly through electrostatic forces between cation and anion point charges located in the ionic array. Unfortunately, while responsible for a stable assembly, these primary electrostatic interactions have little direct influence on properties. Rather, it the effect of ionic interactions on the set of valence electrons that is the key. For elements of the main group this includes p orbitals; for most transition metal ions encountered, these are purely the set of d electrons; in the lanthanoids, we would be dealing with the f-electron set in a similar way.

The CFT developed as the earliest theory from Bethe's work published in 1929, based on a group theory approach to the influence on the energy levels of an ion or atom upon lowering symmetry in a ligand environment. First developed to interpret paramagnetism, it was subsequently applied more widely in the development of the field of coordination chemistry in the 1950s. This led to identification of shortcomings that were addressed by the development of the more versatile ligand field theory (LFT), which includes recognition that both σ and π bonds can occur in complexes as in pure organic compounds. The two are sufficiently closely related in terms of their core operations involving the d orbitals for them to be often used (albeit inappropriately) interchangeably. While deficient in some aspects, the CFT remains a useful theory for qualitative and limited quantitative interpretation. What is perhaps most remarkable about this theory is that it works at all, given the simplicity and level of assumptions – but it does. Further, it has stood the test of time to remain adequate despite the vast changes in the field over the decades since its evolution.

What we are doing in taking this approach to developing a model is mixing a purely ionic model with an atomic orbital model. The valence orbitals of the central atom are assumed to be influenced by the close approach of the ligands acting simply as points of negative charge. We can start considering the way the model operates by employing for illustrative purposes a set of p orbitals, since here we have a simple case where each p orbital lies along one axis of an imposed three-dimensional coordinate system (Figure 3.8). What we know from the simple atomic orbital model is that we can represent the three empty p orbitals as equal in energy. If we surround these bare p orbitals with a symmetrical 'atmosphere' of negative charge associated with the presence of ligand donors, there is no obvious change because our p orbitals carry no electrons yet. Now, consider what happens if we insert an electron into the p-orbital set, and allow it to occupy any orbital. A coulombic repulsion will occur between the inserted electron and the surrounding 'atmosphere' of negative charge that is identical regardless of the orbital it occupies, raising the energy of the set of p orbitals equally. This is the so-called spherical field situation (very occasionally called a steric field). The next step is to introduce some directionality into the interaction, by restricting the negative charge of the ligand to localities along one specific direction, say equidistant from the metal in each direction along the z axis. If we place our p electron in the p_z orbital, the electrostatic interaction will be stronger than if we were to place it in either the p_x or p_y orbital, because of the greater distance between the orbital lobes and the spatially restricted negative charge in the latter cases. The consequence is that an electron in the p_z orbital will feel a stronger interaction than one in the p_x and p_y orbitals, resulting in loss of degeneracy, the latter two orbitals becoming lower in energy that the former orbital; moreover, the latter two are also of equal energy because they are spatially equivalent relative to the fixed negative charge location. The outcome is an energy diagram as shown in Figure 3.8. The p_z orbital has been raised in energy relative to the spherical field case (commonly taken as the zero reference point), whereas the p_x and p_y orbitals are lowered in energy relative to this reference point. Moreover, the model actually leads to the p_z orbital being raised twice as much as the two other orbitals are lowered, so that the overall effect is energy-neutral, in the absence of insertion of electrons into the assembly. The spatially-defined negative charge field introduced has removed or lifted the degeneracy of the p orbitals that existed in a pure spherical field. The situation we have described is not met in reality, since the central atom in a coordination compound does not normally

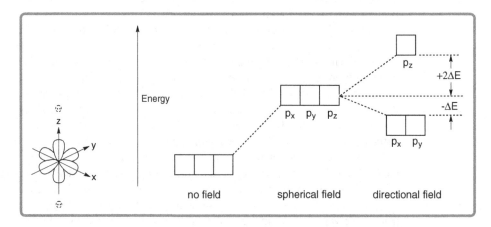

Figure 3.8
The concept of crystal field influences applied for purely illustrative purposes to a set of p orbitals. The directional field here results from imposing a specific interaction along the z axis alone.

employ only a set of p orbitals as its valence orbital set. Usually, we are dealing with central atoms with a set of d orbitals as the highest occupied energy levels employed.

The situation in the case of d orbitals is somewhat greater in complexity because there are now five orbitals involved, and these orbitals have distinctly different characteristics. Whereas all p orbitals look the same and have orbitals directed in an identical manner along an axis, this is not the case with d orbitals. There are a set of 'triplets': the d_{xy}, d_{xz} and d_{yz} orbitals, that each look like a four-leaf clover and whose lobes point between axes of a three-dimensional coordinate set centred on the metal ion. Another orbital ($d_{x^2-y^2}$) looks identical, but locates its orbital lobes along the x and y axes. The final member (d_{z^2}) is unique in appearance, and locates its major lobes along the z axis. Despite these differences, they form, in the absence of any field, a degenerate set – or put simply, they are all of equal energy. In a symmetrical spherical field of charge density the outcome is no different from the initial behaviour for the p orbits; simply, all orbitals are raised in energy equally, remaining degenerate. However, when point charges are introduced in specific locations, this degeneracy is removed. In a situation where six point charges are introduced symmetrically along the x, y and z axes, each equidistant from the central ion at the vertices of an octahedral array (perhaps the most common shape met in coordination chemistry), an entirely different outcome to the spherical field situation results. Because the $d_{x^2-y^2}$ and d_{z^2} orbitals have lobes pointing along the axes, they interact most strongly with the ligand point charges, and their energy is raised as a result of this. The interaction for the other three is significantly less – less, in fact, than they felt in a spherical field since there is no charge density close now – so that they effectively drop in energy versus the spherical field. The three orbitals that point their lobes between axes, d_{xy}, d_{yz} and d_{xz}, are shaped identically and oriented in the same manner relative to a different pair of axes; consequently, it is hardly surprising that they experience identical effects and remain degenerate in the octahedral field. However, it is notable (Figure 3.9) that the $d_{x^2-y^2}$ and d_{z^2} orbitals also end up as degenerate despite their apparent differences in shape. How can this be? While it could be said, simply, that it just works out that way, this is hardly a satisfying outcome. The answer lies deep within the mathematical derivation of the atomic model and orbital functions, and mathematical detail is something we are doing our best to avoid here. However, there is a certain simplicity and logic here that we can inspect without recourse to mathematics. It

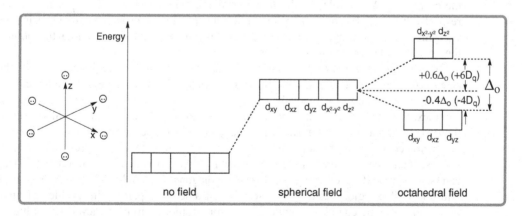

Figure 3.9
Crystal field influences for a set of d orbitals in an octahedral field.

turns out that the d_{z^2} orbital is in reality a linear combination of two orbital functions, $d_{z^2-y^2}$ and $d_{z^2-x^2}$ that are shaped just like the separate $d_{x^2-y^2}$ orbital, but oriented differently. It is then less surprising that the composite we see as the d_{z^2} orbital is energetically identical to the $d_{x^2-y^2}$ orbital. But if you're now wondering why these orbital functions need to be combined in the first place – well, that's a story you simply won't find here.

The outcome of the introduction of a defined spatial arrangement of point charges, here an octahedral field, is that the degeneracy of the five d orbitals is removed, and the orbitals arrange themselves in new sets of differing energy. The d orbitals pointing directly towards the ligands ($d_{x^2-y^2}$ and d_{z^2}), and thus along what are the M–L bonds, can be considered to be involved in a traditional σ bond; they sometimes are referred to as d_σ orbitals. The three remaining and more stable orbitals (d_{xy}, d_{yz} and d_{xz}) point away from the M–L bond direction, but may possibly be capable of involving themselves in π-type bonding, and as a consequence are sometimes called d_π orbitals. However, this view is mixing conventional covalent bonding thinking with a purely ionic model, and thus is problematic, and we shall largely avoid it.

The formation of doubly and triply degenerate sets of orbitals is a characteristic of the octahedral field. Because it is the spatial location of the set of point charges that is significant in generating this outcome, it will hardly come as a surprise to find that *every different shape arrangement of point charges will lead to a different characteristic outcome* – but more on that later. At present, focusing on the octahedral field, the outcome shown in Figure 3.9 applies. The lower energy set of three orbitals (d_{xy}, d_{yz} and d_{xz}) is called the *diagonal* set (or, applying mathematical group theory, which we shall not develop here, also called t_{2g} where *t* stands for triplet degeneracy), and the higher energy set of two orbitals ($d_{x^2-y^2}$ and d_{z^2}) is called the *axial* set (or from group theory, e_g, a doublet level), the names relating to where the orbitals point versus imposed axes. The energy difference between these levels is, compared with the differences between atomic orbital levels generally, relatively small, in line with the influence of the ligand set being a relatively modest one. This energy difference is called the crystal field splitting, represented by a parameter termed Δ_o (where the subscript 'o' is an abbreviation for 'octahedral'). An energy balance between the two sets of orbitals is struck, so that the three lower levels are $-0.4\Delta_o$ lower and the upper two levels $+0.6\Delta_o$ higher than the spherical field position set as the zero reference point. The language of chemistry contains several dialects, and you will find some texts refer to the energy gap Δ_o as $10D_q$ (with the diagonal and axial sets lowered $4D_q$ or raised $6D_q$ respectively); don't be confused, as the same conceptual model is being applied. Using the Δ_o symbolism is more appropriate, since it carries some additional information that defines the type of field operating in the subscript.

One key question that begs an answer at this stage is simply what factors govern the size of Δ_o? Answering this should give us the satisfaction of being able to predict certain spectroscopic properties. Obviously, since we are involving both metal and ligand in our complex, we can anticipate that both have a role to play. If we fix the metal's identity, then we can focus on the ligands, and their particular properties that influence Δ_o. For octahedral complexes of most first-row transition metal ions at least, the presence of colour suggests that somehow part of the visible (white) light spectrum is being removed. This can be envisaged if the energy gap between the diagonal (t_{2g}) and axial (e_g) levels equates with the visible region, leading to absorption of a selected part of the visible light that occurs to cause the complex to undergo electron promotion from the lower to the higher energy level. Simply by monitoring the change in colour as ligands are changed, we can determine the energy gap Δ_o applying for any ligand set. The stronger the crystal field, the larger Δ_o and hence the more energy required to promote an electron, leading to a higher energy transition, seen experimentally as a shift of the absorbance peak maximum to shorter wavelength. This allows us to rank particular ligands in terms of their capacity to separate the diagonal and axial energy levels. This has been done to produce what is called the *spectrochemical series*. For some common ligands that bind as monodentates to a single site, this order in terms of ability to split the diagonal and axial energy levels apart is of the form:

$$I^- < Br^- < \underline{S}CN^- < Cl^- < S^{2-} < NO_3^- < F^- < HO^- < OH_2 < \underline{N}CS^- < NH_3$$
$$< NO_2^- < PR_3 < CN^- < CO.$$

Notably, this order is just about independent of metal ion in any oxidation state or even adopting any common geometry.

The problem is that this experimentally-based trend does not sit comfortably with the crystal field model. For example, this trend indicates that an ion like Br^- is far less effective than a neutral molecule like CO at splitting the d-orbital set, which is at odds with a model based on electrostatic repulsions, in which one might expect a charged ligand to be more effective than a neutral one. Of course, electronegativities, dipole effects and even size may be invoked as contributing to the difference, but these are hard to see as overriding effects, particularly when the O-donor anion HO^- turns out to be a weaker ligand than its larger, neutral O-donor parent water (H_2O), and ammonia is a stronger ligand than water despite having a smaller dipole moment and larger molar volume. Something is not quite right with the CFT world.

The CFT, while useful, simply suffers most from being based on a concept of point charges. Real ligand donors have size – and along with size comes the strong possibility of the donor group or atom undergoing deformation of its electron density distribution simply as a result of being placed near a positive charge centre. In effect, you can think of this as a shift of electron density towards the region between the metal and the donor, or what we would think of as happening when a covalent bond forms. While the outcome here is far from covalent bond formation, there is an introduction of some covalency into the otherwise purely ionic bonding model – think of it as dark grey, rather than purely black. In fact, if we see black and white as the two extremes of ionic and covalent bonding, rarely is one of the extremes applicable; it seems chemistry, like life, is full of compromise. The outcome of allowing some covalency in the model is a fairly minor perturbation, not a drastic change – except the theory morphs into *ligand field theory* to distinguish the changes introduced.

3.3.2 A Covalent Bonding Model – Embracing Molecular Orbital Theory

It's perhaps a bit confusing to assert that such a thing as coordinate covalent bonds exist, and then to introduce a theory based on purely ionic interactions. Even clawing back some of the concept by asserting mixed bonding character seems a grudging concession. So what's so wrong with a purely covalent model? One of the most obvious answers is that, by its localized atom-to-atom nature, covalent models deal poorly with both shape and interpretation of spectroscopic observations. Since the value of any model lies in its applicability in the interpretation of physical observations, not achieving this tends to limit its attractiveness. It's a bit like sun worship compared with space physics; interpretation in the former is dominated by belief, in the latter by scientific facts and models. Sun-worship is fine if you're so disposed, except changing to moon-worship then becomes a much bigger decision than simply applying scientific methodology to a different celestial body. Scientists should be sufficiently flexible that they are able to accept that a system is capable of sustaining several models, whose usefulness depends on the demands placed on them. Thus there is room for a simple covalent model – but it will have limitations.

Let's revise the simple valence bond model discussed earlier. For our main focus of current attention, the octahedral shape, there are several small molecules or molecular ions of this shape that tend to be regarded as simple covalent compound rather than coordination compounds, such as PF_6^-, for which an adequate bonding model exists. We can draw on this as we develop a model for an octahedral metal complex, ML_6. With a pure covalent model, six M—L bonds require six appropriate orbitals on the central metal atom with which a lone pair of electrons on each of six ligands can interact. The valence shell of a d-block metal can be considered to consist of the five nd orbitals, as well as the energetically similar $(n+1)$s and three $(n+1)$p orbitals – a total of nine orbitals. Since physical observations show that all six bonds in an ML_6 system with one ligand type are identical, it is necessary to have involved not only six orbitals on the metal, but six identical orbitals. This is achieved by hybridization, using the one s, three p and two of the five d orbitals to form six equivalent d^2sp^3 hybrids (see Figure 1.2 for an example of this), each capable of accepting a pair of electrons – if empty. Therein lies a problem, as the d orbitals at least may contain electrons, and perhaps too many to reside out of the way in the three nonparticipating d orbitals. One can make a virtue out of a necessity by asserting that the number of d electrons limits the number and type of coordinate covalent bonds that can form, and there is certainly a relationship of sorts here, or else use the next highest set of (empty) d orbitals. But it's all rather unsatisfactory.

A somewhat more sophisticated approach to this description (though still not capable of dealing fully with spectroscopic and magnetic properties) is to look at a more 'molecular' approach, by considering six of the nine available bonding orbitals on the metal interacting or mixing with the six lone pairs on the ligands to form six bonding and six antibonding orbitals. In effect, we can visualize the bonding component in this case essentially as in the simple model discussed above. However, we are moving to a molecular orbital (MO) approach to bonding, which is a holistic approach to bonding. The key to molecular orbital theory is that it involves combinations of orbitals from components to produce a new set of orbitals of different energy from those of the separate components. However, importantly, the number of orbitals in the molecular assembly exactly matches the number in the components. In the simplest case, where we are combining one orbital from atom A and one orbital from atom X, we produce two orbitals; one is stabilized versus the parent orbitals (that is, it is of lower energy), the other destabilized (that is, it is of higher energy). The former is called the bonding orbital, the latter an antibonding orbital; these names reflect their character, as electrons inserted in the former lead to a more stable situation when A and X are linked by a bond, whereas electrons in the latter destabilize the assembly. An excess of electrons in bonding compared with antibonding orbitals leads to a stable compound AX. The model can accommodate σ and π bonding, as well as orbitals in some cases that do not participate directly in bonding (called nonbonding orbitals).

These concepts extend to joining molecular species to form a new bond, and an example for a simple coordination compound F_3B—NH_3 is shown in Figure 3.10, featuring only the relevant two 'frontier' orbitals of the BF_3 and NH_3 molecules involved in formation of the coordinate bond. It is not uncommon, not only simply for ease of visualization but also because the energy levels of frontier bonding orbitals are central to the theory, to restrict a MO diagram to solely depicting 'orbitals of interest'. In this case, the pair of electrons

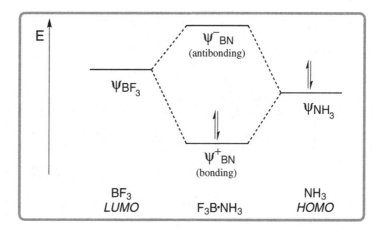

Figure 3.10
A molecular orbital diagram for $F_3B{-}NH_3$, reporting only the orbital interaction of the lowest unoccupied molecular orbital (LUMO) on the precursor molecule BF_3 and the highest occupied molecular orbital (HOMO) on the precursor molecule NH_3. Only if there is useful overlap of the orbital contributions from each precursor are the new orbitals formed.

originally in the N orbital occupy what we can consider as the bonding MO, leading to a more stable situation for the B–N assembly (the electrons occupying a lower energy orbital compared with the energy of the orbital on the original N centre), and hence a stable bonding outcome. In more complicated metal complexes, it is again convenient to consider just the one relevant orbital on each donor molecule or atom (effectively the lone pair in the valence bond description) together with the valence orbitals on the central metal in building the MO diagram.

In our ML_6 model, there were originally nine orbitals of similar energy on the metal considered (five d, one s and three p), yet to form bonds to six ligand orbitals only six are required. This means that three are not involved in forming bonds to the ligands, and are thus designated as nonbonding orbitals (Figure 3.11). While it is not necessarily a good idea to always seek linkages between models, these nonbonding levels effectively correspond to the t_{2g} set of the ionic model already discussed. Overall, 15 'orbitals of interest' appear in the diagram, capable of accommodating up to 30 electrons. Like all MO models, the number of participating orbitals in the components (the metal and six ligand orbitals here) must equal the number of MOs in the assembly (ML_6 here). Moreover, equal numbers of bonding and antibonding orbitals, the former of lower energy and the latter of higher energy than the parent orbitals, must form.

The 12 electrons from the six ligands lone pairs are used to occupy the six bonding molecular σ-orbitals created (defined by their symmetry labels as a_{1g} (singlet), t_{1u} (triplet), and e_g (doublet)), which lie lower but fairly close in energy to the initial ligand orbital energy levels, with the metal d electrons then inserted in the next highest energy levels (t_{2g}, $e_g{}^*$) of the molecular assembly. This means that insertion of d electrons commences in the model at the nonbonding orbitals derived from the original d-orbital set. This set of three nonbonding levels along with the six bonding orbitals provide an upper limit of 18 electrons before antibonding orbitals must be employed, which would then lead to an inherently reduced stability for the system. Moreover, there are some benefits for some types of complexes in having this nonbonding set filled fully, with filling leading to some favourable

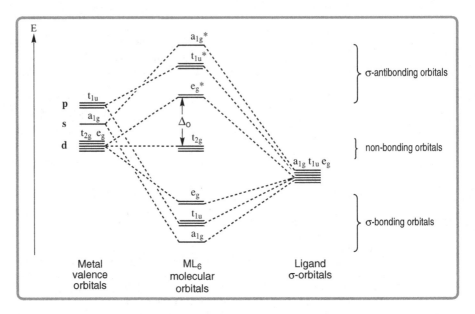

Figure 3.11
Molecular orbital diagram for an octahedral ML_6 complex.

lowering of their energy. This gives rise to the so-called '18-electron rule', met particularly in organometallic chemistry, and discussed earlier in Section 3.3. The model is rather unsophisticated in assuming that sixfold and threefold degeneracy, for the participating ligand and nonbonding d orbitals levels respectively, actually exists, and a higher order treatment is necessary to accommodate this deficiency and allow interpretation of properties.

The molecular orbital theory, as a holistic model, is based on the recognition that it is not essential to limit bonding to linkages between only two atoms at a time. Rather, it allows a MO to spread out over any number of atoms in a molecule, from two to even all atoms. This is a complex and mathematically intensive theory, which is not explored here in depth. It applies mathematical *group theory* to determine the allowed combinations of metal orbitals and ligand orbitals (or orbital overlap) that lead to bonding situations. One example of a combination involving many centres is the overlap of the metal s orbital with all six of the lone pair orbitals from the six ligands at once. This seven-centre combination produces both a bonding molecular orbital (designated as a_{1g} in group theory) and a complementary antibonding orbital (designated as $a_{1g}{}^*$); a representation of the bonding orbital is shown in Figure 3.12.

The three p orbitals differ only in spatial orientation, and it is hardly surprising to find their interactions with donor orbitals are energetically equivalent, leading to a set of three degenerate levels (designated as t_{1u} in Figure 3.11). The five d orbitals interact as two sets due to their spatial arrangement into two groups – those lying along axes and those between axes, as was the case in CFT. The outcome for the $d_{x^2-y^2}$ and d_{z^2} orbitals interacting with ligand orbitals is the formation of a degenerate set of two bonding levels (designated as e_g) and of two degenerate antibonding levels ($e_g{}^*$). A key to the formation of allowed bonding interactions in terms of a pictorial concept is orbital orientation – the lobes of the orbitals need to point along a line joining the atomic centres. This is readily achieved with a $d_{x^2-y^2}$ orbital, for example, but simply cannot occur with an orbital like d_{xy} due to its spatial

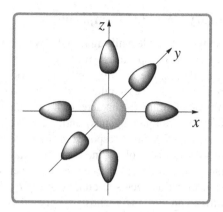

Figure 3.12
Model, for octahedral symmetry, of the components of the a_{1g} bonding molecular orbital formed from overlap of the metal s orbital (centre, of symmetry a_{1g}) with the six a_{1g} ligand group orbital (LGO) lobes formed considering all six ligand lone-pair orbitals. A set of three degenerate t_{1u} LGOs match to the degenerate set of three p orbitals, and a set of two e_g LGOs match to the two degenerate e_g d orbitals. Overall, six LGOs combine with six of the nine available s, p, d metal orbitals, forming six bonding and six antibonding molecular orbitals. The metal's set of three t_{2g} orbitals are nonbonding.

orientation relative to the lone pair orbitals (Figure 3.13). Thus, in this model, the three remaining d orbitals with lobes oriented between the axes become nonbonding orbitals, as a degenerate set (designated as t_{2g}). The overall bonding model is shown in Figure 3.11. This is somewhat similar to that evolved from the valence bond approach in Figure 3.5, but the six bonding and antibonding levels do not form a degenerate set here. Further, note that the t_{2g} and e_g* levels correspond to the t_{2g} and e_g levels of the crystal field model splitting diagram, although arising through quite different concepts. It may be a bit confusing, but may also be comforting to find that there is some common ground between the models. This certainly supports the mixing in of ionic and covalent concepts in 'hybrids' such as LFT.

What has been done above for the octahedral geometry can likewise be applied to other shapes, but it is not the intent to labour the point by pursuing these here. What we have established is another viable model for dealing with metal complexes, albeit based on simple σ-bonding concepts; we will return to look beyond this level later.

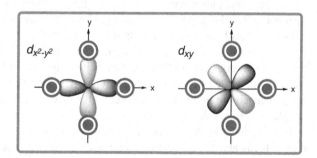

Figure 3.13
Orbital orientation as a contributor to effective molecular orbital formation: effective $d_{x^2-y^2}$-orbital interaction versus ineffective d_{xy}-orbital interaction. The circles represent ligand orbitals located equidistant from the metal centre in each case.

3.3.3   Ligand Field Theory – Making Compromises

A problem we left a little while ago in CFT was how to explain the experimentally observed order of ligands in the spectrochemical series, and issues such as water being a better ligand than hydroxide and the neutral carbon monoxide being such a strong ligand. Molecular orbital theory has given us another perspective on the d-orbital set, with the t_{2g} level seen as dominantly nonbonding in character (which means that its energy should not be influenced particularly by changing ligands) and the $e_g{}^*$ level thus carrying the responsibility for altering the size of Δ_o as it responds to the presence of different ligands. Obviously, with just two energy levels involved, increasing the size of Δ_o can be done in two ways – either by raising the energy of the $e_g{}^*$ level or else by lowering the energy of the t_{2g} level. Any action in the latter direction seems at odds with the t_{2g} level being nonbonding in character, or having little to do with the ligands. The crystal field model focused on raising or lowering of the e_g level through electrostatic interactions between ligands and metal d electrons. Fortunately, the molecular orbital theory can provide us with a means for understanding how the t_{2g} level may also be influenced and its energy manipulated by the ligands. If we return to the spatial orientation of metal orbitals in the t_{2g} set, we are reminded that they point not along the axes associated with ligand positioning, as for the 3d $e_g{}^*$ orbitals or even 4s and 4p orbitals, but between the principal axis directions. Therefore, they can have little to do with traditional σ-bonding (as, by definition, a σ-bond is defined as one where electron density is enhanced in a direct line joining the two atom centres, i.e. metal and donor atom in the present case). The key to any interaction involving t_{2g} d orbitals is whether the ligand donor has p or even π orbitals directed orthogonal (sideways-on) to the metal–donor bond direction. If this does occur, a further interaction between metal t_{2g} d orbitals and ligand donor p (or π) orbitals can arise. In the usual MO manner, interaction of a single d orbital and a single p orbital would lead to two MOs, one bonding (π, or in-phase) and one antibonding (π^*, or out-of-phase) orbital of lower and higher energy than the parent orbitals respectively – a mechanism for manipulating the energy of the t_{2g} set has thus been established, and one that depends on the properties of the ligand. This is developed and exemplified in Section 3.3.4.

This capacity to model and account for the interaction of ligands capable of additional π-bonding with metals provides an enabling mechanism for dealing with the experimental spectrochemical series positioning of ligands. Of course, not all ligands will have the capacity to undergo further π-type interactions with metal orbitals. For example, ammonia has but one lone pair directed to σ-bonding, and otherwise three internal N—H σ-bonds; it cannot undertake any further bonding. However, carbon monoxide (C≡O) has an array of π-bonds resulting from its multiple bond character, and is a clear candidate for further π-type interactions (it is an example of a π-acceptor or π-acid ligand, a group of key importance in organometallic chemistry, with empty orbitals of symmetry appropriate for overlap with a filled d_π, or t_{2g}, metal orbital). The scene has now changed, so that we have greater capacity to understand and predict the effects of ligands on the chemistry of metal complexes – this is the strength of the LFT.

What we have in the end is a reasonably consistent set of models, but ones that differ in their focus and assignment of importance to electrostatic and covalent character. What is the 'real' situation, and how can we effectively assess the contribution of each component? A key and indeed fairly simple approach to this came forward from Pauling, who asserted that metals and ligands would adopt as far as practicable nett charges close to zero, through metal–ligand bond polarization or π-type metal-to-ligand or ligand-to-metal electron density donation. As a consequence of this concept of *electroneutrality*, a metal ion in a high oxidation state with high formal positive charge would seek to involve itself in ligand-to-metal charge donation (i.e. acquisition of negative charge), whereas one in a low oxidation state may do the reverse. Support for this concept, experimentally, is to notice that a high-charged metal ion such as Mn(VII) finds itself dominantly with O^{2-} ligands, whereas lower-charged Mn(II) is satisfied with OH_2 ligands.

What we have seen so far is that bonding between metals placed centrally in an array of atomic or molecular entities that participate in coordinate bonding can be represented by models of eventually high sophistication. Most of our understanding and interpretation of physical observables draws on these models, and it their success in allowing us to interpret what we see that keeps the models in use. But a model is imperfect because it is a model – and so higher levels of sophistication will reinvent or replace these models as time goes by. At present, at least, and for an introduction to the subject, they more than suffice – indeed they have proven remarkably resilient and effective, despite the extraordinary growth in coordination chemistry over the decades since their development. We shall examine bonding in terms of dealing with different geometries in the next section.

3.3.4 Bonding Models Extended

As we have seen above, CFT in its basic form is an ionic bonding model, since it represents ligands simply as negative point charges, which collectively create an electrostatic field around the central metal ion. It links well to the electrostatic VSEPR model that predicts shape. The electrostatic field perturbs the central ion environment, and every d orbital will be raised in energy relative to the prior free-ion situation. In effect, the potential field can be divided into two contributions: a spherically symmetrical component, which affects every d orbital equally, causing an equal increase in energy for each one; and a symmetry-dependent component, where the location of the ligands is paramount, and which may influence each orbital differently, leading to a splitting of the original fivefold degeneracy of the d-orbital set. Conventionally (and conveniently), it is common to represent only the effect of the nonsymmetric component, defined relative to the orbital energy of the degenerate set in the spherical field (see Figure 3.9). Since we are commonly dealing with differences between levels, this is entirely appropriate. The perturbation theory that governs the splitting diagrams has a key requirement which is that, for the set of empty orbitals, the sum of the energies of orbitals when including the presence of a perturbing field (or set of ligands of defined arrangement) must equal that of the orbitals in the spherical field alone. This outcome is achieved by the summed energy of orbitals reduced in energy versus the symmetrical field position needing to equal the summed energy of orbitals increased in energy. Of course, an empty orbital (a virtual orbital) is a rather odd concept to deal with, since we tend to consider orbitals as really only meaningful when they contain electrons. Perhaps, grasping for a macroscopic example, think of a set of orbitals like compartments in a CD rack – the individual compartments only become important to you when they contain a CD, yet you are certainly aware of the full array available, as they define the rack.

Crystal field theory can deal with the observed splitting between the d-orbital sets increasing with increasing oxidation state of the central metal atom. Since the ionic radius of an ion decreases as ionic charge (which equates with oxidation state for a metal ion) increases, the surface charge density increases, metal–ligand bond distances (r) decrease and the splitting Δ_o (which varies with r^{-5}) increases. However, CFT has some obvious limitations, which led to development of alternate models. It cannot easily explain why Δ_o

is invariably larger for 4d and 5d elements versus 3d elements, nor can it provide a really satisfactory interpretation of the variation of Δ_o with ligand type. While there is an array of experimental outcomes from a range of methods that can be, at least qualitatively, fitted to the d-orbital splitting model of CFT, quantitative calculations based on this electrostatic model tend to fail to match experimental values. However, the key failing is that a purely ionic description of bonding in coordination complexes is not consistent with the now abundant experimental evidence. Put simply, electrons in d orbitals, which are fully located therein in the CFT model, in fact spend part of their time in ligand orbitals (they are 'delocalised', or there is covalency in the bonding). The spectroscopic method of electron spin resonance, which 'maps' unpaired electron density, indicates that 'pure' d orbitals in fact can have their electrons appreciably delocalized over the ligands, whereas evidence for donation of ligand electron density to the metal comes from the nephelauxetic effect of electronic spectroscopy. (The term 'nephelauxetic' means 'cloud expanding', and the effect relates to the observation that electron pairing energies are lower in metal complexes than in 'naked' gaseous metal ions. This suggests that interelectronic repulsion has fallen on complexation, which equates with the effective size of metal orbitals increasing. The effect varies with the type of ligand bound to the metal.)

This outcome can hardly be a surprise, since partial covalency has also been invoked to account for the properties of simple ionic solids. Moreover, in the same way that we do not need to throw out the ionic bonding model of the solid state to accommodate these observations, we do not need to reject the CFT wholesale. In particular, the simple orbital splitting model for the five d orbitals can continue to provide useful service, but amendment to accommodate covalency is necessary. This led to the development of ligand field theory (LFT), which, as the name implies, recognizes the role of the ligand more effectively. While CFT and LFT tend to be used interchangeably, they differ in that the former is a classical electrostatic model whereas the latter is more a MO model. As we have seen earlier, LFT leads to metal-centred MOs that correspond to the d-orbital energy levels from CFT, and hence it has become common to use LFT as an encompassing theory that includes elements of CFT, particularly where the focus is on the five d orbitals (Figure 3.14). However, all of our discussion to date has focused on one, albeit common, geometry – octahedral.

One outcome of the earlier discussion, which identified the wealth of shapes adopted by metal complexes, is a recognition that the bonding models we introduced above exclusively for six-coordinate octahedral complexes are clearly limited if they are restricted to this one shape. That different shaped complexes exist is a fact, so our model developed to date must be flexible enough to accommodate other shapes. One of the best ways to understand how this works is to start with our octahedral field, and look at what happens as we distort it. One well known type of structural distortion is where there is one unique axis where the bond distances are longer than along the other two axes; it still has the basic octahedral shape, but is 'stretched' along one axis direction – like a stretched limo still being, at heart, a standard car. In the extreme, these stretched bonds get so long that they effectively don't exist, meaning that there are only four bonds left, in a plane around the central metal. By convention, the 'stretched' axis is defined as the z axis. As such, it is the orbitals which point in the z direction that feel the effect of change most – easy to identify, from their names, as d_{z^2}, d_{xz} and d_{yz}. As we pull the two point charges away from the metal, their influence on these three orbitals diminishes, and the orbitals fall in relative energy (or are stabilized); this removes the degeneracy from both the diagonal set *and* the axial set of orbitals, as at least one of the d_{z^2}, d_{xz} and d_{yz} orbital resides in each set. The outcome is shown in Figure 3.15. This trend continues to the extreme case where they are removed

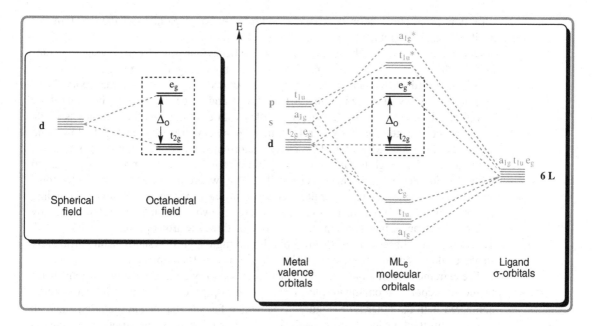

Figure 3.14
Comparison of CFT (ionic; at left) and LFT (molecular orbital; at right) development of the d-orbital splitting diagram for octahedral systems. Both reduce to the equivalent consideration of insertion and location of metal d electrons in two degenerate sets of orbitals separated by a relatively small energy gap (Δ_o).

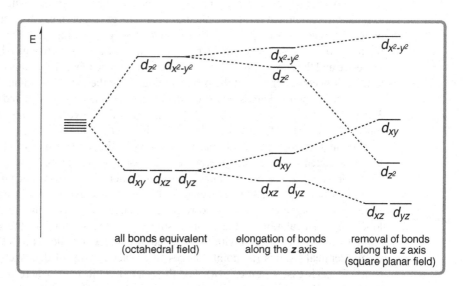

Figure 3.15
Variation in the d-orbital splitting diagram as a result of elongation of bonds along the z axis.

fully (and four-coordination is achieved), with increasing separation of the two types of orbitals in each originally degenerate set.

One classic example of bond elongation is the behaviour of d^9 Cu(II) octahedral complexes, which typically display elongation of M—L bond lengths along one axis direction. This is predicted by the *Jahn–Teller* theory. Without resorting to detail of this theory (inappropriate at this level), it predicts that, where the ground state is degenerate, it is only if the complete orbital set is empty, half-filled or filled with electrons that the complex is stable with respect to distortion that will relieve the degeneracy through splitting the low-lying orbital set into separate orbitals. Otherwise, distortion from regular octahedral geometry, easily achieved through usually axial elongation (or more rarely compression) occurs to relieve orbital degeneracy. The effect is only strong where the e_g level carries electrons, since these have the capacity to split much more strongly than the essentially nonbonding t_{2g} set. Strong Jahn–Teller distortions arise where the two e_g orbitals are differentiated by electron occupancy, that is when there are one or three electrons present in the e_g level. This is routinely observed for d^9, low-spin d^7 and high-spin d^4 systems, with particularly strong evidence coming from structural studies of d^9 Cu(II) complexes.

The extreme case of elongation, complete removal of two ligands along one axis, leads to the four-coordination square-planar shape. The $d_{x^2-y^2}$ orbital is then high in energy compared with the other four (Figure 3.15), which favours the d^8 configuration since all electrons fill the four low-energy levels, leaving the high energy one empty, providing an energetically favourable arrangement. The experimental observation of a large number of square planar d^8 complexes is consistent with this model. The energy gain also increases with increasing ligand field strength; since this occurs for the heavier (4d and 5d) transition elements, the observation in those rows of four-coordinate square planar as a more common geometry is also consistent with the model.

What the above shows is that *the bonding model is responsive to molecular shape*, which is one of its key attributes. In principle, it can be manipulated to yield an energy level arrangement for any shape. For the best known of the four-coordination shapes, tetrahedral, the ligand arrangement is distinguished by none of the point charges lying along the x, y, z, axes that we employed for the octahedral case. The tetrahedron can be visualized as based on a cube, with pairs on point charges on opposite corners of a cube. Thus the tetrahedral ligand field is closely related to the rare cubic ligand field. This cube has the metal atom at its centre and the axes passing through the centres of each face. From this cubic framework at least, it becomes clear that the orbitals lying along the axes ($d_{x^2-y^2}$ and d_{z^2}) will interact least with the point charges, the opposite to the behaviour in an octahedral field (Figure 3.16). Conversely, the d_{xy}, d_{yz} and d_{xz} orbitals that lie between axes are angled out more towards the corners of the cube and will thus interact most.

This interaction hierarchy is the reverse of the situation in the octahedral field, and, as a consequence, the orbital energy sets are also reversed in both the cubic and related tetrahedral fields (Figure 3.17). For the tetrahedral field they also change point group names slightly (now simply e and t_2, associated with the change in overall symmetry).

Geometric arguments relating to relative distances between point charges and orbital lobes allow calculation of the relative strength of the splitting in the tetrahedral case (called, now, Δ_t) compared with the octahedral case (Δ_o), which shows that $\Delta_t = 4/9\Delta_o$ when identical ligands (as point charges) are placed identical distances from the metal centre in each case (Figure 3.18). For the related cubic field, where there are twice as many point charges present, the splitting energy is twice the tetrahedral value, or $\Delta_{cubic} = 8/9\Delta_o$.

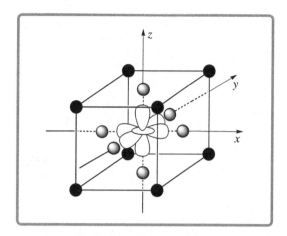

Figure 3.16
A view representing the interaction of the d_{z^2} and $d_{x^2-y^2}$ orbitals with a cubic field (black circles) and an octahedral field (grey circles). The orientation of the d-orbital lobes directly towards the octahedral set contrasts with the orientation for the cubic set.

The derivation of a model that manipulates the d-orbital set is incomplete without consideration of placement of d electrons into the orbitals. Because the model we have developed is ionic in character, one of the advantages is that we do not have to worry about any additional electrons coming from attached groups. The only electrons of concern are the initial d electrons themselves, which will vary from zero (a trivial case) to a maximum of ten (all that the five d orbitals can accommodate) depending on the metal and its oxidation state. At first, we can apply simple rules for filling atomic energy levels – fill in order of increasing energy – and apply Hund's rule (for degenerate levels, electrons add to each orbital separately maintaining parallel spins before pairing up commences). The outcome is the set of electron arrangements shown for the octahedral case in Figure 3.19. What this model provides is support for complexation, since in most cases the energy of the assembly in the presence of the octahedral field is lower than the zero field situation. To clarify this, consider the simple d^1 case. Here, an electron resides only in a t_{2g} orbital, which is more stable by $-0.4\Delta_o$ than the spherical field zero-point. The energy of this stabilization is

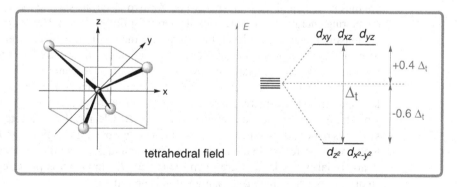

Figure 3.17
The d-orbital splitting diagram for a tetrahedral field.

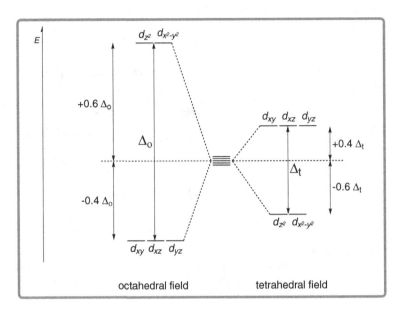

Figure 3.18
Comparison of the d-orbital splitting for octahedral and tetrahedral fields ($\Delta_t = 4/9\,\Delta_o$).

called the crystal field stabilization energy (CFSE). For d^2, with two electrons each in a t_{2g} orbital, stabilization is $2 \times (-0.4\Delta_o) = -0.8\Delta_o$. When the set gets more crowded with electrons in both levels, the situation can still lead to increased stability; for example, d^4 in Figure 3.19 has $3 \times (-0.4\Delta_o)$ for electrons in the t_{2g} set offset here by one electron destabilized by $+0.6\Delta_o$ as a result of residing in the higher e_g set, leading to an overall CFSE of $[3 \times (-0.4\Delta_o) + 0.6\Delta_o] = -0.6\Delta_o$.

This set is only part of the story, however, since we must now consider the consequence of the relatively small energy gap (Δ_o) between the t_{2g} and e_g set (an energy change of $\sim 250\,\text{kJ}\,\text{mol}^{-1}$ equates to absorption of light near the centre of the visible region). We can examine this best with a very simple model of two nondegenerate levels separated by an energy Δ (Figure 3.20), accommodating two electrons. The two levels will then be of energy E and $(E+\Delta)$. If the two electrons occupy the lowest level (of energy E) only and are thus paired up, there is an energy cost for two electrons sharing the same orbital called the pairing energy, P. Thus the overall energy for this system is $2E+P$. If the two electrons are unpaired, one in each level, the energy for the one in the lower level is E, and for the other in the higher level is $E+\Delta$, for an overall energy of $2E+\Delta$. The difference between the former (which we shall call *low spin*; an alternative name is *spin-paired*) case and the latter (which we shall call *high spin*; the alternative name is *spin-free*) case is that the low-spin (that is spin-paired) case is raised in energy above $2E$ by P, whereas the high-spin (spin-free) case is raised in energy above $2E$ by Δ. Clearly, if Δ is greater than P, then the system is better off (that is, of lower overall energy) being low spin; were P greater than Δ, high-spin is preferred. In circumstances where P and Δ are similar in energy, the situation is nontrivial. For a large energy gap (where $\Delta > P$), low-spin is preferred, whereas for a small energy gap (where $\Delta < P$), high-spin is preferred.

For the proper d-electron set in an octahedral field, with its two sets of degenerate levels, there is no case for consideration with d^0, d^1, d^2, d^3, d^8, d^9 and d^{10}, since there is effectively

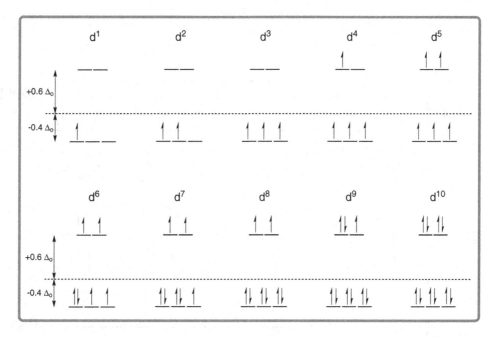

Figure 3.19
Electron arrangements in the d orbitals in an octahedral field with maximum number of unpaired electrons for the $d^1 - d^{10}$ systems.

only one option (assuming Hund's rule applies). However, for d^4–d^7, both low-spin and high-spin options exist, as shown in Figure 3.21, the size of Δ_o influencing outcomes. This is exemplified for d^6 in Figure 3.22. What is important about this model is that it can be used to understand experimental observations (as any viable model should, of course). It allows us to understand how the colour and magnetism of a complex can change even where the central metal ion and its oxidation state is not altered, simply as a result of a changing ligand environment perturbing the size of Δ_o – but more on that later.

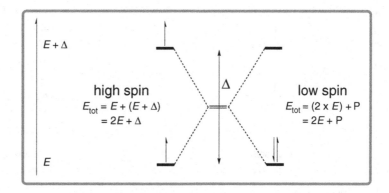

Figure 3.20
Comparing the effects of spin pairing (right) versus separate orbital occupancy (left) for a simple (but artificial) model of two nondegenerate orbitals containing two electrons.

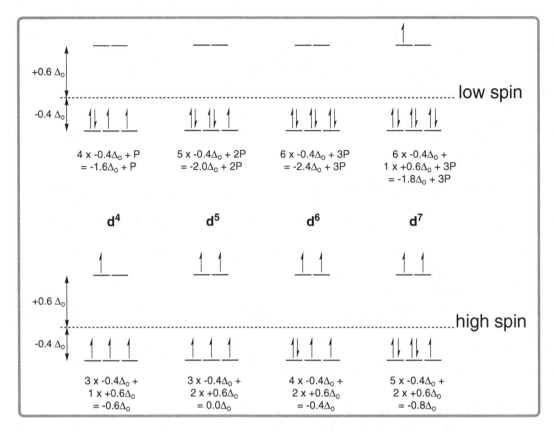

Figure 3.21
Comparing the high spin and low spin arrangements for d^4–d^7 in an octahedral field.

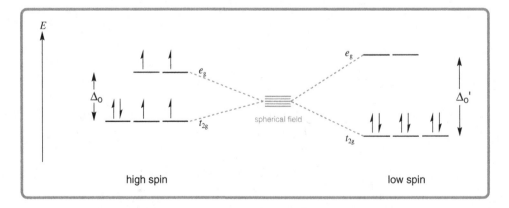

Figure 3.22
How the spin arrangement (high spin or low spin) can vary with the size of Δ_o, exemplified for d^6 in an octahedral field. The type of donors governs the outcome.

The molecular orbital theory recognizes that it is not essential to limit bonding between a metal and its ligands to linkages between only two atoms at a time, but our model developed initially dealt exclusively with what we would formally consider single (and σ) M—L bonds were we applying a covalent bonding approach. So what happens if we next introduce some double- (or π-)bond character into the M—L bonds? Multiple bonds are well known in carbon chemistry, but there is no reason that carbon should hold the patent – and of course it doesn't, as clear from the recognition of S=O bonds, for example. In the case of the M—L bond, at least partial double-bond character involves the consideration of π bonding interactions of the ligand and metal *in addition to* the σ bonding. There are in effect two classes of effects, π-donor and π-acceptor, relating to which way electron density 'flows', as metal centres as well as the donor ligand can be considered as a source of electron density.

We will briefly exemplify the π-donor effect, by considering the interaction of metal-based t_{2g} orbitals with filled π orbitals on a ligand. This interaction involves 'sideways' type overlap somewhat like the simple covalent bonding concept, but multi-centred here. If empty t_{2g} metal-based orbital interact with filled π ligand orbitals, the usual bonding and antibonding style outcome occurs, except the resultant bonding and antibonding levels are very close in energy to their ligand- and metal-based parents respectively. This means that the former has dominantly the character of the ligand, and the latter that of the metal – we can think of this as simply the metal levels being pushed up a little in energy and the ligand orbitals down a little, through a partial transfer (or donation) of electron density from ligand orbitals to the metal – an effect we term π-donation. There is one obvious consequence for the metal as a result of the energy of the metal e_g^* orbitals remaining unaltered – the energy gap Δ_o is decreased, as seen in Figure 3.23, an effect which is reflected in the physical properties of the complex. It is ligands such as halide or O^{2-} anions that are π-donors, because they have available extra lone pairs after one lone pair has been used for σ bonding. These electron pairs, involved in what are in effect repulsive interactions with filled metal orbitals, interact to lower Δ_o. Not surprisingly, metals with no or few d electrons tend to

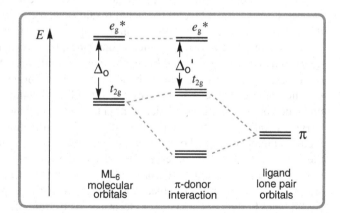

Figure 3.23
The π-donor concept in metal–ligand bonding, and its influence on energy levels. Occupied lone pair (π) orbitals on the ligand are more stable than the metal nonbonding t_{2g} orbitals. A repulsive-type interaction between these orbitals leads to a rise (destabilization) of the t_{2g} and a fall in energy of the ligand lone pair orbitals, and a related fall in the size of Δ_o.

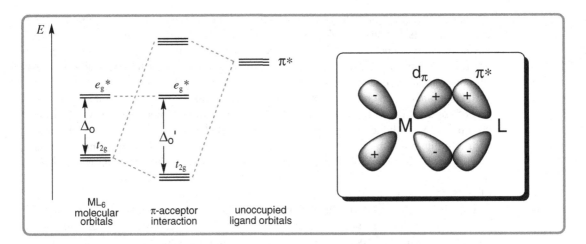

Figure 3.24
The π-acceptor concept in metal–ligand bonding, and its influence on energy levels. Unoccupied ligand π^* orbitals are less stable than the metal nonbonding t_{2g} orbitals. A repulsive-type interaction between these orbitals leads to a fall (stabilization) of the t_{2g} and a rise in energy of the ligand π^* orbitals, and a related increase in the size of Δ_o. An orbital interaction view of the process in terms of overlap of a metal d_π and empty ligand π^* orbital (with the $+$ and $-$ signs relating here to orbital symmetry, not charge) is also given at right.

bind π-donor ligands best, as exemplified by d^0 Ti(IV) in $[TiF_6]^{2-}$, since the empty t_{2g} (or d_π) orbitals are most suitable for π-bonding.

The opposite effect is the π-acceptor situation, which involves electron density being donated back from the metal to the ligand, into empty orbitals of appropriate symmetry that interact with metal orbitals (such as empty π^* orbitals on carbon monoxide with metal t_{2g} orbitals, Figure 3.24). Here, the ligand antibonding orbitals are higher in energy than the metal t_{2g} levels, and the interaction leads effectively to a lowering of energy of the metal t_{2g} set and an increase in Δ_o, and an increase in the system's total bonding provided electrons occupy these levels. The metal is said to be participating in 'back bonding' with the ligand, because electron 'flow' is in the opposite direction to usual. We shall return to these concepts, in terms of experimental outcomes, later. Although the metal d_π (or t_{2g}) set does not match with ligand σ orbitals, the metal orbitals can interact with empty ligand π^* orbitals. One might ask how an antibonding orbital can form a bond. The key to this is that the antibonding character is with respect to the ligand component atoms, which does not preclude bonding of the ligand donor atom to the metal, which is 'external' to the ligand.

Back-bonding is a common feature of metal coordination to a molecular ligand that includes double or triple bonds. The outcome, as seen in Figure 3.24, is a stabilization of the t_{2g} (or d_π) set and an increase in the size of the splitting (or of Δ_o). The experimental observation that organometallic compounds tend to be low-spin is consistent with the larger Δ_o predicted in the model, which also equates with π-acceptor ligands having a strong ligand field. Further, the transfer of electron density from the metal to a π-acceptor ligand such as carbon monoxide acts to stabilize complexes of formally low-valent and even zero-valent metals. Low-valent metals have little capacity for accepting electrons from simple σ-donor ligands such as ammonia, however. In concert, these concepts explain why $[W(NH_3)_6]$ is not stable, whereas $[W(CO)_6]$ is stable even to oxidation by air as a result of highly efficient back-bonding.

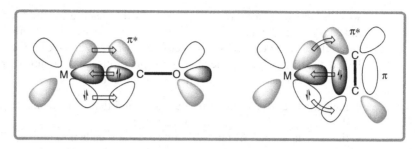

Figure 3.25
Bonding interactions in organometallic complexes featuring carbon monoxide and ethylene ligands, illustrating the synergy involved. Filled CO or C_2H_4 to empty M σ-donation as well as back-donation from filled M d orbitals to empty CO or C_2H_4 π^* orbitals occurs. The only difference is that the type of ligand orbital involved in the M—L σ-bonding varies.

There is a form of synergy operating in bonding of ligands like carbon monoxide, as electron density 'flows' in each direction. There is a donation from filled ligand orbitals to empty metal σ orbitals, and a back-donation from filled metal d orbitals to empty ligand π^* orbitals. This is also seen with alkenes, which take the unusual arrangement of 'side-on' bonding, introduced in Chapter 2. This is illustrated in an orbital picture in Figure 3.25.

Simple spectroscopic evidence for back-bonding comes from infrared (IR) spectroscopy of carbon monoxide compounds, where the strength of the carbon-oxygen bond can be related to the carbon–oxygen IR stretching vibration, $\nu(CO)$. When CO forms an adduct with BH_3, which is required to be σ-bonded in character, the $\nu(CO)$ shifts just slightly higher in energy from 2149 cm^{-1} to 2178 cm^{-1}. By contrast, coordination to metals cause a substantial shift to lower energy (to ≤ 2000 cm^{-1}), interpreted as a weakening of the carbon–oxygen bond as a result of electron density being pushed back into the π^* antibonding orbital. The size of the shift can actually be used as a guide to the electron richness of the central metal. Any ligand with a filled orbital that can act as a σ-donor and an empty orbital that can act as a π-acceptor can be consider a π-acceptor ligand, which includes a wide array of carbon-bonding molecules such as alkenes, arenes and carbenes, as well as other molecules such as dinitrogen. Usually, these ligands employ their highest occupied molecular orbital (termed the *HOMO*) for σ-bonding and lowest unoccupied molecular orbital (termed the *LUMO*) as π-acceptor. These orbitals are often referred to as *frontier orbitals*, being dominant in bonding in these systems.

Notably, the spectrochemical series introduced in Section 3.3.1 is also rationalized partly by the π-donor/acceptor concept, since ligands on the left-hand side tend to be π-donors, whose interaction with the octahedral t_{2g} set tends to raise the energy of this set and decrease Δ_o, whereas ligands on the right-hand side tend to be π-acceptors, whose interaction lowers the t_{2g} set and increases Δ_o. Ligands which are pure σ-donors (like ammonia, which offers no additional lone pair after the σ-bonding pair, or vacant orbitals) do not participate in π overlap at all, and thus tend to lie in the middle of the spectrochemical series.

3.4 Coupling – Polymetallic Complexes

All of the examples met so far are based on simple monomeric complexes, containing just one metal centre. This might suggest that these are the sole or dominant type – not so. It is not uncommon to meet complexes that feature two or even more metal centres linked together. Complexes may be monomeric, dimeric (two metals involved), oligomeric (several metals involved), or polymeric (many metals involved). The metal centres may be linked directly, yielding a metal–metal bond, or else may be linked or bridged by a ligand or ligands. This latter class can arise in a number of different ways:

- linkage via a small neutral or anionic ligand that binds simultaneously to both metals through a common donor atom (e.g. Cl^-);
- linkage via a small ligand that binds to both metals through two different donor atoms (e.g. NCS^-, using both the N and S atoms);
- linkage via a polydentate ligand that spans across two (or more) metals and provides some donor groups for each (e.g. $^-OOC-CH_2-COO^-$, using the two carboxylate groups);
- incorporation of two or more metal ions in a large cavity in a polydentate ligand, each metal binding to several of a large set of potential donors (a process called encapsulation).

Moreover, the complex may involve: all metals of the one type and in the same oxidation states; all metals of the same type in several oxidation states; two or more different types of metals in a common oxidation state; or two or more different types of metals in a range of oxidation states – that is, all combinations are feasible!

Complexes that are oligomeric and roughly spherical are often called clusters. Clusters may involve several layers of metal ions from a core outwards and, perhaps not surprisingly, metals in different environments (core or surface, for example) behave differently, even if inherently the same type and in the same oxidation state – in other words, 'environmental' effects contribute to the metal's behaviour.

We shall very briefly examine some of these classes a little more carefully, with some examples. One of the simplest families of related compounds is the halo-metals. Many metal ions can exist with a set of halogen anions as common donors in monomers of general formula MX_n^{q-}; an example is the $[CuCl_4]^{2-}$ ion. In some cases, oligomers can form, of which the first is the dimer $M_2X_n^{q-}$; an example is $[Fe_2Cl_6]^{2-}$, distinguished by having two bridging halides in addition to four terminal halides (Figure 3.26). Higher oligomeric clusters are also known; trimers, such as $[Re_3Cl_{12}]$, have the three metal centres disposed in an equilateral triangle arrangement, with both bridging and terminal halides,

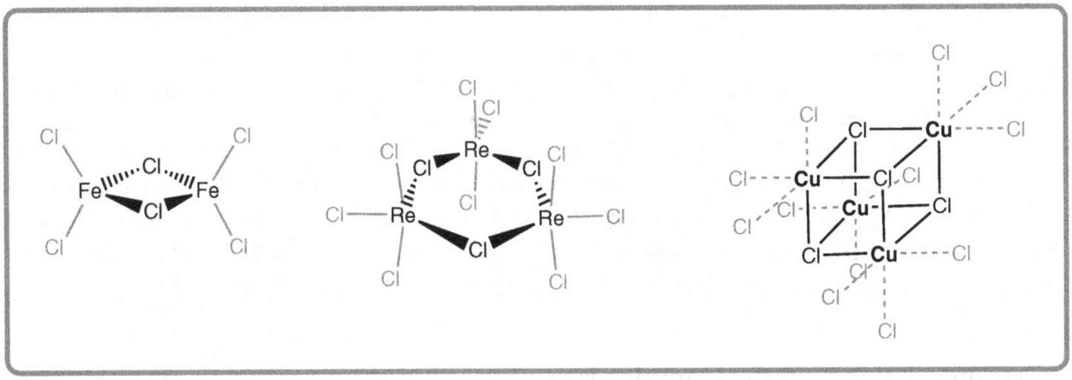

Figure 3.26
Examples of oligomeric complexes of metal ions with halides. Both bridging and terminal halide can be identified in the figures; terminal halides are shown in grey.

whereas tetramers, such as $[Cu_4Cl_{16}]^{4-}$ adopt a slightly distorted box-like shape. These may also involve metal–metal bonding; for example, $[Re_3Cl_{12}]$ has bonds between the Re atoms additional to those formed by bridging chloride ions.

The above examples are relatively simple, since they contain just two types of atom and adopt symmetrical shapes. It should be anticipated that many examples will prove much more complicated in structure. Although polymeric coordination compounds are growing in number and importance, we shall limit our exploration of these, leaving this to more advanced textbooks.

3.5 Making Choices

3.5.1 Selectivity – Of all the Molecules in all the World, Why This One?

The heading above takes us back to a movie analogy again, but we're not going to take it too far. Suffice to say that, in the same way there is a vast number of bars or cafes in the world to choose from, there is an even vaster range of potential ligands for a metal ion to select. Why, then, does selection happen (or choice arise) at all?

What we do know from experimental evidence is that metal ions do not interact with potential ligands purely on a statistical basis. A simple example will suffice. Water, the most common solvent, is $\sim$55 M. If ammonia is dissolved in water to a concentration of $\sim$0.5 M, there is a 100-fold excess of water molecules over ammonia molecules. If a small concentration of copper ion is introduced into the solution, the vast majority of the copper ions bind to the ammonia – even though each cation 'meets' many more water molecules. Why is there this clear preference for ammonia over water as a ligand in this case? This is but one example of a general observation.

3.5.2 Preferences – Do You Like What I Like?

Ligand preference for and affinities of metal ions present themselves as experimentally observable behaviour that is not easily reconciled. As a general rule, when a metal (M) is mixed with equimolar amounts of ligands A and B, the result is *not* usually equimolar amounts of MA and MB. Both metal ions (*Lewis acids*) and ligands (*Lewis bases*) show preferences.

In seeking an understanding of this phenomenon, we can sort ligands according to their preference for forming coordinate bonds to metal ions that exhibit more ionic rather than purely covalent character. Every coordinate covalent bond between a metal ion and a donor atom will display some polarity, since the two atoms joined are not equivalent. The extreme case of a polar bond is the ionic bond, where formal electron separation rather than electron sharing occurs; coordinate bonds show differing amounts of ionic character. Small, highly charged entities will have a high surface charge density and a tendency towards ionic character. In a sense, we can think of these ions or atoms as 'hard' spheres, since their electron clouds tend to be drawn inward more towards the core and thus are more compressed and less efficient at orbital overlap. A large, low-charged entity tends to have more diffuse and expanded electron clouds, making it less dense or 'soft', and better suited to orbital overlap. This concept, applied to donor atoms and groups, leads to us defining 'hard' donors as those with a preference for ionic bonding, and 'soft' donors as those preferring covalent bonding. An important observation is that the 'hard' donor atoms are also the most electronegative. Therefore, arranging donor ions or groups in terms of increasing electronegativity provides us with a trend from 'soft' to 'hard' ligands:

'SOFT' – *increasing electronegativity* → 'HARD'									
CN^-	S^{2-}	RS^-	I^-	Br^-	Cl^-	NH_3	^-OH	OH_2	F^-

The ligands at each 'end' show distinct preferences for different types of metal ions. The F^- ligand binds strongly to Ti^{4+}, whereas CN^- binds strongly to Au^+, and it is clear these metal ions differ in terms of both ionic radius and charge. Any individual metal ion will display preferential binding when presented with a range of different ligands. This means that metal ions can be graded and assigned 'hard–soft' character like ligands; however, by convention, we define the least electronegative as 'hardest' and the most electronegative as 'softest'. For the metals, they grow increasingly 'soft' from left to right across the Periodic Table, and also increase in softness down any column of the table. This definition for metals allows us to apply a simple 'like prefers like' concept – *'hard' ligands (bases) prefer 'hard' metals (acids), 'soft' ligands (bases) prefer 'soft' metals (acids)*. This principle of hard and soft acids and bases (HSAB), developed by Pearson, is a simple but surprisingly effective way of looking at experimentally observable metal–ligand preference. To see the concept in action, we should look at a few examples of how it allows 'interpretation' of experimental observations; yet, without a strong theoretical basis, the concept is somehow unsatisfying.

One example involves the long-established reaction of Co^{2+} and Hg^{2+} together in solution with SCN^-. The ligand has both a 'soft' donor (S) and a 'hard' donor (N) available; although like-charged, Hg^{2+} is larger than Co^{2+} and further across and down the Periodic Table, and thus 'softer'. This reaction results in a crystalline solid $[(NCS)_2Hg(\mu\text{-}SCN)_2Co(\mu\text{-}NCS)_2Hg(SCN)_2]$, with each metal in a square plane of thiocyanate donors, with four S atoms bound to each Hg^{2+} and four N atoms bound to the central Co^{2+}. Only this thermodynamically stable product, where the softer S bonds to softer Hg^{2+}, and harder N bonds to harder Co^{2+}, is known. If the 'wrong' end of a ligand like thiocyanate binds initially to a mismatched metal ion, it will usually undergo a rearrangement reaction to reach the stable, preferred form. Where SCN^- is forced initially to bond to the 'hard' Co^{3+} through the 'soft' S atom, it undergoes rearrangement to the form with the 'hard' N atom bound readily. It is possible to apply the concepts exemplified above to reaction outcomes generally. In all cases it is a comparative issue; for example, a ligand defined as 'harder' versus one other donor type (such as OH_2 versus Cl^-) may be considered 'softer' if compared with another type of ligand (such as NH_3 versus OH_2); this aspect will also be true for metal ions.

Definition of hard/soft character is the result of empirical observations and trends in measured stability of complexes. For example, hard acids (such as Fe^{3+}) tend to bind the halides in the order of complex strength of $F^- > Cl^- > Br^- > I^-$, and soft acids (such as Hg^{2+}) in the reverse order of stability. However, as with any model with just two categories, there will be a 'grey' area in the middle where borderline character is exhibited. This is the case for both Lewis acids and Lewis bases. Selected examples are collected in Table 3.2 below; a more complete table appears later in Chapter 5.

The 'hard'–'soft' concept is, from a perspective of the metal ion, often recast in terms of two classes of metal ions as follows:

- *'Class A'* **Metal Ions ('hard')** – These are small, compact and not very polarizable; this group includes alkali metal ions, alkaline earth metal ions, and lighter and more highly charged metal ions such as Ti^{4+}, Fe^{3+}, Co^{3+}, Al^{3+}. They show a preference for ligands (bases) also small and less polarizable.

Table 3.2 Selected examples of hard/soft Lewis acids and bases.

Character	Lewis acids	Lewis bases
Hard	H^+, Li^+, Na^+, Mg^{2+}, Cr^{3+}, Ti^{4+}	F^-, HO^-, H_2O, H_3N, CO_3^{2-}, PO_4^{3-}
Intermediate	Fe^{2+}, Co^{2+}, Ni^{2+}, Cu^{2+}, Zn^{2+}	Br^-, NO_2^-, SCN^-
Soft	Cu^+, Ag^+, Au^+, Hg^+, Cd^{2+}, Pt^{2+}	I^-, CN^-, CO, H^-, $\underline{S}CN^-$, R_3P, R_2S

- **'Class B' Metal Ions ('soft')** – These are larger and more polarizable; this group includes heavier transition metal ion such as Hg^{2+}, Pt^{2+}, Ag^+, as well as low-valent metal ions including formally $M(0)$ centres in organometallic compounds. They exhibit a preference for larger, polarizable ligands. This leads to a preference pattern outlined in Figure 3.27.

There is also a changing preference order seen across the rows of the Periodic Table, with *Class A* showing a trend from weaker to stronger from left to right, and *Class B* showing the opposite trend of stronger to weaker, defined in terms of the measured stability constant of ML complexes formed.

3.5.3 Complex Lifetimes – Together, Forever?

Metal complexes are not usually unchanging, everlasting entities. It is true that many, once formed, do not react readily. However, the general observation is that coordinate bonds are able to be broken with more ease than cleaving a covalent carbon–carbon bond. The strength of metal–donor bonds are typically less than most bonds in organic molecules, but much greater than the next most stable type, hydrogen bonds. We recognize carbon–carbon bonds as strong and usually unchanging and hydrogen bonds as weak and easily broken; the coordinate covalent bond lies in the middle ground, albeit nearer the carbon–carbon end of the park. Breaking a metal–ligand bond is almost invariably tied to a follow-up process of making another metal–ligand bond in its place, so as to preserve the coordination number and shape of the complex. When the same type of ligand is involved in each sequential process, *ligand exchange* is said to have occurred. Where the ligands and solvent are one and the same, there is no opportunity for any alternative reaction. Even at equilibrium, metal complexes in solution display a continuous process of ligand exchange, where the rate of the exchange process is driven by the type of metal ion, ligands and/or solvent involved.

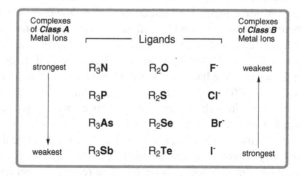

Figure 3.27
Metal–ligand preferences for key ligands. In any column, a *Class A* metal ion prefers ligands from the top whereas a *Class B* metal ion prefers ligands from the bottom.

From the perspective of the central metal, ligand exchange can vary with metal ion from extremely fast (what we refer to as '*labile*' complexes) to extremely slow (termed '*inert*' complexes). Whereas direct exchange of one ligand by another of exactly the same type is the simplest process inherently, the fact that such exchanges can occur suggests that, if other potential ligands of a different type are present, they may intercept the process and be inserted in the place of the original type – ligand exchange has become *ligand substitution*.

3.5.3.1 Lability – Party Animals

Chemistry is full of 'opposites' – resulting from often setting two extremes as the limits for defining behaviour (the 'black or white' approach). Of course, these limits are not isolated options, but are usually the extremes of a continuum of behaviour (the chemical equivalent of saying that something is rarely ever black or white, but more a shade of grey). How rapidly metal ions take up or lose ligands is a good example of this characteristic. There are two extreme positions; very fast reactions of labile compounds or very slow reactions of inert compounds. Labile systems are the party animals of metal complexes – making and breaking relationships rapidly. We can measure the rate at which ligand exchange with the same type of ligand occurs, even though it appears that nothing changes, by using radioactive isotopes to allow the rate of the process to be monitored, provided the process is not too rapid. It's a bit like painting a white house with fluorescent white paint – nothing appears to have changed, until you see it at night. At the molecular level, we simply measure the uptake of radioactive ligand into the coordination sphere of the metal over time as it replaces nonradioactive ligand, which allows us to define the rate of ligand exchange.

3.5.3.2 Inertness – Lasting Relationships

An inert system is like modern marriage – maybe not joined for ever, but willing to give it a good try. These are complexes which, once formed, undergo any subsequent reaction very slowly. Inertness can be so great that it overcomes thermodynamic instability. This means that a complex may be pre-disposed towards decomposition, but this will happen so very, very slowly that to all intents and purposes the complex is unreactive, or inert. Cobalt(III) amine complexes are the classic example of this; thermodynamically they are unstable in aqueous solution, but they are so inert to ligand substitution reactions that they can exist in solution with negligible decomposition for years.

Even for a metal ion regarded as inert, however, the rate of substitution or replacement of particular ligands will differ, and may differ significantly. For example, the half-life for replacement of the coordinated perchlorate ion from the cobalt(III) complex $[Co(NH_3)_5(OClO_3)]^{2+}$ by a water molecule in acidic aqueous solution is about seven seconds, whereas the half-life for replacement of an ammonia from the related $[Co(NH_3)_6]^{3+}$ is about 3 800 years! Chemistry is astounding in its diversity, if nothing else.

Note one important aspect of the above discussion – lability and inertness are *kinetic* terms, and all about the rate at which something reacts. A species that is inert is kinetically stable. This does not require that it be thermodynamically stable, however (although it may be). Thermodynamic stability is about being in a form which has no other readily accessible species lower in energy. The reason we can have kinetic stability in a system that is thermodynamically unstable is that the two differ. Kinetics is about transition *to* equilibrium; thermodynamics is about the situation *at* equilibrium. For a molecule to convert from one form to another, by any process, it is considered necessary for it to overcome an activation energy barrier, whereby it must proceed through a higher energy transition

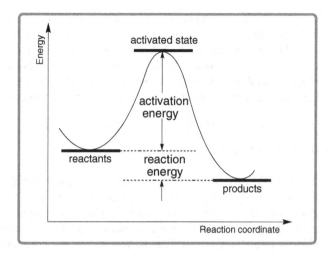

Figure 3.28
A reaction coordinate, defining the barrier to reaction (activation energy) as well as the energy difference between reactants (initial state) and products (final state) or reaction energy.

state (represented in Figure 3.28). Reactants convert to products through this transition or activated state, which is a short-lived 'half-way house' between the two forms.

A simple macroscopic example may illustrate this. Consider a thin flat piece of board painted white on one side and red on the other. Now arrange to turn this over on a tabletop with the board touching the table throughout. You can do this, of course, by turning the board onto its edge, then continuing the motion until it lies flat on its other side. In doing this transformation, the position where it is precariously standing on its edge can be considered a transition (or activated) state, and if you let go of it in that position it may fall back so the white side is up (equivalent to no reaction) or fall forward so the red side is up (equivalent to a completed reaction). To get it from lying flat to on its side costs you effort, or energy – what we would call the activation energy at the molecular level. Molecules acquire this activation energy mainly through collision as a result of their motion; if the collision energy suffices to take the reactants to an arrangement where they can proceed to products without further energy, they have achieved the status of an activated or transition state species. In our macroscopic example, your physical effort takes the board to the upright activated state. For a very heavy piece of board, the amount of effort is significant and you may not get it into an upright position every time you try. This amounts to a high activation barrier or large activation energy, which equates with a slower rate of reaction. If the board is small and light, the effort required is small and the task easy and rapidly completed; this amounts to a low activation barrier or small activation energy, consistent with a fast rate of reaction.

If you next consider your board starting on a different level to where it finishes (table to floor, for example), it is obvious that one is the lower level; the board could fall from table to floor, but not easily the reverse way. On the molecular level, being on the 'floor' would be a thermodynamically more stable arrangement for the molecule than being on the 'table'. The difference in energy between the reactants and products is the reaction energy, and the stability of a complex depends on the size of this energy difference. This additional consideration (thermodynamics) doesn't change the process by which you turn the board over (kinetics), although there is a relationship between them at the molecular level, which we won't explore here. At the molecular level, however, raising temperature increases molecular velocities and increases the probability of a collision achieving the transition state, so the rate of a reaction increases with increasing temperature.

Suffice to say that kinetics and thermodynamics, in combination, govern all of our chemical reactions, but obviously tempered by what exactly is available chemically to react. We shall examine stability and reactions of coordination complexes in more detail in Chapter 5, and the application of these in chemical synthesis is met in Chapter 6.

3.6   Complexation Consequences

What are the observable physical consequences of complex formation? We are bringing together usually highly positively charged ionic central atoms and negatively-charged or polar and electron-rich donor atoms or groups to assemble a tightly-bound complex. Inevitably, the ligands affect the core element (commonly a metal) and the central or core element (metal) affects the ligands. We can see this in experimentally measurable effects, some of which are mentioned below.

It is the properties of the donor groups of a ligand that are changed most substantially upon coordination to a metal ion. This can be illustrated by looking at molecules where the donor group carries protons. Acidity of coordinated water groups increases substantially on coordination – from a pK_a of $\sim$14 for free bulk water to a pK_a of 5–9 typically for coordinated water molecules. Thus aqua metal complexes are often only fully protonated in acidic solution. As charge on the metal ion increases, acidity increases to the point where deprotonation occurs readily, to form the hydroxide ($^-$OH), and to even in some cases to form the oxide (O^{2-}):

$$M^{n+}-OH_2 \leftrightharpoons M^{n+}-OH^- + H^+$$
$$M^{n+}-OH^- \leftrightharpoons M=O^{2-} + H^+$$

High-valent metal ions tend to promote deprotonation to form oxo-metal complexes and further desolvation to eventually yield simple metal oxides. Thus pure aqua complexes are found mainly with metals in oxidation states III and below.

The influence of the central metal ion on the ligand falls off rapidly with distance from the central metal. For example, in a molecule featuring coordination of a diol through one of the two alcohol groups, such as $M^{n+}-OH-CH_2-CH_2-OH$, the alcohol group bound directly to the metal will have a pK_a of $\sim$7, whereas the unbound alcohol group on the end of the pendant chain will have a pK_a of $>$14, essentially the same as for the free ligand. Even where two heteroatoms are joined directly, as in hydrazine ($M^{n+}-NH_2-NH_2$), it is the metal-bound amine group that is more acidic.

Ligand substitution leads to clear changes in a complex, as seen when commencing with aqua metal ions. When the water ligand is substituted by other ligands (common donors are halide ions and O-donor, N-donor, S-donor and P-donor groups), in monodentate or polydentate ligands, observable physical changes occur. *Colour* and *redox potential* change with ligand substitution – for example $[Co(OH_2)_6]^{3+}$ is light blue, with a reduction potential E_o of $+1.8$ V (making it unstable in solution, as it oxidizes the solvent slowly); $[Co(NH_3)_6]^{3+}$ is yellow, and E_o is shifted substantially to 0.0 V (making it stable to reduction in solution). Sometimes, the change of ligand can lead to a gross change in molecular shape also; light pink, octahedral $[Co(OH_2)_6]^{2+}$ changes to purple, tetrahedral $[CoCl_4]^{2-}$ when water ligands are replaced completely by chloride ions. Changes due to complete or even partial ligand exchange in the shape, colour, redox potential, magnetic properties and wider physical properties are characteristic behaviour for coordination complexes. These are explored in detail in following chapters.

That the position of an element in the Periodic Table influences its chemistry is inevitable, and a consequence of the Periodic Table reflecting the electronic configuration of the element. However, it is less clearly recognized that elements within the same column (and hence with the same valence shell electron set) differ in their chemistry. It is instructive to overview the chemical impact of the central atom to illustrate both similarities and differences. These, along with specific examples of synthetic coordination chemistry, appear in Chapter 6, whereas shape and stability aspects are detailed in the next two chapters.

Concept Keys

Complexes form a limited number of basic shapes associated with each coordination number (or number of ligand donor groups attached). Six-coordinate octahedral is a particularly common shape.

A simple covalent σ-bonding model employing five d orbitals and the next available s and three p orbitals, with appropriate hybridization to match a particular shape, allows a limited interpretation of bonding in coordination complexes.

The crystal field theory (CFT), an ionic bonding model, is focused on the d-orbital set and the way this degenerate set of five orbitals on the bare metal ion is split in the presence of a set of ligands into different energy levels. It provides a fair understanding of spectroscopic and magnetic properties.

A holistic molecular orbital theory description of bonding in complexes provides a more sophisticated model of bonding in complexes, leading to ligand field theory (LFT), which deals better with ligand influences. Both CFT and LFT reduce to equivalent consideration of d electron location in a set of five core d orbitals.

Variation in the splitting of the d-orbital set into different energy levels is dependent on complex shape.

Arrangement of a set of electrons in the d-orbital energy levels, as a result of relatively small energy gaps, can in some cases occur with different arrangements where unpaired electrons are either minimized (low-spin) or maximized (high-spin). Properties differ as a result.

Certain ligands can employ their π orbitals for additional interaction with the central metal, either for increasing (π-donor) or decreasing (π-acceptor) electron density on the central metal. This alters the energies of levels associated with the d orbitals, influencing physical properties.

Some ligands are capable of binding to more than one metal ion at once, acting as bridging groups in polynuclear assemblies.

Preference exists between ligands and metals undertaking complexation. Ligands and metals can be categorized as 'hard' or 'soft' bases/acids; a like-prefers-like situation operates.

The properties of both the ligand and the metal play a role in complex formation. Moreover, the physical properties of both the metal ion and the ligand are altered as a result of complex formation.

Further Reading

Cotton, F.A., Wilkinson, G., Bochmann, M. and Murillo, C. (1999) *Advanced Inorganic Chemistry*, 6th edn, Wiley Interscience, New York, USA. A classical advanced textbook in the broad inorganic field that includes a detailed coverage of d-block chemistry, with a leaning towards organometallic chemistry.

Gispert, J.R. (2008) *Coordination Chemistry*, Wiley-VCH Verlag GmbH, Weinheim, Germany. This is a large advanced text for graduate students, but undergraduates seeking an extended view of aspects of the field should find it useful.

Kauffman, G.B. (ed.) (1994) *Coordination Chemistry: A Century of Progress*, American Chemical Society. This is a readable account of the rise of coordination chemistry, celebrating the centenary of Werner's key work, for those with a historical bent; puts coordination chemistry in its historical context.

McCleverty, J.A. and Meyer, T.J. (eds) (2004) *Comprehensive Coordination Chemistry: From Biology to Nanotechnology*, Elsevier, Pergamon. Wilkinson, G., Gillard, R.D. and McCleverty, J.A. (eds) (1987) *Comprehensive Coordination Chemistry: The Synthesis, Reactions, Properties, and Applications of Coordination Compounds*, Pergamon, Oxford, UK. These two multi-volume sets provide coverage at an advanced level from ligand preparation through their detailed coordination chemistry and beyond; they provide, through independent chapters, a fine though sometimes daunting in-depth resource.

4 Shape

4.1 Getting in Shape

Coordination complexes adopt a limited number of basic shapes. We have developed in Chapter 3 a predictive set of molecular shapes evolving from an electrostatic model of the distributions predicted for from two to six point charges dispersed on a spherical surface. These shapes, evolving from a modification of the valence shell electron pair repulsion (VSEPR) model that was itself developed initially for main group compounds, are satisfactory as models for many of the basic shapes met experimentally for complexes throughout the Periodic Table. The original VSEPR model and its electron counting rules have limited predictive value for shape in complexes of d-block elements compared with its application for p-block elements. This relates to the defined directional properties of lone pairs in p-block elements, whereas in transition elements nonbonding electrons play a much reduced role in defining shape. Rather, it is simply the number of donor groups bound about the metal that is the key to shape in transition metal complexes. This is recognized in the Kepert model, which is a variation of the VSEPR concept developed for transition elements that ignores nonbonding electrons and considers only the set of donor groups represented as point charges on a surface. This essentially electrostatic model has limitations, as shape is influenced by other factors such as inherent ligand shape and steric interactions between ligands, as well as the size and valence electron set of the central metal ion.

The actual shape of complexes is now able to be determined readily in many cases. The advent of X-ray crystallography at a level where highly automated instruments allow rapid determination of accurate and absolute three-dimensional structures of coordination complexes in the crystalline form has been a boon for the chemist. Provided a complex can be crystallized, its structure in the solid state can be accurately defined. Although it is important to recognize that, for a coordination complex, solid state structure and structure in solution can differ, it is nevertheless true that they often are essentially the same, so we have at our fingertips an exceptionally fine method for structural characterization. This is a technique that can define angles with an error approaching $0.01°$ and bond distances (which are typically between 100 and 300 pm) with an error as low as 0.1 pm. It is least successful at detecting the very light hydrogen atoms, although the closely related neutron diffraction method provides greater resolution of these. An example structure is shown in Figure 4.1, in which all atom locations were accurately defined except for hydrogen atoms, which have been placed at calculated positions.

What crystallography has now shown clearly is that the predicted shapes we developed earlier are often observed but usually not achieved ideally – we notice that bond angles are often not exactly those anticipated, and, at times, geometries occur that are clearly better

Introduction to Coordination Chemistry   Geoffrey A. Lawrance
© 2010 John Wiley & Sons, Ltd

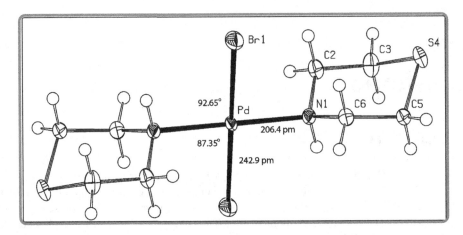

Figure 4.1
A view of the structure determined by X-ray crystallography of the simple neutral complex [PdBr$_2$(thiomorpholine)$_2$]. The atom locations are represented by probability surfaces, with the smaller the size the better defined is the atom. (Reprinted with permission from Australian Journal of Chemistry, 'Complexation of constrained ligands piperazine, N-substituted piperazines and thiomorpholine' by Sarah E. Clifford, Geoffrey A. Lawrance, et al., 62, 12 Copyright © (2009) CSIRO Publishing)

described in terms of a different basic shape. There are obviously several factors influencing the outcomes we see experimentally.

There are some basic ideas we can immediately introduce to explain the experimental observations. Consider the simple cation [Co(NH$_3$)$_6$]$^{3+}$. As an ML$_6$ compound, we would predict initially (see Figure 3.3) that this complex cation will be octahedral in shape. From X-ray crystallography, this is exactly what is found; all N—Co—N angles between neighbouring ammonia groups are (or at least are very close to) 90°, and in additional all M—N bond distances are identical within very small margins of error. Experiment has justified use of our simple point-charge model. Now consider what happens if just one ammonia ligand is replaced by a bromide ion, to form [CoBr(NH$_3$)$_5$]$^{2+}$. The basic shape is still octahedral, but there are some changes found. First, the Co—N distances are different from the Co—Br distance. This is reasonable, since we are, after all, linking different species of different sizes, charges and shapes. However, the intraligand angles also change, with N–Co–Br angles opening out a little to be greater than 90° and some N—Co—N angles closing up a little to be less than 90°. This implies that the interactions 'sideways' between types of ligands differ – an ammonia and a halide interact in a nonbonding manner (or push against each other) differently than do two neighbouring ammonia molecules. We can define these effects between neighbouring ligands as *ligand–ligand repulsion forces*, called generally a *steric effect*. Simply, two ligands cannot occupy the same space, and compromises must be reached which involve bond angle distortion and bond length variation. In addition, there are other effects that are electronic in character. It is apparent that the Co—N distance for the ammonia bound directly opposite the bromide ion (*trans*) differs from those of the four bound adjacent to the bromide ion (*cis*). This could arguably be related to steric effects for *cis* groups differing from *trans* groups, but may also be related to electronic effects; in the simplest view, consider the latter as reflecting the way different ligands compete differently for d orbitals, or 'push' or 'pull' electron density to or from the metal centre.

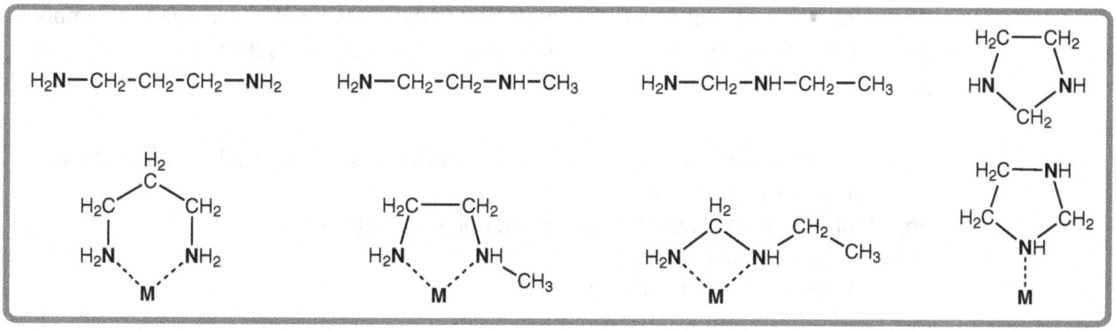

Figure 4.2
Chelation modes for simple linear diamines of common empirical formula $H_{10}C_3N_2$. Also included for contrast is a related molecule where the linear diamine to its immediate left has the five atoms in the chain confined in an organic ring; it can adopt only monodentate coordination to the metal.

Regardless of the identification of the effects, one obvious outcome is that the 'octahedral' $[CoBr(NH_3)_5]^{2+}$ ion is not an ideal octahedron.

If we replace the cobalt(III) centre in $[Co(NH_3)_6]^{3+}$ by a different metal ion such as Rh(III) or Ni(II), what we discover is that the octahedral shape is retained, but the average M—N distance changes. Obviously, each metal has a preferred distance between its centre and the donor atom. We can understand this in terms of both the size of the metal ion – the larger, the longer the bonds – and in terms of the charge on the metal ion. With variation in charge (or oxidation state) two factors operate; first, the number of d electrons differ, which can influence both shape preference and repulsive terms related to the number of valence shell electrons; second, the charge changes, influencing simple electrostatic interactions. The effects are exemplified by examining one metal in two oxidation states; for example, Co(III)—N bonds are invariably shorter than Co(II)—N bonds. Because a coordinate bond links two different atomic centres, it is reasonable to expect that preferred metal–donor distance changes with donor also. We have seen already how a Co(III)—N bond distance differs from a Co(III)—Br$^-$ bond distance. This is a universal observation – the preferred metal–donor distance varies with the type of donor even when the metal is fixed. In fact, the effect is quite subtle, as the Co(III)—NH_3 distance differs from that found for Co(III)—NH_2CH_3, even though they both form a Co—N bond. This can be accounted for by two factors; first, ammonia is a smaller, less bulky molecule than methylamine; second, the basicity of ammonia and methylamine differ, affecting their relative capacities to act as a lone pair donor (Lewis base) in forming a coordinate covalent bond to the metal ion (Lewis acid). The size and also the shape of ligands can force much more dramatic effects upon their complexes, as we shall see, so it is apparent that ligand geometry and rigidity are important. We can illustrate this in part for three simple diamines, all of formula $H_{10}C_3N_2$. Chelation of each of these in turn would produce a six, five and four-membered ring (Figure 4.2). With five-membered chelate rings usually being more stable than six- or four-membered rings, it is not surprising to find that shapes of complexes with the ligand forming a five-membered chelate ring are less distorted than those with the other ligands. Further, if we join the terminal carbon and nitrogen atoms in the latter example to form a small five-membered heterocyclic ring (Figure 4.2), the ligand shape is now such that only one of the two amine groups can bind to a single metal centre at any one time, the other having its lone pair directed in an inappropriate direction for chelation.

There are in fact several effects that contribute to the outcome of metal–ligand assembly. Overall, stereochemistry and coordination number in complexes appear to depend on four key factors:

1. central metal–ligand electronic interactions, particularly influenced by the number of d (or f) electrons of the metal ion;
2. metal ion size and preferred metal–ligand donor bond lengths;
3. ligand–ligand repulsion forces;
4. inherent ligand geometry and rigidity.

The simple amended VSEPR point-charge model discussed in Chapter 3 is based in effect on the third of the above, but despite this provides a basis for predicting shape. However, it must, because of its limitations, be deficient in predicting shape in metal complexes. We shall see as we explore actual shapes below that it is, nevertheless, a good starting point.

4.2   Forms of Complex Life – Coordination Number and Shape

Molecules are certainly more varied than life forms. Even carbon-based compounds can be considered as unlimited in number, despite the fact that they almost exclusively involve four bonds around each carbon centre in a very limited number of shapes. When we move to coordination compounds, the range of coordination numbers and shapes is expanded considerably, so that coordination complexes live up to their name – they are inherently complex molecular forms. Fortunately, we can identify a number of basic shapes and even some system that governs outcomes – that is, there *is* some predictive aspect to shape in coordination complexes. We shall examine complexes from the perspective of coordination number below.

4.2.1   One Coordination (ML)

This unlikely coordination number suffers from the fact that a single donor bound to the metal would still leave the metal highly exposed, a situation that would most likely lead to additional ligands adding and thus increasing the coordination number. It is nevertheless prudent to describe it as extremely rare, because there is a small possibility that a suitably bulky and appropriately shaped ligand may achieve one-coordination. It may be more practicable in the gas phase under high dilution conditions, where metal–ligand encounters are limited.

As a consequence, it is not surprising that there appears to be only one isolated structure claimed. This is of the indium(I) and thallium(I) complexes with a single M—C bond from a σ-bonded benzene anion that carries two bulky tri-substituted benzene substituents in *ortho* positions; these partially block approach of other potential donors to the metal cation (Figure 4.3). An X-ray crystal structure has been determined, defining the shape. It is a rare observation, as other complexes (such as Mn(I) and Fe(I)) of the same ligand bind another ligand at the 'open' side of the molecule, as expected. In any case, this coordination number is trivial in the sense that it can have only one shape – a linear M—L arrangement.

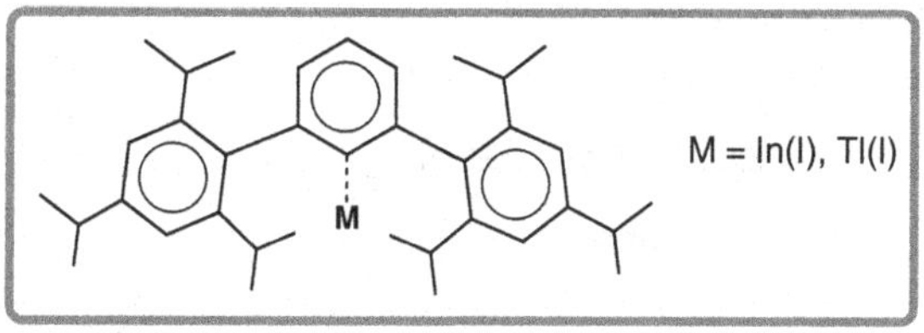

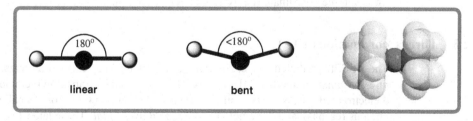

Figure 4.3
An extremely rare one-coordinate complex. Rotation about the single bonds linking the aromatic rings relieves the steric clashing of *iso*-propyl substituents (apparent in this drawing), and potentially opens up the other side of the metal for addition of another ligand.

4.2.2 Two Coordination (ML_2)

Two coordination is the lowest stable coordination number that is well reported. It also presents the first opportunity for variation from the shape predicted in Figure 3.3. We expect a ML_2 molecule to be linear, with the two donor groups disposed as far away from each other as possible on opposite ends of a line joining them and passing through the metal centre. Deviation from prediction may arise in this case simply by *bond angle deformation*, with the usual 180° L—M—L bond angle reduced to <180° through bending (Figure 4.4).

Experimentally, ML_2 complexes are overwhelmingly *linear*. Both electron pair repulsion and simple steric arguments favour this shape. If bending occurs, it brings the two ligands closer towards each other, providing greater opportunity for repulsive interaction between the ligands; this would seem both unreasonable and unlikely, yet bent molecules do occur. 'Bent' geometries are well known in p-block chemistry, of course, where lone pairs play an important directional role. Water is the classical example, with its two lone pairs and two bond pairs around the oxygen centre leading to a bent H—O—H as a result of its inherently tetrahedral shape (on including lone pairs) as well as additional effects due to differing repulsions between the lone pairs and bond pairs. Where such bending is seen in metal complexes, it can often be assigned to a higher pseudo-coordination number shape with nonbonding orbitals present contributing and occupying some region of space. X-ray crystallography defines atomic centres but cannot readily detect regions of electron density that do not involve atoms. Lone pairs are effectively not observable directly; the presence of directed nonbonding electron density can only be inferred, as a result of an influence on structure through repulsive terms, seen experimentally as changes in bond angles.

The ML_2 geometry is rare for all but metal ions rich in d electrons, particularly d^{10} and d^9 metal ions. This is a recurring theme in coordination number for transition metal

Figure 4.4
Possible shapes for two-coordination, and (at right) an example of a linear complex cation, $[Au(PR_3)_2]^+$ (where R = CH_3).

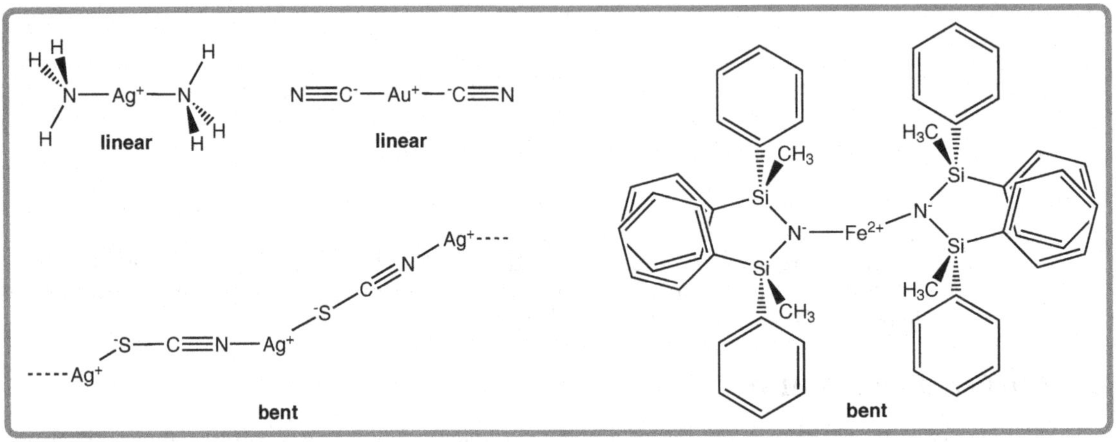

Figure 4.5
Examples of complexes with two-coordination, including both linear and bent species.

complexes – as a very rough rule, the more electrons in the valence shell, the lower the coordination number. Complexes which are two-coordinate include those of the d^{10} cations Ag(I) and Au(I), for example the $[Ag(NH_3)_2]^+$ and $[Au(CN)_2]^-$ complexes, which have linear N—Ag—N and C—Au—C cores respectively; a related example with a bulkier PR_3 ligand is drawn in Figure 4.4. It is also believed that the simple dihalides (MX_2) of most d-block metal ions in the gas phase tend to be linear two-coordinate species, although this is not generally the case in the solid or solution state. Examples of bent complexes are few in number. One of the simplest is Ag(SCN), which in the solid state forms a polymer with . . . SCN—Ag—SCN . . . character, and a S—Ag—N angle of 165°. The Ag—N≡CS linkage is driven towards linearity by the multiply-bonded nitrogen, with a lone pair directed towards the metal directly opposite the multiple bond, whereas for the Ag—S—CN linkage the set of lone pairs on the pseudo-tetrahedral S atom influence the coordination and lead to a bent arrangement (Figure 4.5).

Occasionally, very bulky ligands enforce two-coordination on metals with fewer d electrons that would usually prefer other higher coordination numbers. One of very few such compounds is the d^6 Fe(II) complex of the amido anion $^-N(SiR_3)_2$. The particular two-coordinate example $[Fe\{N(Si(CH_3)(C_6H_5)_2)_2\}_2]$ approaches linear shape (Figure 4.5), but the N—Fe—N angle is only ~170°. As the bulk of the substituents on the Si atoms decrease, there is a switch to higher coordination number; the anion $^-N(Si(CH_3)_3)_2$, for example, forms a three-coordinate $[Fe\{N(Si(CH_3)_3)_2\}_3]$ complex.

4.2.3 Three Coordination (ML_3)

The VSEPR-predicted shape, *trigonal planar*, is well represented amongst this relatively rare coordination number. ML_3 is (like ML_2) favoured by transition metal ions with lots of d electrons (d^8, d^9, d^{10}). Two other shapes are known, however; one is called *T-shape* (for obvious reasons), and the other called *trigonal pyramidal*. These latter two can be seen to arise from distortions of the 'parent' trigonal planar shape, as depicted in Figure 4.6. It is usual for different shapes for a particular coordination number to be able to interconvert without any bond breaking, simply through rearrangements such as those exemplified in

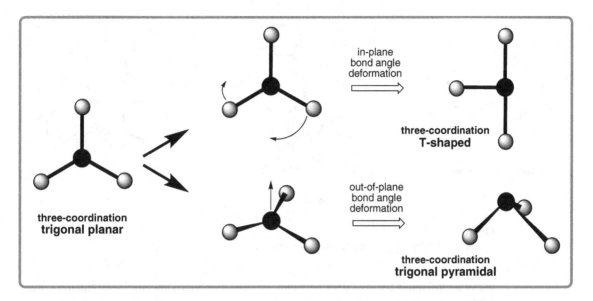

Figure 4.6
Trigonal planar, the parent shape of three-coordination, and the way it can transform to other known geometries.

Figure 4.6. There is an obvious outcome of this. These various shapes reported in this chapter should be considered as limiting shapes – they are the extremes or termini of change, and as a consequence molecules may adopt a shape that is intermediate or part-way along the process of changing from one basic shape to another. This is something that three-dimensional structural studies have clearly identified. However, it is both convenient and essential if we are to have any pattern to our family of shapes to identify a coordination complex in terms of its nearest basic shape – and in most cases the deviation from a basic shape is small, suggesting that they do have inherent stability.

A simple trigonal planar complex ion is $[HgI_3]^-$. Two more elaborate examples of trigonal planar geometry, both involving bulky ligands, are $[Cu(SR_2)_3]^+$ (where $R = CH_3$) and $[Pt(PR_3)_3]^+$ (where $R = C_6H_5$) for which all metal–donor bonds are equivalent (and, coincidentally, both Cu—S and Pt—P distances are 226 pm) and L—M—L angles all very close to the expected 120° (Figure 4.7). Whereas bond distances are usually identical also for the three-coordinate trigonal pyramidal geometry, the L—M—L angles are all <120°, and diminish the further the metal is raised above the plane of the donors.

A rare example of T-shaped geometry is $[Rh(PR_3)_3]^+$ (where $R = C_6H_5$), which (unlike its trigonal planar Pt analogue) has one P—Rh—P angle close to 180°, and the others near 90°, with Rh—P distances also not identical. Overall, trigonal planar is the dominant shape of three-coordination, and this is the shape predicted by Kepert's amended VSEPR model.

4.2.4 Four Coordination (ML$_4$)

Coordination number four (ML$_4$) is common and has two major forms, *tetrahedral* and *square planar*. The former is the shape predicted by the electron pair repulsion model; the latter is a different shape observed experimentally, with many examples known. These are ideal or limiting structures, in the sense that they represent the perfect shapes which lie at

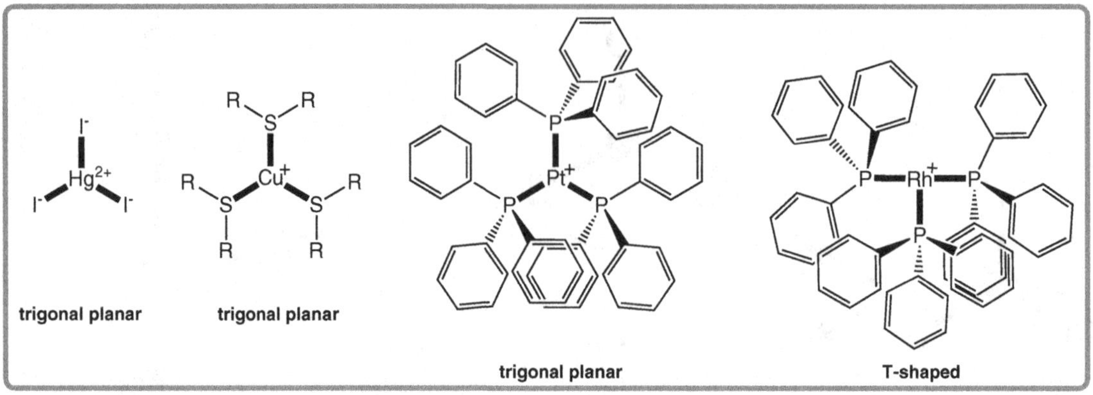

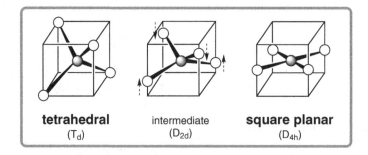

Figure 4.7
Examples of complexes with trigonal planar or the rarer T-shaped geometry.

the structural limits for this coordination number; as mentioned already, ideal structures are relatively rare in coordination chemistry, and distorted or *intermediate* geometries are more likely met, so named since they can be achieved by distorting one or other shape partially towards the other class. The two limiting geometries can be converted one into the other by displacement of groups without any bond breaking being involved, which is energetically less demanding, since bond angle deformations with their lower energy demand than bond breaking are then the dominant energy 'cost'.

The shapes of the two limiting and an intermediate geometry are shown in Figure 4.8, both based on a cubic box frame. The tetrahedral shape is defined by placing two donors on opposite corners at the top of the box, and the other two donors on opposite corners at the bottom of the box, but bottom corners that do not lie directly beneath the corners occupied at the top of the box. This geometry can be converted to the square planar shape by simply 'sliding' the top two donors down two edges of the cubic box and 'sliding' the other two bottom donors up the other two edges of the box until they are all half-way along the edges. When this occurs, they are all co-planar, and also co-planar with the metal ion placed at the centre of the box; this is, in effect, the square planar shape. If the 'sliding' is stopped part-way, then an intermediate distorted shape is achieved. In reality, all types are well known;

tetrahedral   intermediate   **square planar**
(T_d) (D_{2d}) (D_{4h})

Figure 4.8
The two limiting shapes for four-coordination, tetrahedral and square planar, along with an intermediate geometry formed in transition from one limiting shape to the other.

for example, for simple $[MCl_4]^{2-}$ ions, d^7 Co(II) is tetrahedral, d^8 Ni(II) is an intermediate geometry, and d^9 Cu(II) is square planar in shape. Many complexes described as square planar display small tetrahedral distortions that place the two pairs of donors slightly above or below the average plane including the metal ion; likewise, many tetrahedral complexes display small distortions towards square planarity. Where these distortions are minor, it is convenient to ignore them in defining the basic shape.

It is also often convenient to represent shapes in terms of their actual symmetry, expressed as the appropriate mathematical 'label' – T_d for tetrahedral, D_{4h} for square planar, and D_{2d} for intermediate geometries in this case – as this defines the shape succinctly and is appropriate for application in spectroscopy. There are simple rules for deciding the symmetry of a complex, and these are described and exemplified in Appendix B.

Tetrahedral or distorted tetrahedral geometries are, from experimental observations, dominantly found in complexes that are overall *neutral* or *anionic*. Simple examples include $[CuX_4]^{2-}$, $[FeX_4]^{2-}$ and $[CoX_4]^{2-}$ (X = halogen anion). Otherwise, it is d^0 compounds (such as $[TiCl_4]$) or d^{10} compounds (such as $[Ni(PF_3)_4]$ and $[Ni(CO)_4]$) that lean towards tetrahedral geometry. Other d^n configurations (except d^3) exhibit limited examples, and then only usually for the first row transition metals. Ligand arrangement in the tetrahedral geometry minimizes inter-ligand repulsions, so negatively charged ligands prefer this shape because of the more favourable charge-based repulsions when adopting this shape. However, this shape does lead to a weak ligand field when compared to what occurs in square planar systems, which can make this geometry less effective. With a balance between favourable repulsions and unfavourable ligand field effects, it is not surprising to find that steric effects, or ligand size, are an important consideration in this geometry.

Square planar complexes mainly used as examples are those having a d^8 metal ion, such as Rh(I), Ir(I), Pd(II), Pt(II) and Au(III). Several examples are shown below (Figure 4.9). One consequence of square planarity is clear in these examples – there are structural isomers possible. The neutral $[PtCl_2(NH_3)_2]$ exists in two geometric isomers, *trans* (where each pair of groups is as far apart as possible on opposite sides of the molecule) and *cis* (where the two pairs occupy adjacent sites). Like all geometric isomers, they display distinct chemical and physical properties, including biological properties; the *cis* isomer is otherwise known as the anti-cancer drug cisplatin (see Chapter 8), but the *trans* isomer is not effective as a drug.

That square-planarity is common for d^8 complexes does not mean that this shape is not seen for other metal ions, and indeed it is reasonably common amongst metals with from d^6 to d^9 configurations, It is also important to note that not all d^8 systems are square planar. Some metal ions are ambivalent, with the ligand field playing a clear role in the outcome;

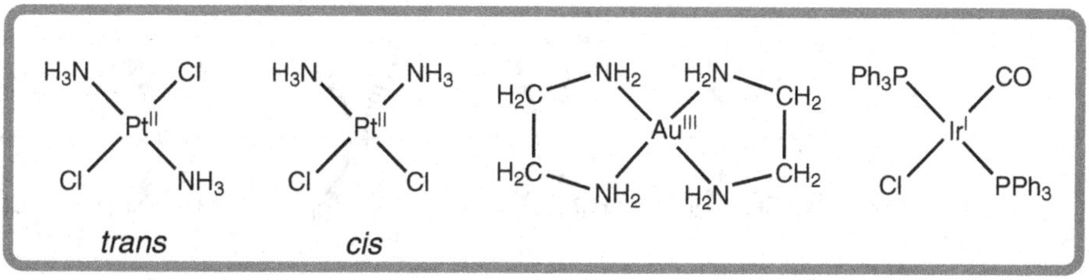

Figure 4.9
Examples of four-coordinate square planar complexes.

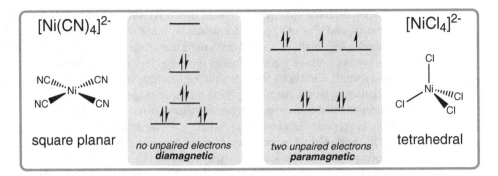

Figure 4.10
Examples of d^8 nickel(II) complexes adopting square planar or tetrahedral geometry, depending on the type of ligand. The square planar complex has no unpaired electrons (diamagnetic) whereas the tetrahedral complex has two unpaired electrons (paramagnetic), allowing easy identification (energy splitting diagrams not to scale).

the d^8 Ni(II) ion can form octahedral, tetrahedral and square-planar complexes, with strong-field ligands tending to favour square-planarity. They are distinguished readily from their colours and spectroscopic properties; for example, octahedral NiN$_6$ complexes are purple high-spin species and square-planar NiN$_4$ complexes yellow low-spin species (here, N represents an N-donor ligand, not an isolated atom). This ambivalence is a reminder that the square planar geometry can be reached by removal of two *trans*-disposed ligands from around an octahedron, as well as by bond angle distortion of a tetrahedron. The structural ambivalence of d^8 Ni(II) is also exemplified in Figure 4.10, where different ligands lead to different geometry; apart from spectroscopic differences, magnetic properties of square planar and tetrahedral complexes differ, the latter alone having unpaired electrons.

Since square-planar complexes are essentially two-dimensional compounds, it is not a surprise to find other molecules coming into reasonably close but still nonbonding positions (distances >300 pm) above and below the plane in the solid state and in solution, including metal–solvent and even some cases of metal–metal interaction. The square planar cation–anion pair $[Pt(NH_3)_4]^{2+}[PtCl_4]^{2-}$ (called, in earlier times, Magnus' green salt) for example, stack cationic and anionic complexes in alternating positions in their crystalline lattice, with a M . . . M separation of $\sim$325 pm (Figure 4.11).

The separation is close enough to have an effect on physical properties, with the green colour indicating some weak metal–metal interaction, since another form of the salt with a very long separation of over 500 pm and no semblance of interaction is pink, colour arising

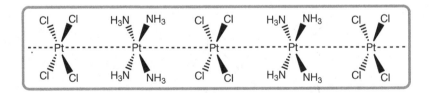

Figure 4.11
Stacking of a d^8 platinum(II) complex in the solid state leads to weak Pt· · ·Pt interaction, influencing physical properties.

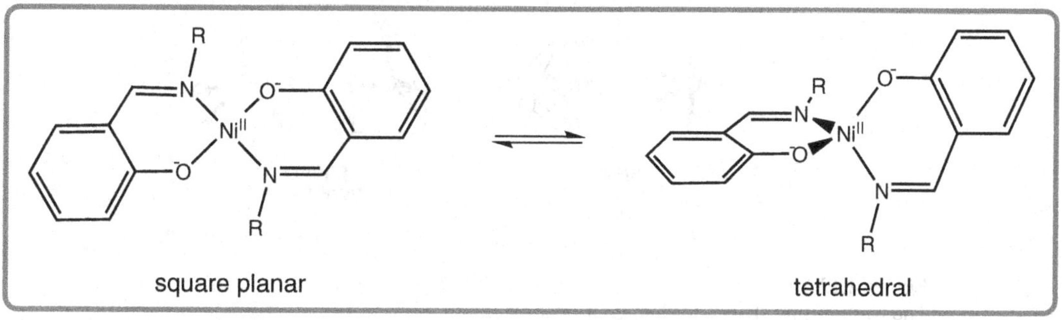

square planar   tetrahedral

Figure 4.12
An N,O-chelate ligand whose Ni(II) complex displays interconversion between different geometries depending on conditions.

in that case simply from combination of the colours of the two independent ions. When metal ions come very close together, this interaction will also affect the magnetic properties of their assembly significantly.

In describing the relationship between tetrahedral and square planar complexes above, we used a model where interconversion could occur without bond breaking. If this is a fair representation of reality, then it should be possible to find some systems that exist either as a mixture of the two forms in equilibrium or can convert between the two forms when a change in conditions is applied. Fortunately, there are indeed some compounds that undergo conversion between tetrahedral and square planar forms in solution, a situation which implies that the stabilities of the two forms are very similar. One now classical example involved the Ni(II) complex of a chelated N,O-donor ligand shown in Figure 4.12, where conversion between dominantly tetrahedral and dominantly square planar forms depends on the temperature, solvent and the type of R-group attached to the coordinated imine nitrogen. Change between the two forms can be readily monitored, since their colours and absorption band positions and intensities are different.

4.2.5   Five Coordination (ML₅)

Examples of ML$_5$ are found for all of the first row transition metal ions, as well as some other metal ions. Although once considered rare, growth in coordination chemistry has led to five-coordination becoming met almost as frequently as four-coordination. This exhibits one of the limitations of making comparisons of this type; rarity may not be a result of any inherent restriction, but may simply reflect limited experimental development. Given that four-coordination is common and six-coordination very common, it is perhaps not surprising to find five-coordination also having matching status, at least for lighter, smaller metal ions. Five-coordination is not commonly met in complexes of the heavier transition metals, however.

The amended VSEPR model predicts two forms of five-coordination, and experimental chemistry has clearly identified many examples of both forms. These limiting structures are square-based pyramidal (or, simply, square pyramidal) and trigonal bipyramidal (Figure 4.13). The classical square-based pyramidal shape is formed simply by cleaving off one bond from an octahedral shape, which leaves the metal in the same plane as the four square-based ligands. In reality, almost no complexes exhibit this shape, but rather adopt a distorted

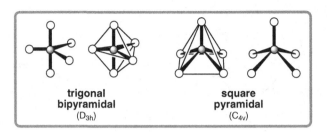

Figure 4.13
The two limiting shapes for five-coordination.

square pyramidal shape where the metal lies above the basal square plane, with typically the angle from the apical donor through the metal to each donor in the base around 105° instead of 90°. Considering electron pair repulsion alone, this distorted shape (distorted only in terms of the metal location; the Egyptian pyramid shape created by the donor groups is otherwise unaltered) is actually more stable than the form created by simply truncating an octahedron, and is only slightly less stable than the trigonal bipyramidal geometry. As a consequence, it has become usual to regard the square pyramidal shape as that with the metal above the pyramidal base plane, and you will see it represented in this way almost exclusively.

Once again, as described for three- and four-coordination, it is possible to convert from one form to the other through bond angle changes without any bond-breaking. Both geometries are common, but in practice there are many structures that are intermediate between these two. The two limiting structures are of similar energy and as predicted, some complexes display an equilibrium between the two; $[Ni(CN)_5]^{3-}$, for example, crystallizes with both structural forms of the anion present in the crystals.

If we examine the two five-coordinate shapes from a crystal field perspective, the d orbitals split in a different way to that found for octahedral, tetrahedral and square planar shapes since the d orbitals find the ligands in clearly different locations in space. The crystal field splitting pattern for the two is shown in Figure 4.14. From this pattern, crystal field stabilization energies can be calculated, and favour the square pyramidal geometry in all cases (apart from the trivial situations d^0 and d^{10}) except for high spin d^5. This prediction differs from the outcome from the electron pair repulsion model.

Although electron pair repulsion and crystal field stabilization energy (CFSE) influences operate, it appears nevertheless that the steric or shape demands of at least polydentate ligands play a dominant influence on complex shape. This is exemplified in Figure 4.15, where an example of a 'three-legged' ligand shape fits best to the trigonal bipyramidal geometry, occupying the top four positions of the complex shape, with a fifth simple ligand then occupying the underside – the whole assembly looks a little like an open umbrella. The ligand is pre-disposed to this shape, with limited steric interaction when bound. Likewise, the cyclic tetraamine represented in Figure 4.15 is predisposed to binding in the square pyramidal shape.

Overall, trigonal bipyramidal is found rather than square pyramidal shape except where steric requirements of polydentate ligands are important or where π bonding occurs (as in vanadyl complexes $[VOL_4]$ where the V=O bond occupies the axial site with four other ligands in the basal plane). However, exceptions abound, reflecting the similar energies of the two forms. Examples of complexes with trigonal bipyramidal geometry are

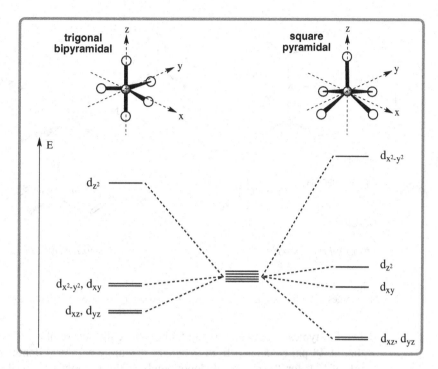

Figure 4.14
Crystal field splitting pattern for trigonal bipyramidal (left) and square pyramidal (right) ML$_5$ complexes.

[Fe(CO)$_5$], [Co(NCCH$_3$)$_5$]$^+$, [NiBr$_3$(PEt$_3$)$_2$] and [CuCl$_5$]$^{3-}$; square pyramidal shapes occur for [Ni(CN)$_5$]$^{3-}$, [VO(SCN)$_4$], [ReBr$_4$O]$^-$ and [Fe(NO)(S$_2$CNEt$_2$)$_2$] (Figure 4.16).

Whereas a limited number of complexes display equivalent lengths for all bonds, it is more common for some distortions from the regular stereochemistry to be found, both in terms of bond distances and angles. For example, in trigonal bipyramidal [Co(NCCH$_3$)$_5$]$^+$ the two axial Co—N bonds are 5 pm longer than the three equatorial bonds, and in [CuCl$_5$]$^{3-}$ axial bonds are 95 pm shorter than equatorial bonds. For square pyramidal complexes, the

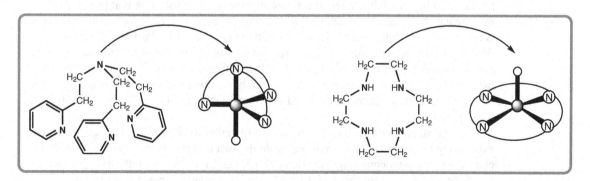

Figure 4.15
Ligand shape directing complex shape in five-coordination.

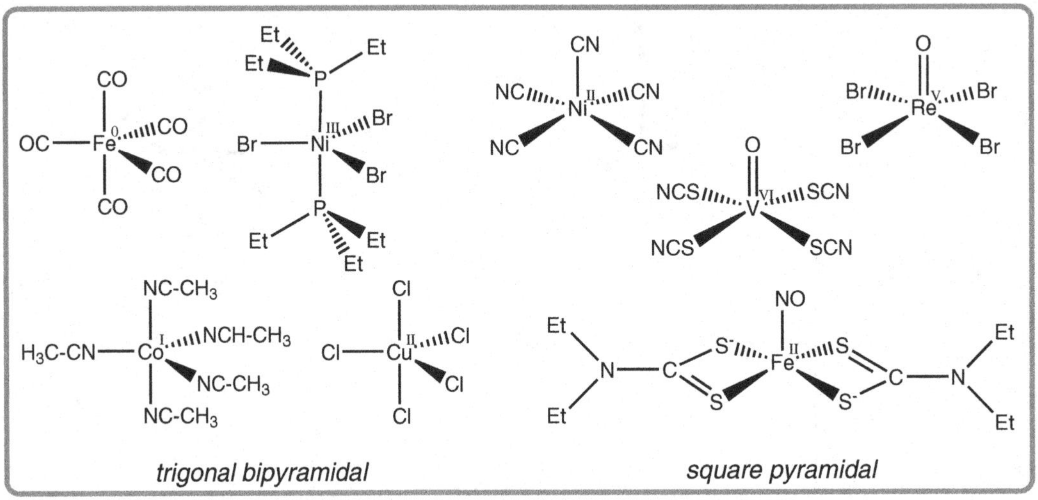

Figure 4.16
Examples of complexes adopting one of the two shapes of five-coordination.

metal lies typically between 30 and 50 pm above the square plane of donors, close to the value of 48 pm calculated from geometry assuming an apical donor–metal–basal donor angle of 104° for 200 pm metal–donor bonds. In this stereochemistry, the single axial bond tends to be longer than the four equatorial bonds; for $[Ni(CN)_5]^{3-}$, the former is 217 pm and the latter are 187 pm.

4.2.6 Six Coordination (ML$_6$)

ML_6 is the most common coordination type by far that is met for transition metal elements (seen for all configurations from d^0 to d^{10}), and also is often met for complexes of metal ions from s and p blocks of the Periodic Table. Of the two limiting shapes (Figure 4.17), the *octahedral* geometry is by far the most common, though a few examples of the other limiting shape, *trigonal prismatic*, exist. Because the six donor atoms come into closer contact in the trigonal prismatic than in the octahedral geometry, trigonal prismatic is predicted to be less stable. However, many structures show distortion that places them as intermediate between ideal octahedral and the ideal trigonal prismatic form. This distortion is best viewed in terms of the orientations of two opposite triangular faces of sets of three donors. For octahedral (which is also a special case of trigonal antiprismatic, where edge and face length uniformity applies), the 'top' face is twisted around 60° versus the 'bottom' face; for trigonal prismatic, the two faces exactly superimpose (the twist angle has been reduced to 0°). Intermediate or distorted structures show a twist angle of between 60° and 0° (Figure 4.17).

The trigonal prismatic geometry is usually enforced by the ligand. Simple ligands almost exclusively form octahedral complexes, as do the vast majority of polydentate ligands. Of complexes of monodentate ligands, $[Re(CH_3)_6]$, $[Zr(CH_3)_6]^{2-}$ and $[Hf(CH_3)_6]^{2-}$ (which are d^0 or d^1 systems) are three of a very few with simple σ-bonded ligands that display trigonal prismatic geometry. A fairly rigid chelate with a short separation between the two donor atoms (the so-called 'bite' of the ligand, the preferred separation of donors from each

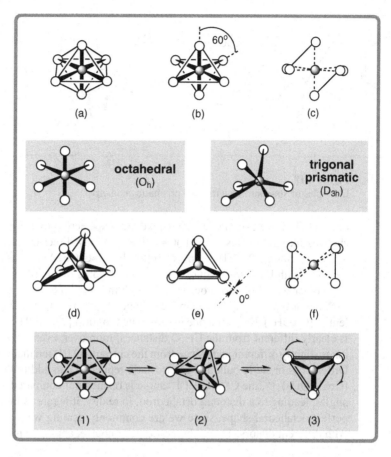

Figure 4.17

Six-coordinate geometries. Views of the common octahedral and rare trigonal prismatic shapes with (a), (d) all faces defined; with the trigonal faces carrying sets of donor groups defined through views perpendicular into the face (b), (e); and side on, showing the 'sandwich' character of the L_3ML_3 sets (c), (f). Note how the angle between the triangular faces projected onto a common plane varies from 60° in the octahedral case to 0° in the trigonal prismatic case. Twisting of one octahedral face (1) with the other fixed in position through an intermediate (2) to the trigonal prismatic geometry (3) provides a mechanism for interconversion without bond breaking.

other, influenced by their ligand framework) may also direct the shape to trigonal prismatic. Transformation between the two limiting geometries of octahedral and trigonal prismatic can occur simply by rotation of one triangular face relative to the other, as illustrated in Figure 4.17.

The dominance of octahedral geometry may be assigned to a number of factors: this arrangement is favoured by the amended VSEPR concept, is an ideal shape for minimizing steric clashing between donors, promotes good metal–ligand orbital overlap, and leads to favourable ligand field stabilization energies. Distortions from pure octahedral geometry can arise from twisting of faces, as discussed above, and this effect is most likely met with chelate ligands, where the 'bite' of the chelate donor groups may be satisfied by a twisting distortion. Most chelates do not lead to this outcome, and octahedral predominates, except that chelates with non-ideal 'bite' are those that often show some distortion away from ideal

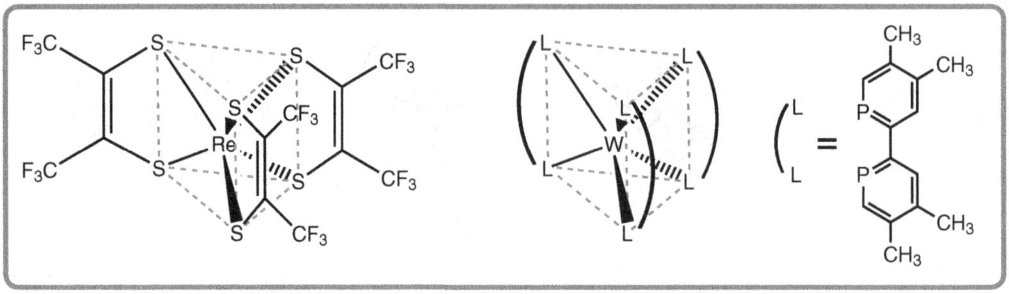

Figure 4.18
Examples of six-coordinate trigonal prismatic geometry.

octahedral. In the extreme, this may take the shape across to trigonal prismatic; fairly rigid chelates like dithiolates ($^-S-(R)C=C(R)-S^-$) may direct this outcome, as found in the trigonal prismatic Re(VI) complex [Re(S(CF$_3$)C=C(CF$_3$)S)$_3$]. The tungsten(0) complex of a chelate with P-donors (Figure 4.18) also is trigonal prismatic.

Another way distortions occur is through the attachment of a mixture of ligands, since each donor type has a preferred bond distance. For example, although all bonds are equivalent in [Cr(OH$_2$)$_6$]$^{3+}$, when one is substituted to form [CrCl(OH$_2$)$_5$]$^{2+}$, the Cr—Cl distance is clearly different from the Cr—O distances; moreover, even the O-donor opposite the Cl ion is slightly different in distance from the chromium(III) ion than the remainder O-donors. Differences in steric bulk and electrostatic repulsion can add to these distortions; for *cis*-[CrCl$_2$(OH$_2$)$_4$]$^+$, the Cl$^-$—Cr—Cl$^-$ angle is opened out compared with the H$_2$O—Cr—OH$_2$ angles, leading to a distorted octahedron. In reality, it is rare to find a complex that adopts perfect octahedral shape, since we are commonly dealing with ligands or sets of ligands with different donors. These observations, of course, apply to any basic stereochemistry.

4.2.7   Higher Coordination Numbers (ML$_7$ to ML$_9$)

Beyond ML$_6$ lies a range of higher coordination numbers that reach as high as ML$_{14}$. While we will not dwell on these at any great length at this level, it is appropriate to be aware of some of these, and how they may arise. Thus we will briefly examine ML$_7$ through to ML$_9$. For these, a point-charge model of the distributions predicted for charges dispersed on a spherical surface can still be applied, and predicted shapes are found experimentally. Rather than expand on this aspect, however, we will examine another approach to understanding how some of the structures arise by relating them to our lower coordination numbers already described.

If we examine selectively some shapes for two- to six-coordination already discussed, we can arrange them in such a manner where they are related by an increase in the number of groups dispersed in a symmetrical fashion around a plane including the metal, along with an additional group in each of the two axial sites (Figure 4.19). Following this trend beyond six-coordinate octahedral, where there are four donor groups around the centre in a square arrangement, it is possible to suggest that one seven-coordinate shape could arise through placing five groups around the centre (the symmetrical arrangement for which is a pentagon) or else, for eight-coordination, six groups (in a hexagonal arrangement). These would lead to a pentagonal bipyramid and a hexagonal bipyramid respectively (Figure 4.19). As it happens, both these are known shapes for seven- and eight-coordination respectively.

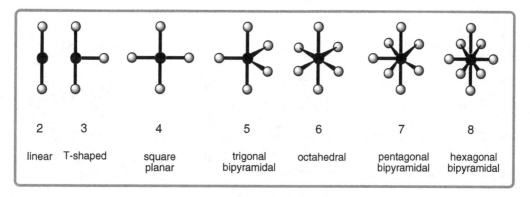

Figure 4.19
Shapes for coordination numbers from 2 to 8 which reflect a trend involving a stepwise addition of groups arranged around the central plane that also includes the metal.

How far this approach can be extended depends on the size of the metal ion and the size of the donor groups, as clearly steric factors will become important as more and more groups are packed into the plane around a metal ion. This also introduces one of the general observations regarding complexes with high coordination numbers – they are found with larger metal ions, or else with metal ions that exhibit long metal–donor distances, as both these aspects contribute to a reduction in steric 'crowding' of the metal centre.

Another approach to expansion of the coordination number is through expanding the layers (or planes) of donors around a metal ion. This concept is best explained with some illustrations. If we consider the six-coordinate trigonal prismatic shape, we can visualize this as a metal ion in a central layer and two layers of donors above and below the central layer, as illustrated (Figure 4.20). We can then expand the coordination number in two ways: addition of extra donors to the upper and lower planes of donors; or addition of donors to the central plane containing the metal, analogous to that already described in Figure 4.19 except here we are starting with more than one group in 'axial' locations. Converting the two donor layers from three to four donors each, effectively converts each trigonal plane of donors to a square plane of donors, leading to eight-coordination *cubic*. In reality, steric clashing between the layers is significantly reduced by a rotation of the top square through 45° to produce a *square antiprismatic* (or Archimedean antiprismatic) shape (Figure 4.20). This is analogous to twisting the six-coordinate trigonal prismatic shape through 60° to produce the preferred octahedral shape, discussed earlier. In line with the expectations for relative stability, the square antiprismatic shape is experimentally much more common than cubic, although both are known; for example, the $[MF_8]^{5-}$ complex ion for M = Pr(III) is cubic but for M = Ta(III) is square antiprismatic.

Alternatively, adding additional donors into the central plane including the metal, best achieved by making a new bond through the centre of a square face of the six-coordinate trigonal prism, leads to different geometries (Figure 4.20). With one insertion, a seven-coordinate *one face-centred* (or mono-capped) *trigonal prismatic* structure results, whereas, if an addition is made to all three faces, the nine-coordinate *three-face centred* (or tricapped) *trigonal prismatic* form is obtained. There are additional shapes for these coordination numbers, such as a *one face-centred octahedral* (seven-coordinate) and *dodecahedral* (eight-coordinate), but we shall not extend the story. What has been established is that higher

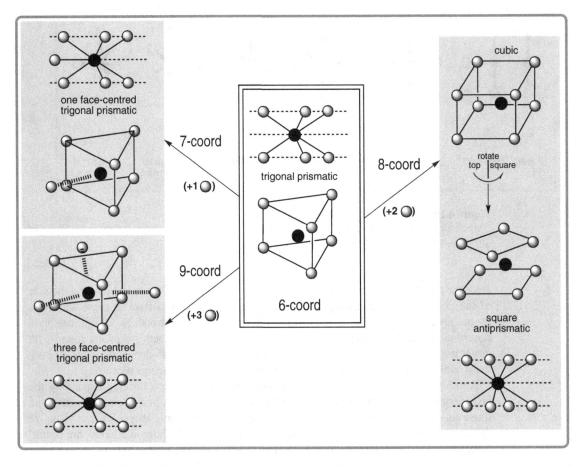

Figure 4.20
Methods of converting the six-coordinate trigonal prismatic shape, where the metal is 'sandwiched' between two layers of donors, into various seven-, eight- and nine-coordinate shapes through addition of other groups either in the plane of the metal to form another layer of donors or else in the two existing layers of donors to expand the set of donors in those layers.

coordination numbers can be evolved or understood through extrapolation from the known lower coordination shapes. Again, distortions from these limiting shapes are common in actual complexes.

Because there is a 'crowding' of ligands around the metal ion in these higher coordination number species, repulsive interactions between adjacent ligands become more important than in lower coordination number complexes. Thus it is smaller ligands that tend to occupy sites in the coordination sphere of such d-block metal ions. For example, the small fluoride ion forms with zirconium(IV) the $[ZrF_7]^{3-}$ complex ion, which has the pentagonal bipyramidal geometry, whereas the tiny hydride ion forms with rhenium(VII) the tricapped trigonal prismatic $[ReH_9]^{2-}$. Further, larger metal ions, with longer metal–ligand bond lengths, also tend to support higher coordination numbers. Thus molybdenum(V) exists as either square antiprismatic or dodecahedral $[Mo(CN)_8]^{3-}$ ion, the shape dependant on the cation involved. However, it is with the f block that the higher coordination numbers are dominant, with examples of complexes with coordination numbers below seven much less common than those with seven or higher coordination. For example, for the aqua ions $[M(OH_2)_n]^{3+}$, $n \leq 6$ for the d-block metal ions, whereas $n \geq 7$ for the f-block metal ions.

Idealized structures up to nine-coordination are summarized in Figure 4.21. These do not represent all of the shapes met, since, apart from all these idealized structures, it is necessary to remember that bond angle and bond length distortions of these structures can occur; some of the shapes resulting from these effects are themselves common enough to be represented as named shapes, and we have discussed some examples of these earlier. Further, beyond nine-coordination, an array of additional shapes can be found, of which perhaps the best known are the *bicapped square antiprism* (for ten-coordination), the *octadecahedron* (for eleven-coordination) and the *icosahedron* (for twelve-coordination). Clearly, the options are extensive, so it may be time to find out what directs a complex to take a particular shape.

4.3 Influencing Shape

4.3.1 Metallic Genetics – Metal Ion Influences

Although a coordination complex is an assembly of central atom and donor molecules or atoms, and as such is best viewed holistically, it is still instructive to identify factors influencing shape that depend on each component. If we focus first on the central atom or ion, there are two of the four key factors mentioned earlier that we can consider metal-centric. These are:

- the number of d electrons on the metal ion; and
- metal ion size and preferred metal ion–ligand donor group bond length.

Each metal in a particular oxidation state brings a unique character, almost like a gene, to play in its complexes. Examples of the way the size and bond distances vary across the d block are given in Table 4.1, reporting data for N-donor, O-donor and Cl^- ligands. The M—L distances are averages only, as distances vary over a range of at least 20 pm, influenced by the specific type of donor group for a particular type of donor atom, the ligand shape and associated strain energy, as well as influences of other donors in the coordination sphere (through what are *trans* or *cis* effects of an electronic nature). Moreover, the spin state of the central metal plays a role (e.g. M—O distances for high spin Mn(III) are typically 20 pm longer than for low-spin compounds).

As mentioned above, distances vary for any particular metal ion depending on the character of the donor (carboxylate-O or alcohol-O, for example, or else amine-N versus amido-N), influences from the ligand framework itself, and influences of other ligands bound to the same complex. Some modest trends are apparent, such as the fall in M—L distance with increasing charge on central metal ions with the same number of d electrons. There are some more consistent trends, nevertheless; for example, the variation in bond length M—Cl > M—N > M—O is almost universally observed. Further, there is a modest relationship between bond distances and the size of the metal ion. The increase in metal cation size from the first to the second and third row of the Periodic Table is accompanied by usually longer M—L distances, as exemplified for Co(III), Rh(III) and Ir(III) in Table 4.1. Overall, metal–donor distances fall within a range of ~160–260 pm, with the smaller distances found where highly charged metal ions, small anionic ligands and/or multiple bonding operate. For example, the $V^{IV}{=}O$ distance of ~160 pm is markedly shorter than usual $V^{IV}{-}O$ distances of ~185 pm. Experimentally, every metal ion–donor group unit

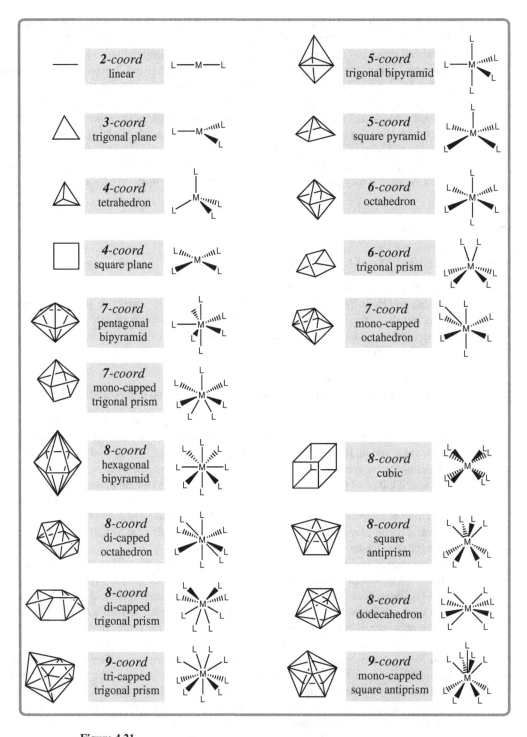

Figure 4.21
A summary of ideal polyhedral shapes for the limiting structures for coordination numbers 2 to 9.
More practical metal bonding representations for complexes are also shown.

Table 4.1 Variation of ion size and average bond distances found for first-row d-block metal ions in common oxidation states and with six-coordinate octahedral geometry. Entries for a second and third row element, to exemplify differences down a column of the Periodic Table d block, appear in italics.

Metal ion	d-Electron configuration	Free ion radius (pm)	Typical M—O (pm)	Typical M—N (pm)	Typical M—Cl (pm)
Sc(III)	d^0	74.5	210	—	245
Ti(IV)	d^0	60.5	195	210	230
Ti(III)	d^1	67	185	215	235
V(IV)	d^1	58	185	205	215
V(III)	d^2	64	215	225	235
Cr(III)	d^3	61.5	195	210	235
Mn(IV)	d^3	53	185	210	230
Cr(II)	d^4	80	200	215	245
Mn(III)	d^4	64.5	195	205	230
Mn(II)	d^5	83	215	240	250
Fe(III)	d^5	64.5	190	205	230
Fe(II)	d^6	78	205	215	240
Co(III)	d^6	61	185	195	225
Rh(III)	*d^6*	*66.5*	*195*	*205*	*235*
Ir(III)	*d^6*	*68*	*210*	*215*	*240*
Co(II)	d^7	74.5	205	220	240
Ni(II)	d^8	69	205	210	235
Cu(II)	d^9	73	200	200	225
Zn(II)	d^{10}	74	205	210	230

displays bond distances that vary slightly across a usually large number of ligand systems examined, indicating that each assembly does have a preferred metal–donor distance.

4.3.2 Moulding a Relationship – Ligand Influences

If we focus next on the bound donor molecules or ions, there are also two of the four key factors mentioned earlier that we can consider ligand-centric. These are:

- ligand–ligand repulsion forces; and
- ligand rigidity or geometry.

Obviously, each ligand is unique in its shape and size. The effect of repulsion between ligands, which we can term *nonbonding interactions*, can usually be recognized readily in the solid state from distortions in shape as defined by crystal structure analysis. Two PH_3 molecules may bind in a square planar shaped complex with little preference for *trans* over *cis* geometry, whereas two very bulky $P(C_6H_5)_3$ molecules may exhibit strong preference for coordination in a *trans* geometry, where they are much further apart. Another way that two *cis*-disposed bulky ligands can relieve ligand–ligand repulsion is for the complex to undergo distortion from square planar towards tetrahedral, which leads to the two ligands moving further apart in space; this is also observable in the solid state and, from spectroscopic behaviour, in solution. However, there are also weak structure-making effects arising from favourable hydrogen bonding interactions that can assist in stabilizing a particular shape. This might occur where a carboxylate group oxygen participates in a hydrogen bond with an amine hydrogen atom, arising from the interaction

$$R_2N–H^{\delta+} \cdots ^{\delta-} O–CO–R.$$

Such directed close contacts can actually be observed in the crystal structure also, but distances are longer than formal covalent bonding distances and the bonds much weaker.

Some ligands are structurally so rigid that they can bind to a metal ion in only one manner. The aromatic porphyrin molecule is completely flat, and large amounts of energy are required to distort it. Therefore, when it binds to a metal, it seeks to retain this shape, and will simply use the square-shaped array of four N-donors to wrap around a metal ion in a planar manner. Not only aromatic ligands are rigid; some polycyclic fused-ring aliphatic molecules may be sufficiently rigid and require a particular shape. An example of a cyclic rigid four nitrogen donor ligand that can bind effectively only with the square-shaped array of four N-donors in the plane about the metal ion (Figure 4.22) contrasts with the flexible aliphatic ligand also shown, which can bind in a flat or 'bent' arrangement. Even ligands where the donors do not form part of a ring are affected by rigidity; the tridentate ligand at the right in the figure is conjugated and must preferentially remain flat, limiting the way

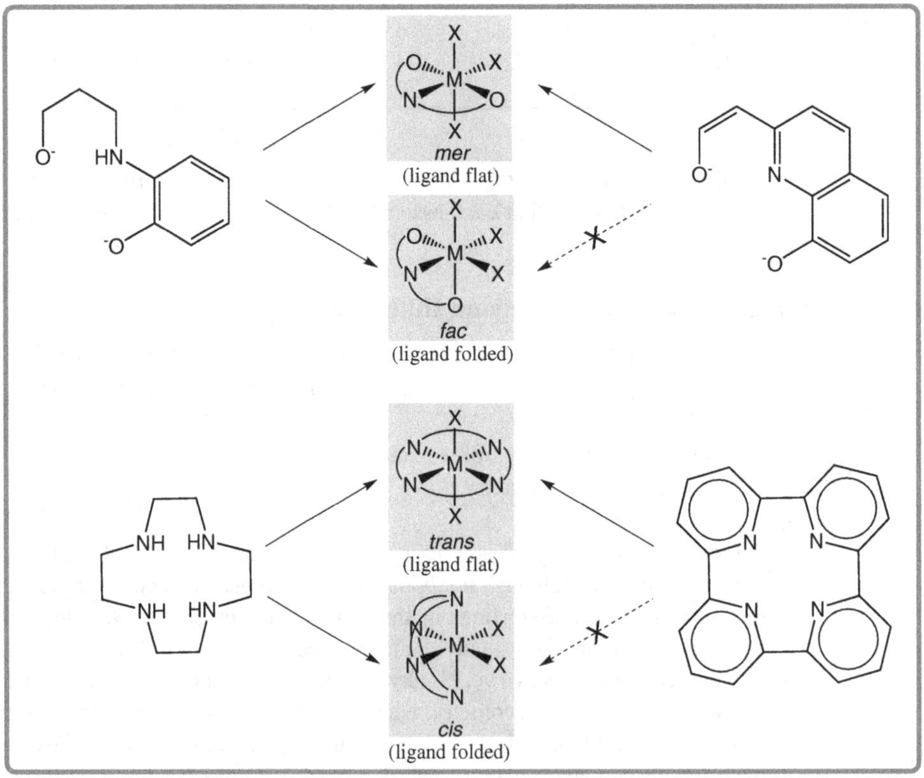

Figure 4.22
Rigid and flexible ligands influence the way they bind to metal ions. The more rigid ligands on the right do not permit folding, whereas the more flexible ones on the left can accommodate folding and thus offer options when coordinating to an octahedral metal ion.

it can bind all donors, whereas the analogue on the left, with a flexible saturated pendant chain, can adjust to flat or bent coordination modes.

4.3.3 Chameleon Complexes

Because complexes usually arrive at their particular three-dimensional shape because it is the thermodynamically most stable form, we tend to expect that they cannot change this outcome for the particular set of ligands they carry – each time they are made, the same result is achieved. While this is dominantly the case, it should be remembered that the thermodynamic stability relates to a particular set of conditions, such as temperature, solvent and counter ions or electrolyte. Change the conditions, and the system may be perturbed enough to undergo a physical change. Some changes may be readily reversible, so that reversing the change allows the complex to revert to its original form – genuine chameleon-like behaviour. The most obvious example of this is where a change in conditions creates a change in geometry. We have already noted that for some coordination numbers two different shapes may be similar in stability; providing the barrier to conversion between them is not too great, then interconversion may occur as a result of, for example, changing the temperature. The tetrahedral/square planar interconversion shown in Figure 4.12 is an example of this. Of course, conversion from one isomer to another irreversibly to achieve the more thermodynamically stable form is common, and we shall deal with this later in this chapter.

The tendency to show flexibility in shape varies with coordination number, because ligand–ligand repulsion energy differences between various shapes vary, as do the heights of energy barriers in reaching transition intermediates. As a consequence, five-, seven-, eight- and nine-coordination tend to be substantially less rigid than four- and six-coordination, and display more examples of chameleon-like behaviour. For example, not only is the ligand–ligand repulsion energy difference between the trigonal bipyramidal and square pyramidal geometries close to zero and definitely much less (perhaps as much as 50-fold) than between the two common four- and six-coordination shapes, but also the transition states for rearrangement are less disfavoured, so five-coordination tends to be less rigid.

4.4 Isomerism – Real 3D Effects

Given a large pile of bricks, there exists an almost infinite number of ways in which they can be assembled into a three-dimensional shape. Likewise, on the molecular scale, one of the inevitable consequences of assembling three-dimensional molecules with components arranged around a central core atom is the existence of options for arrangement of those components – a phenomenon that we call *isomerism*. Because of the designation of a core shape for the complex, limitations on ligand shapes, and connectivity and bonding rules, the number of options is far from infinite – yet not much less concerning for that. Defining the basic shape of the complex is the first stage, with next the task of defining options for ligand attachment needing to be examined. In very few cases is the exercise trivial, because we need to think in three dimensions to identify possibilities. This can be illustrated in the case of a tetrahedral MA_3B complex, where there is only one possible outcome, yet this is not immediately obvious (Figure 4.23).

Only in the case where all ligands are identical monodentates with a single common donor atom, is the situation simple and the prospect of positional isomers removed for

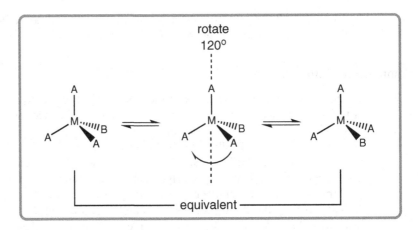

Figure 4.23
Despite first appearances, the molecules at the left and right are identical, since simple rotation by 120° around the marked axis as shown in the central view 'converts' left view into right view. They are not isomers, merely different orientations of the same molecule in space. The need to consider orientation issues in three dimensions complicates decisions on isomers.

a particular stereochemistry; this occurs for $[CoCl_4]^{2-}$ and $[Ni(OH_2)_6]^{2+}$, for example. Wherever more than one type of ligand is bound, and even where either one type of ligand with a set of more donor atoms than required, or a number of identical chelating ligands bind, the possibility of isomers needs to be considered, although it is not the case that all those possible will always exist.

4.4.1   Introducing Stereoisomers

Characteristically, metal ions form complexes that can exist as several isomers. This is a consequence of the stereochemistry, resulting from high numbers of ligands, adopted by most metal complexes. The best known examples of isomerism in complexes are geometrical isomers (such as *cis* and *trans* isomers), but these are not the sole type. We can, following a traditional approach, divide the area into two classes: constitutional (or structural) isomerism and stereoisomerism.

4.4.2   Constitutional (Structural) Isomerism

Some of the forms of isomerism have little more than historic interest now, as their significance has diminished with the rise in physical methods that makes their identification and origins routine, and no longer involves the demanding experiments of an earlier era to probe their form. Nevertheless, some remain important, and others at least give a flavour of the historical development of the field, and this deserves a brief discussion. Constitutional isomers are characterized by species of the same empirical formula (which was able to be determined at an early date in the development of the field) but clearly different physical properties associated with different atom connectivity.

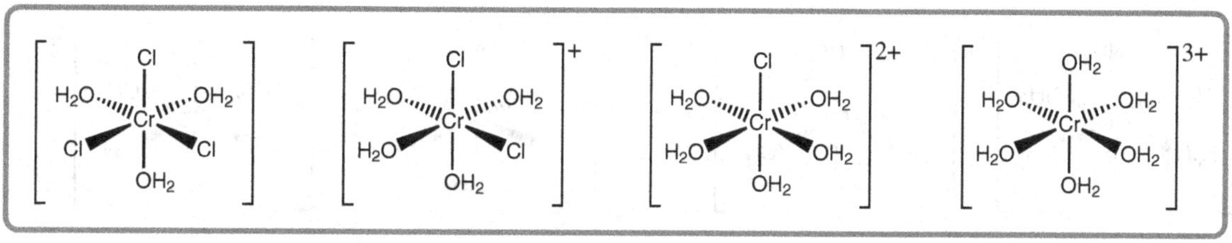

Figure 4.24
Hydrate isomers: the complexes possible for the empirical formula $CrCl_3 \cdot 6H_2O$. In addition, the neutral and 1+ compounds on the left may in principle exist as two geometric isomers, of which only one is shown.

4.4.2.1 Hydrate Isomerism

Hydrate isomerism (sometimes called solvate isomerism) was named to identify an at first puzzling observation, which was that some hydrated compounds with the same empirical formula were obviously different, in colour, charge and number of ions. A classic example is the inert compound $CrCl_3 \cdot 6H_2O$, for which three forms were detected. We know these now as the compounds $[CrCl_2(OH_2)_4]Cl \cdot 2H_2O$, $[CrCl(OH_2)_5]Cl_2 \cdot H_2O$ and $[Cr(OH_2)_6]Cl_3$, being distinguished by the groups bound as ligands to the chromium(III) ion, as shown in the views of the complex cations (Figure 4.24); a fourth option, the neutral complex $[CrCl_3(OH_2)_3] \cdot 3H_2O$ shown at left in the figure, is not detected due to its reactivity in solution, converting readily to other species. The commercial form is the dark green $[CrCl_2(OH_2)_4]Cl \cdot 2H_2O$ (sufficiently venerable to have once been called Recoura's green chloride), formed by crystallization from a concentrated hydrochloric acid solution. Upon redissolution in water, substitution reactions to release additional coordinated chloride ions commence.

4.4.2.2 Ionization Isomerism

Ionization isomerism is another case defined by recognizing that an empirical formula allows some options for the coordination sphere of the metal. It is essentially the same situation as hydrate isomerism, but involves ligands other than water. For example, consider the inert cobalt(III) compound $CoBr(SO_4) \cdot 5(NH_3)$, which forms two different compounds, one violet, the other red. We know these now as $[CoBr(NH_3)_5](SO_4)$ and $[Co(NH_3)_5(SO_4)]Br$, which differ in the choice of which anion occupies the coordination sphere, the other remaining as the counter-ion (Figure 4.25).

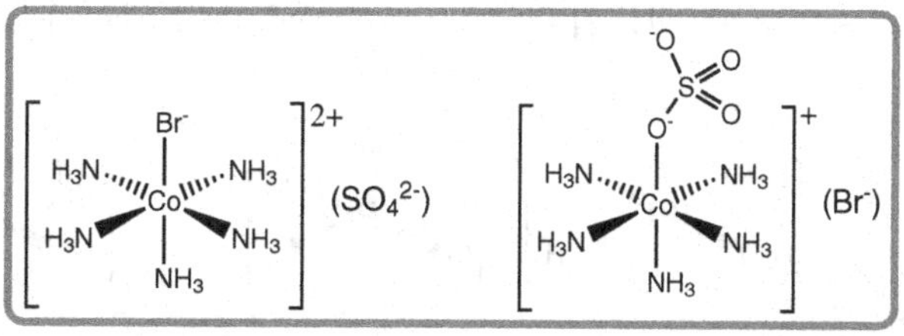

Figure 4.25
Ionization isomers: the complexes exemplified differ in which anion of the two present is coordinated to the metal.

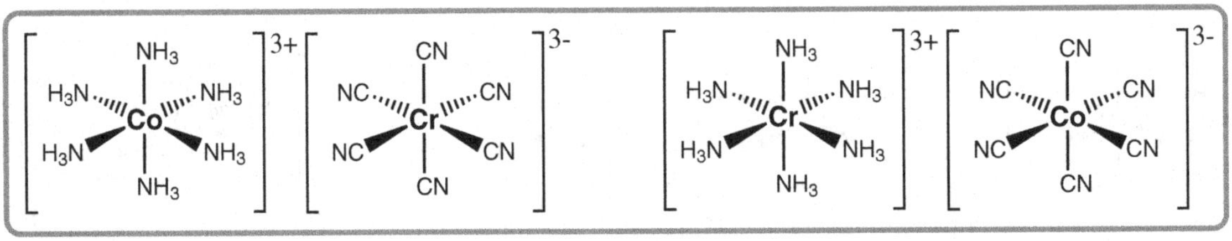

Figure 4.26
Coordination isomers: the complexes exemplified differ in which of the two sets of ligands present is coordinated to which metal.

Early experimentalists identified their character through simple chemical reactions. For example, with silver ion, it is only [Co(NH$_3$)$_5$(SO$_4$)]Br that produces an immediate precipitate of AgBr, as the Co—Br bond in the other form is too strong to permit reaction readily. Likewise, reaction with barium ion causes an immediate precipitate of BaSO$_4$ only with the [CoBr(NH$_3$)$_5$](SO$_4$) form, again because the Co—OSO$_3$ bond in the other isomer is too strong to allow reaction. These experiments allowed identification of differing ionic character in the two compounds.

4.4.2.3   Coordination Isomerism

For complex salts where there are metal ions present in both the cation and the anion, both functioning as a complex ion, there are, with two types of ligands, a number of possible coordination forms. The simplest is where all of one type of ligand associate with one or the other metal centre; for example, CoCr(CN)$_6$·6NH$_3$ can be either [Co(NH$_3$)$_6$][Cr(CN)$_6$] or [Cr(NH$_3$)$_6$][Co(CN)$_6$], where, effectively, metal ions 'swap' ligands, both being capable of binding to either (Figure 4.26). Mixed-ligand assemblies on each metal offer more options.

4.4.2.4   Polymerization Isomerism

The empirical formula obtained from elemental analysis identifies the ratio of components, not their actual number. Thus an empirical formula MA$_2$B$_2$ could be considered as any of [MA$_2$B$_2$]$_n$ (for $n = 1, 2, 3, \ldots$), a series of compounds with the same empirical formula, and with the molecular formula of each some multiple of the simplest formula. In an era where we can determine the molecular mass and/or three-dimensional structure fairly readily, this is not much of an issue, but was of concern in earlier times. A classical example is for the empirical formula Pt(NH$_3$)$_2$Cl$_2$, which exists not only as the $n = 1$ form [PtCl$_2$(NH$_3$)$_2$], but also as [Pt(NH$_3$)$_4$][PtCl$_4$] ($n = 2$) and [PtCl(NH$_3$)$_3$]$_2$[PtCl$_4$] ($n = 3$) (Figure 4.27).

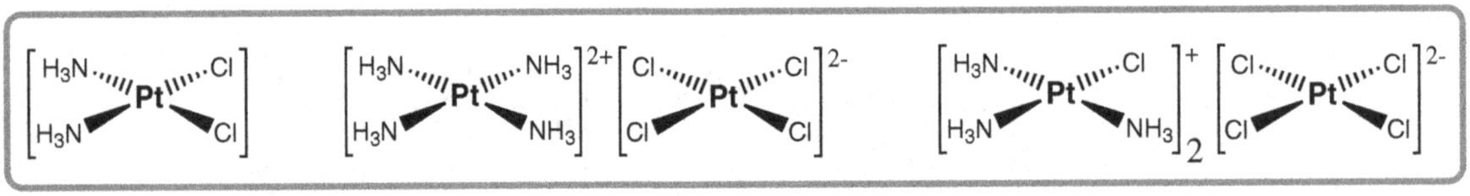

Figure 4.27
Polymerization isomers: the complexes exemplified have identical empirical formulae but differ in the number of replications of the empirical formula, {Pt Cl$_2$(NH$_3$)$_2$}$_n$.

A more unusual version of the $n = 2$ form possible is the dimeric complex $[(NH_3)_2Pt(\mu\text{-}Cl)_2Pt(NH_3)_2]Cl_2$, where two chloride ions bridge between and link the two metal centres with coordinate covalent bonds. Since the syntheses of the two $n = 2$ species differ, this is more an intellectual exercise than a difficult case. However, this example does serve to remind us that oligomers (small polymers) are frequently met in coordination chemistry.

4.4.2.5 *Linkage Isomerism*

There are a number of molecular ligands that contain two different atoms carrying lone pairs, both capable of coordination to a metal ion. These are called *ambidentate* or *ambivalent* ligands, distinguished by their capability for binding a metal ion through either of the different donor groups (Figure 4.28). A simple example is the thiocyanate anion (SCN^-), which offers either a N atom (NSC-*N* isomer) or a S atom (NCS-*S* isomer) to metal ions; which one a metal ion selects depends on metal–ligand preferences discussed earlier. For example, the 'hard' Co(III) ion prefers to form Co—NCS complexes with the 'hard' N-donor, whereas the 'soft' Pd(II) ion prefers the 'soft' S-donor and forms Pd—SCN complexes.

Another classical example is nitrite ion, which offers N or O atoms as donors. This example has been deeply studied, and the way it behaves is fairly well understood. The O-bound isomer converts (isomerizes) to the thermodynamically stable N-bound isomer, sometimes even in the solid state, by an intramolecular process (without the ligand departing the coordination sphere) in inert complexes. Another feature of ambidentate ligands is that they can display a tendency to 'bridge' between two metal ions, with each of the two different donor atoms attached to one of two metal ions.

4.4.3 Stereoisomerism: in Place – Positional Isomers; in Space – Optical Isomers

Stereoisomerism is the name given to cover the more general cases of isomerism in coordination complexes. *Stereoisomers* are molecules of the same empirical formula that have identical coordination number and basic shape, and display the same atom-to-atom bonding sequence throughout, but in which the ligand atoms differ in their locations in space. Put simply, everything is the same except where the ligands are positioned. Changing atomic locations in space changes physical and usually chemical properties, so the stereoisomers

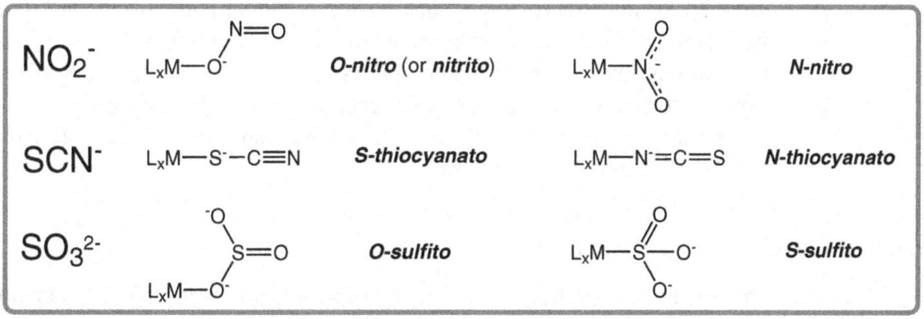

Figure 4.28
Linkage isomers: coordination modes for coordinated nitrite ion, thiocyanate ion and sulfite ion.

display at least some differences; these properties can be measured and used to identify the isomer under examination.

There are two gross classes or types with which you should become familiar. *Diastereomers* are the general class of stereoisomers, and includes geometric isomers (such as *cis*, *trans* isomeric forms) as a sub-class. Any particular diastereomer may have an *enantiomer*, which is a stereoisomer that is a *non-superimposable mirror image* of the original diastereomer (like your left hand is a non-superimposable mirror image of your right hand). Not all compounds will exist as two mirror image forms (the enantiomers), as it is a property related to the symmetry of the compound; consequently, diastereomers are not the same thing as enantiomers. For complexes to have enantiomers, they must be either *asymmetric* (that is be a molecule totally lacking in symmetry apart from the identity operation) or *dissymmetric* (that is lacking an S_n axis, which is a rotation-reflection axis – without an S_n axis, it can have *neither* a plane of symmetry *nor* a centre of symmetry; however, a dissymmetric molecule *can* have a proper axis of rotation, C_n ($n > 1$)). Definition of symmetry operations and the rules of symmetry are discussed in Appendix B.

A molecule that is asymmetric or dissymmetric (and therefore not superimposable on its mirror image) is called a *chiral* compound; this means that all enantiomers are chiral. Such a compound will display *optical activity* as an individual enantiomer, which is the ability to rotate the plane of plane-polarized light (measured using a polarimeter), which is one way that we can detect the presence of an enantiomer and define its optical purity. Whereas *diastereomers* usually differ appreciably in their chemical and physical properties, *enantiomers* differ only in their ability to rotate polarized light and related optical properties. Normally, when a diastereomer that has an enantiomer is synthesized, a 50 : 50 mixture of the two enantiomeric forms of the compound is produced, and thus no optical activity is observed. However, if the compound is separated into its two enantiomers (or *resolved*), each enantiomer will show optical activity; in a polarimeter; the responses of the two enantiomers will differ only in the sign of rotation of plane-polarized light.

Any synthetic procedure that produces more of one enantiomer than the other is termed a *stereoselective* reaction, and, in the extreme of only one enantiomer forming, is called a *stereospecific* reaction. Most simple synthetic reactions lead to equal amounts of enantiomers, called a *racemic* mixture.

Whereas chirality in tetrahedral compounds of carbon requires four different groups to be bonded around the tetrahedral carbon atom, this is not necessarily the case for other central atoms with other stereochemistries. For example, octahedral complexes have more relaxed rules. Whereas a chiral tetrahedral organic compound is, as a consequence of the rule for chirality, asymmetric (or totally lacking in symmetry), chiral octahedral complexes need not be asymmetric, but may have axes of rotation (they are then dissymmetric). The common rule for chirality is simple – a compound must have non-superimposable mirror images. For an octahedral complex, this can occur even when three different pairs of monodentate ligands are coordinated, as discussed later. We shall look a little more closely at four- and six-coordinate complexes below.

4.4.3.1 *Four-Coordinate Complexes*

One of the two limiting forms of four-coordination, square planar, involves at least one plane of symmetry (the plane including the metal and four donor groups), and as a result square-planar complexes cannot form enantiomers (unless a ligand itself is chiral). It can form geometric isomers (*cis* and *trans*), however, as defined in Figure 4.29. Tetrahedral

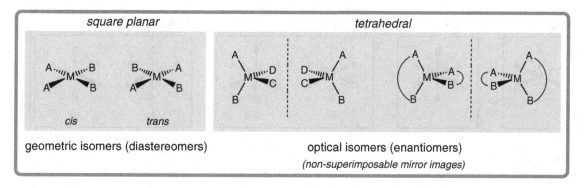

Figure 4.29

Isomers possible with four-coordination. The square planar geometry, with a plane of symmetry, cannot exhibit optical isomerism but can display geometric isomerism, whereas tetrahedral geometry, with its symmetrical disposition of bonds, cannot exhibit geometric isomerism but may display optical isomerism.

complexes can have only enantiomers; they cannot, as a result of their shape, have *cis* and *trans* isomers. Chirality will occur for complexes with four nonequivalent ligands (although there appear to be no known examples isolated yet) *or* with two unsymmetrical didentate ligands (Figure 4.29).

4.4.3.2   Six-Coordinate Octahedral Complexes

Six-coordination presents greater options for location of ligands around the coordination sphere than occurs in four-coordination, as a result of the greater number of donor sites. Whereas there is only one form of MA_6 and of MA_5B possible, for MA_4B_2 there arise two geometric isomers, *cis* and *trans* (Figure 4.30), although there are no enantiomers as both diastereomers have planes of symmetry. When we extend to MA_3B_3, there are two geometric isomers, *facial* (abbreviated as *fac*) and *meridional* (abbreviated as *mer*), neither of which has an enantiomer. If we add additional ligand types, however, the number of isomers can increase rapidly. For $MA_2B_2C_2$ there are now five diastereomers, one of which has an enantiomer (Figure 4.30). With more diverse sets of ligands, the number of possible isomers can grow even greater.

Simple examples of MA_4B_2 and MA_3B_3 complexes are those of cobalt(III) formed with ammonia and chloride ion as ligands, namely the cations *cis*- and *trans*-$[CoCl_2(NH_3)_4]^+$ and the neutral molecules *fac*- and *mer*-$[CoCl_3(NH_3)_3]$. These are illustrated with ball-and-stick drawings, based on X-ray crystal structures of the complexes, in Figure 4.31. What should be clear from these drawings is that the *fac* isomer carries three of each type of ligand in an equilateral triangular arrangement on a separate face of the octahedron; hence the origin of the name. Conversely, the *mer* isomer has each of one type of ligand lying in a plane that includes the metal, at right-angles to the plane of the other type of ligand.

The introduction of chelate ligands usually acts to limit the number of geometric isomers, or at least to not extend the number. For example, the coordination of three symmetrical didentate chelates can occur in only one way, in the same sense that six identical monodentate ligands can coordinate in only one way. Likewise, $M(AA)_2B_2$ compounds (where (AA) refers to a symmetrical chelate) has precisely the same number and type of diastereomers as

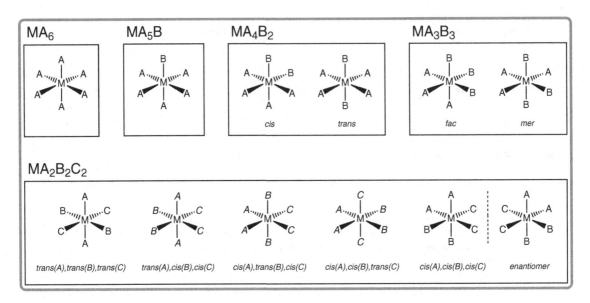

Figure 4.30
Diastereomers and enantiomers possible for octahedral complexes with various sets of monodentate
ligands.

the all-monodentate analogue MA_4B_2. The difference is that introduction of chelates usually
leads to lower symmetry, such that enantiomers can exist. The presence of at least two
chelates in *cis* dispositions always leads to chiral (optically active) octahedral compounds,
as the molecules become dissymmetric. A simple example is the octahedral complex *cis*-
$[CoCl_2(en)_2]^+$ that has two ethane-1,2-diamine chelates coordinated, and which (unlike the

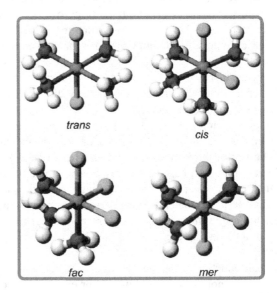

Figure 4.31
Diastereomers of the octahedral MA_4B_2 complex $[CoCl_2(NH_3)_4]^+$ and the MA_3B_3 complex
$[CoCl_3(NH_3)_3]$.

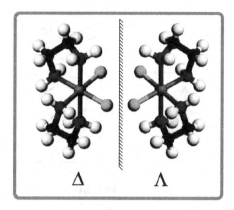

Figure 4.32
Non-superimposable mirror image forms (enantiomers) of the diastereomer *cis*-[CoCl$_2$(en)$_2$]$^+$. The enantiomers are named Δ and Λ respectively.

trans diastereomer) has a mirror image that is non-superimposable on the original, a key requirement for optical activity (Figure 4.32).

So far, we have introduced the concept of optical activity (or chirality) in a complex arising as a result of the way ligands are arranged around the central metal. For six-coordination octahedral, there are actually several ways in which dissymmetry (chirality) can arise, namely from:

1. the distribution of monodentate ligands about the metal (which occurs with at least three pairs of different ligands, as exemplified for MA$_2$B$_2$C$_2$ in Figure 4.30);
2. the distribution of chelate rings about the metal (as occurs when at least two didentate chelates are coordinated, as exemplified in Figure 4.32);
3. coordination of unsymmetrical polydentate ligands (where the assembly of different donors and linkages resulting means the complex becomes dissymmetric);
4. conformations of chelate rings (where tetrahedral carbon and other atoms enforce their geometry in the chain linking donor groups, causing chelate rings to be nonplanar and able to adopt conformations that have δ and λ mirror images forms – see Figure 2.7);
5. coordination of an asymmetric and hence chiral organic ligand (whereby the chirality of the part is assigned to the whole) – this chirality in the ligand may be conventional (arising from asymmetric carbon centres of D or L form), or arise from binding to a helical ligand (which has *P* and *M* isomers associated with opposite helicities);
6. coordination of a donor atom that is asymmetric (which leads to induced chirality in the whole complex, similar to the situation in (5) above, but with the asymmetric centre actually attached to the metal in this case).

Examples of several of these cases appear in Figure 4.33, but detailed discussion will not be pursued here, this being a task for more advanced texts.

4.4.4   What's Best? – Isomer Preferences

Polydentate ligands present particular problems, as a range of geometric isomers, often with enantiomers, may exist in principle. This is illustrated in Figure 4.34 for a simple system where a symmetrical tetradentate is bound. This can occur in three ways, two of which

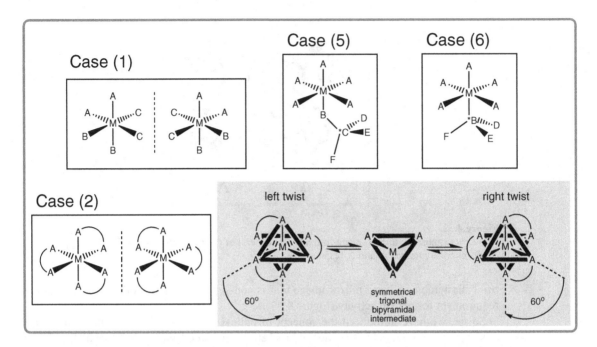

Figure 4.33
Selected examples of ways in which dissymmetry is introduced into octahedral complexes, with case numbers equating with those in the text immediately above. Enantiomers are shown for two cases only [(1) and (2)]. A process by which conversion in case (2) from one enantiomer to the other (*optical isomerization*) may occur is illustrated, based on twisting of one octahedral face by 120° while keeping the other fixed. The two isomers must twist in opposite directions to avoid breaking the chelate chains, to generate a symmetrical intermediate, from which it may continue twisting to yield the opposite enantiomer. The 'left-twisting' isomer is called Λ and the 'right-twisting' one Δ, the symbols for the two enantiomers or optical isomers.

lead to dissymmetric complexes. What should be appreciated is that these three isomers are not equal in thermodynamic stability, meaning that the percentage of each formed is not based purely on a statistical ratio, but related to their relative stabilities. This means, in effect, that in some cases only one geometric isomer could be observed experimentally, if it is significantly more stable than others, or that at least only several of a set of options

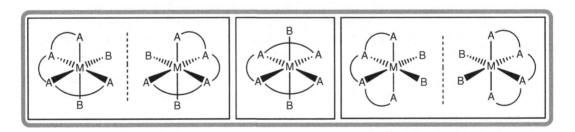

Figure 4.34
The three geometric isomers (diastereomers) for $M(AAAA)B_2$ complexes, one of which is of *trans* geometry with respect to the two simple ligand B (centre), and two of which are of *cis* geometry (*cis*-β at left, *cis*-α at right) and in those cases have enantiomers, also drawn.

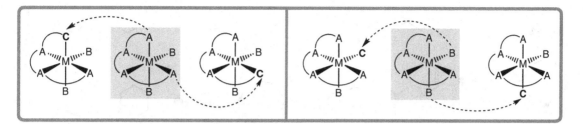

Figure 4.35

Left: One of the geometric isomers of a M(AAAA)B$_2$ complex is transformed into two options when a mixed donor ligand replaces the common donor ligand in forming a M(AAAC)B$_2$ complex. Consider the two arising from replacement of two different terminal donors, so the product has the introduced C group either opposite a B ligand (left) or opposite an A ligand (right). In reality, of course, the mixed donor ligand is preformed and coordinates as a single entity. *Right*: One of the geometric isomers of a M(AAAA)B$_2$ complex is also transformed into two options when two different monodentate ligands replace the common monodentate ligands in forming a M(AAAA)BC complex. Consider the two arising from replacement of two different monodentate donors, so the product has the introduced C monodentate either opposite a central A group of the tetradentate (left) or opposite a terminal A group (right).

may occur experimentally. The isomers usually display different total strain energy, and it is where there are large energy differences that there is observed one thermodynamically stable form or at most a limited number of forms. Thus only some, and not all, will be found experimentally, as some may be too strained relative to others to exist.

If we introduce mixed donors into the simple tetradentate AAAA of Figure 4.34, for example, to form AAAC, the number of isomers will increase. For example, now there are two different forms of the isomer shown at the left in Figure 4.34, as illustrated in Figure 4.35. The two new isomers have the same spatial arrangement of the chelate chains, but the unique terminal groups (A and C) are located differently. An alternative reaction that increases geometric isomers is where the two common monodentate donor groups are replaced by two different monodentate donor groups (Figure 4.35, right); there are again different outcomes depending on which of two groups is replaced. There is adequate nomenclature to deal with naming these difference diastereomers, but this is not required to comprehend the outcomes when pictorial representations are available; some aspects of naming complexes appears in Appendix A. These two simple examples suffice to illustrate the array of geometric isomers that can result when polydentate and mixed donor ligands bind to a central metal ion.

Geometric isomers are energetically inequivalent because of a suite of effects that contribute additively to strain in each complex. Sources of strain in complexes are: bond length deformation (enforced contraction or extension); valence bond angle deformation (enforced opening/closing up); torsion angle deformations (for chelate rings); out-of-plane deformations (for unsaturated groups); nonbonded interactions; electrostatic interactions (of charged groups); and hydrogen bonding interactions. Models that allow estimation of the strain energies in isomers exist, providing prediction of and/or interpretation of experimental behaviour. This is addressed later in Chapter 8.3.

4.5 Sophisticated Shapes

The shapes we have described have employed, in all but the last section, simple ligands that bind at one or two sites around a metal ion. However, most ligands are more complicated

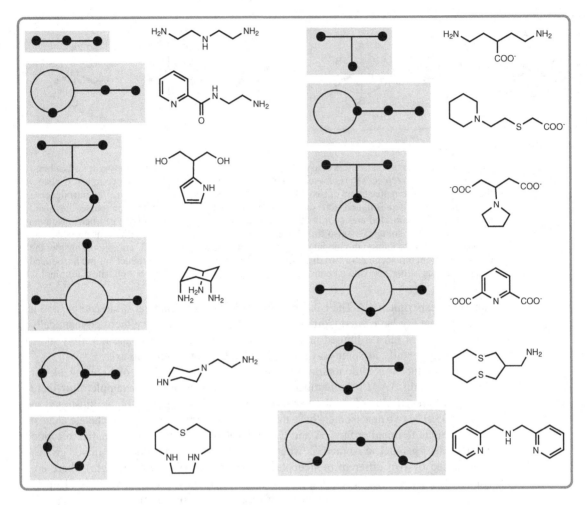

Figure 4.36
Some possible shapes for three-donor ligands, with examples.

in both their potential donor set and basic shape. The ultimate in complicated ligands are the natural ligands met in Chapter 8, but it is appropriate to examine briefly some examples of synthetic ligands, in order to illustrate the way in which they coordinate and form complexes.

4.5.1   Compounds of Polydentate Ligands

Immediately above, we have introduced some of the issues that arise when a polydentate ligand is bound rather than simple monodentate or didentate ligands. Even stepping from two to three donors increases the options in terms of ligand shape (or *topology*), and this shape will affect the way a molecule may bind to a metal ion. Some shapes for potentially tridentate ligands appear in Figure 4.36.

The shape of a ligand influences the way it can bind to a metal ion, as introduced in Chapter 2. This is illustrated for potentially tridentate ligands in Figure 4.37, comparing

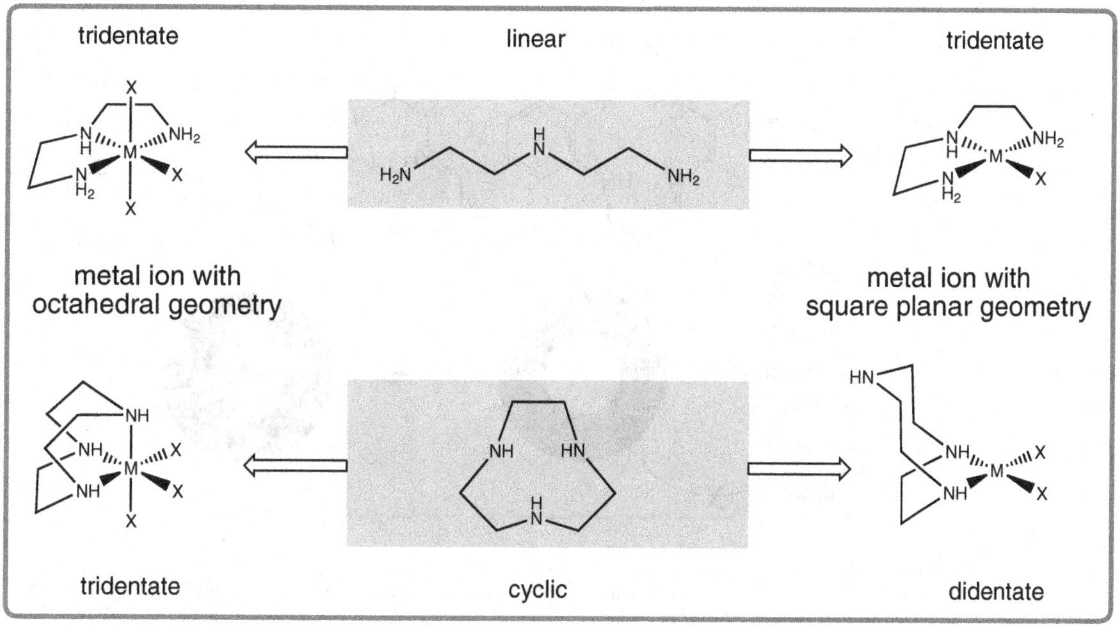

Figure 4.37
Possible coordination for three-donor linear and cyclic ligands to square planar and octahedral metal complexes. The cyclic ligand is unable to bind all three donors in the square planar geometry.

a linear triamine and a cyclic triamine binding to a square planar or octahedral metal ion. Whereas the former can bind to a square planar centre with all three donors attached, the latter cannot, as the ring is too small to permit the third donor to bind in the plane of the metal after the first two bind. This means that the cyclic ligand acts not as a tridentate but as a didentate ligand with one group not coordinated (or pendant). Only if there is a change to octahedral or tetrahedral geometry, where the ligand can bind to a triangular face, do all three amine groups bind. This simple example serves to highlight some of the considerations that apply when ligand topology comes into play.

As the number of donors increase, so does the diversity in shape that may be met. We tend to classify ligands in terms of their appearance or basic framework – for example, they may be described as linear, branched, cyclic and cyclic with pendant chains. Nevertheless, despite the variations possible, these types of ligands are still regarded as basic or simple in character; there exists an array of more complicated and specific shapes that we shall touch upon briefly below.

4.5.2 Encapsulation Compounds

Increased sophistication in chemical synthesis has led to the development of a wide range of molecules which can 'wrap up' or 'encapsulate' metal ions (as well as, in some cases, even whole small molecules). A range of topologies (or three-dimensional shapes) has been devised for accommodation of either metal ions with different preferred coordination numbers or small molecules. The former usually involve covalent bonds, the latter weaker noncovalent interactions (such as hydrogen bonds). Developing functionality for encapsulated

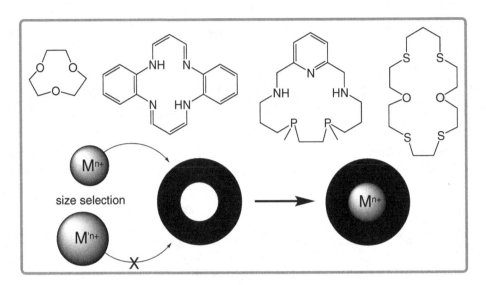

Figure 4.38
Examples of three-, four-, five- and six-donor macrocycles. Metal ion selection and binding strength is based in part on ion – hole 'fit'.

or structurally confined systems has been a growing interest, directed towards the development of molecular machines.

4.5.2.1 Macrocycles

Macrocycles are a diverse family of large ring organic molecules (defined as having nine-membered or larger rings) characterized by a strong metal binding capacity when several heteroatoms are present in the ring. Several simple examples have been introduced earlier. This class includes the so-called 'pigments of life' – nature's aromatic macrocycles built for important biochemical use in plants (the chlorophylls) and animals (the hemes). Visible light is absorbed in these systems by the presence of a sequence of alternating double and single bonds to give characteristic colours. *Macromonocyclic* (single large ring) molecules including several heteroatoms represent the simplest members of the family of macrocycles. They can bind metal ions and even small molecules reasonably efficiently, depending on size. A range of types, both aliphatic and aromatic are known, as exemplified in Figure 4.38. In a simple sense, what is important in these cyclic systems is the matching of cavity hole size to metal ion size; a good 'fit' means a stronger complex.

One well-known group of macrocyclic ligands are polyamines. Many may not be capable of 'wrapping up' the metal fully, through not having sufficient donor groups to satisfy fully its coordination sphere. However, with sufficient donor groups this can be achieved, and in a very large ring even more than one metal ion may be accommodated. Coordination of a saturated polyamine macrocycle, however, introduces subtle stereochemical consequences relating to the disposition of amine (R–NH–R$'$) hydrogen atoms on complexation. This is illustrated for a tetraamine bound to a square-planar metal ion in Figure 4.39; different amine proton dispositions possible are highlighted.

At a simple level, we can see this as 'four up', 'three up and one down', and two types of 'two up and two down'; other options (such as 'four down' compared with 'four up') are

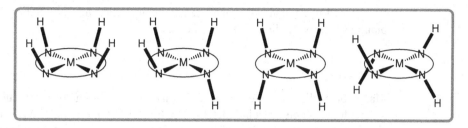

Figure 4.39
Possible dispositions of the secondary amine hydrogen atoms in a coordinated macrocyclic tetraamine. Each isomer, formed only on coordination, exhibits slightly different physical properties.

equivalent to one of those shown, as they are equated simply by inverting the macrocycle. In effect, these isomers equate to a different chirality set for nitrogen donors, fixed upon coordination; each secondary amine RR′HN group, when coordinated, forms RR′HN—M, and then, with four nonequivalent groups covalently bound around the tetrahedral nitrogen, the N centre is chiral. They exist in addition to any other sources of isomerism and chirality in the molecule. Although the N-based isomers may have slightly different physical properties, interconversion between these isomers can be readily achieved by raising the solution pH, which promotes N-deprotonation and exchange, leading to formation of the thermodynamically most stable N-based isomer. Usually, this subtle N-based isomerism tends to be ignored, as it is a level of complication too far for most.

Macrocycles carrying *pendant groups* also capable of binding metal ions produce the opportunity to 'wrap up' metal ions better (Figure 4.40). These 'molecular wrappers' have pendant groups that can come on or off, so they behave as 'hinged lids'. The pendant groups may be of any type, and carry any form of potential binding group – amine, carboxylic acid, thiol, alcohol, pyridine and others. These groups may themselves be further elaborated

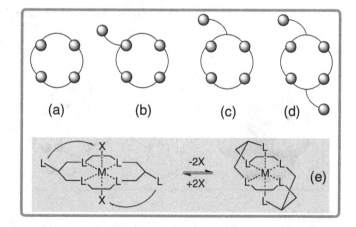

Figure 4.40
Simple macrocycles (a) may be augmented with pendant group(s) attached to either a heteroatom (b) or a carbon atom (c) of the ring. Those with two pendant groups (d) offer better opportunities for 'wrapping up' octahedral metal ions, as the two pendant arms can supply additional donors that, by being linked to the ring, enhance entrapment by 'capping' the metal, as shown in (e).

or extended using standard organic reactions, including forming dimers, linking them to biomolecules, or binding them to surfaces.

4.5.2.2 Macropolycycles

Macropolycyclic molecules (with several large organic rings fused together) that include several heteroatoms can bind metal ions efficiently, and a range of types are known. One simple family is the *sarcophagines*, bicyclic (two fused large ring) amine molecules with the capacity to 'cage' metal ions. They have a cavity which offers six donors, usually saturated nitrogen groups, to metal ions. This cavity is too small for anything but single atoms or ions; the metal ion 'fit' in the cavity is excellent for small metal ions, less efficient for large metal ions. As their name implies, they are molecular 'graves' – once a metal ion is interred in the cavity it is trapped, and can only be removed with difficulty (usually requiring very strong acid or cyanide solution). Metal ions are effectively in a prison with the frame of the macrobicycle as the bars. A typical example is shown in Figure 4.41.

4.5.2.3 Cryptands

Cryptands are cyclic (mainly) polyether molecules with usually three chains linked at nitrogen 'caps' at each end of the molecule (Figure 4.42), much like the sarcophagines but with a different capping atom and different donors. They can, depending on host cavity size, bind metal ions (alkali or alkaline earth ions preferred) or small molecules. A wide range of molecules of this family have been prepared. They can be effective in metal ion selection from a group of ions, useful in both analysis and separation of mixtures. They also help solubilize metal ions in aprotic solvents.

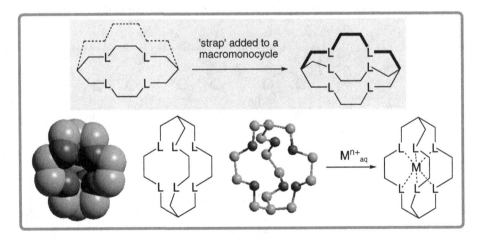

Figure 4.41
A simple macrobicycle can be considered to arise by addition of another chain or 'strap' of atoms to a macromonocycle, to provide three chains joined at two capping carbon atoms (top), effectively two fused macromonocycles. This example offers six donor groups to a metal ion, as exemplified with ball and stick and space-filling models; the latter shows the small central cavity able to accommodate only a single cation. These readily form very stable octahedral complexes with a range of metal ions; the best-known example has L = NH (called 'sarcophagine').

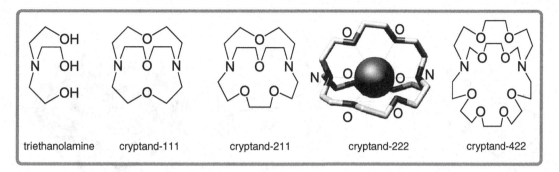

Figure 4.42
The simple noncyclic analogue triethanolamine, and macrobicyclic cryptands, which have three chains joined at two capping nitrogen atoms. The cryptand trivial names reflect the number of O-atoms in each linking chain. (Cryptand-222 is shown as a model of the K^+ complex, based on X-ray structural data; all six O atoms bind to the cation.)

4.5.2.4   Catenanes

A particularly unusual class of macrocyclic molecules are the *catenanes*, where, instead of two side-by-side fused rings as in the cryptands above, two separate and interlinked rings each offer donors to a single metal ion (Figure 4.43). With appropriate components in the ring, a change in oxidation state of the metal ion can lead to a ring rotating into a different position, a process that can be reversed if the metal ion oxidation state is turned back to its original. This process of switching position may provide an electrochemically-driven molecular 'switch'. Synthesis of these interlinked systems is not simple, so their potential application may be limited by this aspect.

4.5.3   Host–Guest Molecular Assemblies

A close relative of simple coordination chemistry is *supramolecular chemistry*. Whereas coordinate covalent bonds link the ligand host and the metal guest in coordination complexes, in supramolecular compounds the guest is a whole molecule, not a metal ion, and as such can only involve itself in weaker nonbonding interactions with the guest, typically involving the most potent of these forces, hydrogen bonding. Although hydrogen bonds are relatively strong compared with other nonbonding effects, they are still much weaker that

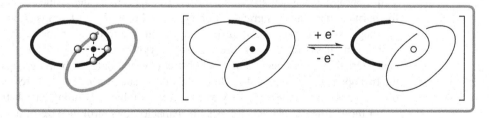

Figure 4.43
A schematic representation of a catenane complex with donors from each of two interlinked rings binding to a central metal ion, and (at right) the process by which a system with differing ring components may operate as an electrochemically-driven molecular switch, involving ring rotation.

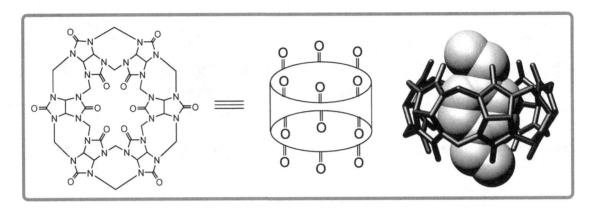

Figure 4.44
A cucurbituril macropolycycle composed of six units, and a side-on view of the cucurbituril acting
as a host to an organic guest cation.

coordinate covalent bonds. However, in certain cases, the shape and outcome is reminiscent
of coordination chemistry, and it may be of value to exemplify it here in passing.

Host–guest interactions occur where a cavity in one molecule (the host) permits selected
entry of another molecule (the guest) into the cavity space; it is, like the macrocycles with
metal ions, a case of 'best-fit', although here it is a whole molecule that is fitted into the
cavity. The host–guest terminology is usually applied to molecule–molecule interactions,
not to metal ion binding, however. The assembly which forms in solution (with measurable
stability, as a result of spectroscopic changes that occur when the guest occupies the different
environment of the host), is held together by noncovalent bonding forces – hydrogen
bonding and other weaker interactions. The 'container' molecules (hosts) are now many
and varied, but all have the common character of acting as hosts for guest molecules or
molecular ions.

Cyclodextrins, calixarenes and cucurbiturils are well established examples of host
molecules. *Cyclodextrins* are large-ring molecules made of linked sugar units. They are
made with typically at least four units in the ring, and are rich in polar ether and alcohol groups. There are several related structural types that differ depending on the class
of monomer units and linkages adopted in macrocyclic ring formation. *Calixarenes* (from
the Latin calix = bowl), are large cyclic molecules made up of typically four to eight
phenol units linked via $-CH_2-$ groups and containing large bowl-like cavities of dimensions appropriate for sequestering a variety of small molecules, which is of great interest
for purification, chromatography, storage, and slow release of drugs *in vivo*. Hydrogen
bonding involving alcohol groups plays a crucial role in the supramolecular chemistry of
such assemblies. *Cucurbiturils* are a family of cyclic host molecules, with their common
characteristic features being a hydrophobic cavity, and polar carbonyl groups surrounding
the two open ends (Figure 4.44). Better defined as cucurbit[n]urils (where n = the number
of monomer units in the ring), they can be water-soluble, depending on substituent groups.

For these different classes of hosts, the capacity to control the size of the rings and the
introduction of various functional groups make it possible to 'tailor' them for a variety of
chemical applications. Their sequestering properties can be exceptional, while appropriate
substitution renders the cavities shape-selective and thus suitable for molecular recognition.
Changes in a range of spectroscopic properties with increasing formation of the host–guest

adduct in water allow us to identify and quantify host–guest formation. The stability of host–guest assemblies in water can be high, approaching the strength of some metal complexes. Aquated metal ions and other small metal complexes can also act as guests, depending on the ring size.

The array and shapes of molecules that can act as ligands for metal ions and even as hosts for small molecules is truly extensive, and the field continues to grow. It is inappropriate to dwell on this diversity here. Rather, the reader is directed to specialist textbooks for a more extensive coverage of this fascinating field. For us, it is sufficient to recognize that the simple ligands which we employ largely at the introductory level are no more than the tip of the iceberg, and a wealth of chemistry remains hidden. Perhaps this isn't such a bad thing, as the field is demanding enough for a student new to its wonders.

4.6 Defining Shape

The above discussion of shape relies on our ability to actually identify and prove a shape. We have seen some examples of how this has been done in the past, using simple but revealing experiments that allow inferences to be drawn, on which base further experimental evidence has allowed structures to be well defined. Nowadays, we have moved into an era where we have available methods that allow us to define structure and shape with startling clarity and certainty in many cases. It is appropriate that we identify aspects of this methodology.

Three-dimensional structure can be probed by a wide variety of spectroscopic methods. Because most techniques tend to give limited or highly focussed outcomes, subject to some interpretation, they have sometimes been called 'sporting methods' – in other words, no certainty guaranteed! Where a crystalline solid can be isolated, the technique of single crystal crystallography offers a usually highly accurate method of defining atom-by-atom location in three dimensions – a very unsporting, but popular and revealing physical method. These and other physical methods are discussed in some detail in Chapter 6.

It is also possible to employ pure computational methods (sometimes called *in silico* methods) to predict shape, or at least isomer preferences, for complexes. The simplest approach to modelling employs molecular mechanics; this relies on a classical model that treats atoms as hard spheres with the bonds as springs. This is introduced in Chapter 8.3. More sophisticated approaches, such as density functional theory (DFT) are growing in popularity and capacity. Molecular modelling promises to provide excellent predictive capacity in the future without the need for laboratory synthesis, at least in the initial stages. However, laboratory-free chemistry is still far off, and synthesis and product identification remains the essence of chemistry.

Concept Keys

Coordination number and shape of complexes is influenced by the number of valence electrons on the metal ion, metal ion size and preferred coordinate bond lengths, inter-ligand repulsions, and the shape and rigidity of the ligand.

Coordination numbers vary from one up to nine, and in some cases reach into the mid-teens. For d-block metals, coordination numbers from two to nine may be met, with four to six most common; for f-block metals coordination numbers above six and up to fourteen may be observed, with seven to nine most common.

Shapes predicted by a simple amended VSEPR model are observed, but some others are found as well. Inter-conversion between shapes in a particular coordination number can occur, and many complexes are non-ideal in shape, showing distortion away from one limiting shape towards another.

Isomerism – the presence in a particular complex of a number of different spatial arrangements of atoms – is common in complexes.

The two key types of stereoisomers are positional or geometric isomers (such as *cis* and *trans*) and, where a complex is asymmetric, optical isomers (such as Δ and Λ). The general class of isomers is diastereomers; where a complex is asymmetric or dissymmetric (optically active), a diastereomer will have a non-superimposable mirror image, called an enantiomer.

The absence of any plane of symmetry is a key requirement for a diastereomer having an enantiomer (and hence being chiral). Thus 'flat' molecules, such as those of square-planar shape, do not exist as optical isomers; octahedral complexes may be chiral, depending on the type and arrangement of ligands.

The possible existence of a set of geometric isomers does not mean that all are seen experimentally. They will differ in strain energy and thermodynamic stability, and as few as a single one may be isolable.

Ligand shape has an impact on the resultant complex shape, the number of possible isomers that could form, and the thermodynamic stability of complexes formed.

Apart from formation of coordination complexes, some molecules bind other molecules or complexes by weaker noncovalent means, such as strong hydrogen-bonding, forming outer-sphere (host–guest) complexes.

Further Reading

Atkins, P., Overton, T., Rourke, J. *et al.* (2006) *Shriver and Atkins Inorganic Chemistry*, 4th edn, Oxford University Press. This popular but lengthy general text for advanced students contains some clearly-presented sections on shape and stereochemistry of coordination complexes.

Clare, B.W. and Kepert, D.L. (1994) Coordination numbers and geometries, in *Encyclopaedia of Inorganic Chemistry*, vol. 2 (ed. R.B. King), John Wiley & Sons, Ltd, Chichester, UK, p. 795. A detailed and readable review, replete with examples.

Fergusson, J.E. (1974) *Stereochemistry and Bonding in Inorganic Chemistry*, Prentice-Hall, New Jersey, USA. An old but valuable resource book for advanced students, but the depth and detail may concern others.

Steed, J.W. and Atwood, J.L. (2009) *Supramolecular Chemistry*, 2nd edn, John Wiley & Sons, Ltd, Chichester, UK. This up-to-date but more advanced and lengthy text provides a comprehensive coverage of the field, for those who wish to stray beyond the introductory level.

Vögtle, F. (1993) *Supramolecular Chemistry: An Introduction*, John Wiley & Sons, Ltd, Chichester, UK. An ageing but still very readable account for the student of supramolecular compounds, as distinct from coordination compounds, that will assist student understanding of this field.

von Zelewsky, A. (1996) *Stereochemistry of Coordination Compounds*, John Wiley & Sons, Inc., New York, USA. A mature but detailed coverage of topology in coordination chemistry, with a good many clear illustrations; a useful, though more advanced, resource book.

5 Stability

5.1 The Makings of a Stable Relationship

Stability is something we all seek to achieve – molecules included. For molecules, stability is a relative term, since it depends on their environment. Consider an aquated metal ion in aqueous solution; in the absence of any added competing ligands or reaction with water itself, it is perfectly stable. However, when ligands that bind strongly to the metal are added, a reaction may occur leading to new complex species. How much of each complex species exists in solution depends on ligand preference, and the stability of the various species. Two key aspects are involved in the above situation. The mixture changes to achieve the energetically most favourable products, and this process does not occur instantaneously. The first aspect is governed by *thermodynamics*, the second by *kinetics*. Thermodynamic stability is a view of the process at equilibrium, once the formation chemistry is completed. Kinetic stability is a measure of the rate of change during formation leading to equilibrium.

5.1.1 Bedded Down – Thermodynamic Stability

A system in the process of achieving thermodynamic stability can be recognized readily in many cases. For example, addition of ammonia to a solution containing Cu^{2+}_{aq} causes a rapid change in colour. This is a signal that a new complex species is forming, and it is apparent from the observation that new complexes between Cu(II) and ammonia form to a great extent, even for low concentrations of ammonia – evidently, the Cu(II)–ammonia complex is more stable than the Cu(II)–water complex. We have discussed some of the reasons for this type of outcome in Chapter 3. Now, we need to understand the processes involved, so that we can quantify such qualitative observations. One important point arising from earlier discussion was that the amount of a complex species existing cannot be predicted simply from the ratio of added ligand to solvent molecules, because ligand preferences override purely statistical aspects. Establishing the actual composition in solution experimentally, and expressing it in terms of some general parameter, is thus important.

The process we are seeing in the copper(II)–ammonia solution is a ligand substitution process, where one ligand is replacing another. In general, we can represent this, for reaction in aqueous solution with a neutral monodentate ligand at this stage, by Equation (5.1):

$$M(OH_2)_x^{n+} + L \rightleftharpoons M(OH_2)_{(x-1)}L^{n+} + H_2O \qquad (5.1)$$

Introduction to Coordination Chemistry   Geoffrey A. Lawrance
© 2010 John Wiley & Sons, Ltd

Since we have written this reaction as an equilibrium, it is possible to write an equilibrium constant for this reaction (5.2), namely:

$$K = \frac{[M(OH_2)_{(x-1)}L^{n+}][H_2O]}{[M(OH_2)^{n+}_x][L]} \tag{5.2}$$

in which the sets of square brackets in this case refer to concentration of the species. In fact, thermodynamics tells us that the equilibrium constant above should be written in terms of activities (a), not concentrations. The relationship between these two is (5.3):

$$a_S = [S]\gamma_S \tag{5.3}$$

where γ_S is the activity coefficient, which has a value $\gamma_S = 1$ in extremely dilute solutions but for practical solutions has a value of less than 1, caused by the influence of other solute species present on the behaviour of a particular solute molecule. Because activities are difficult to determine, and vary with concentration and composition of the solution, it is convenient to work with concentrations by assuming that $\gamma_S = 1$ (pure water has this value itself). It is then important to quote the experimental conditions when reporting equilibrium constants, as the value will change with conditions. Fortunately, the size of the change with conditions usually met, where concentrations of complexes may vary between 0.001 and 0.1 M, are not large in the context of what we seek to determine. Of more concern is devising processes that permit accurate determination of the concentrations of species present, which is not a trivial task by any means.

Because water has a concentration of approximately 55 M, it varies by only trivial amounts for reactions of dilute species. As a result, it is convenient, and traditional, to leave out the solvent term and usually also to ignore coordinated water molecules. We shall adopt this representation henceforth – but do not assume we are dealing with 'bare' metal ions! This reduces Equation (5.1) to Equation (5.4):

$$M^{n+} + L \rightleftharpoons ML^{n+} \tag{5.4}$$

and thus Equation (5.2) to Equation (5.5):

$$K = \frac{[ML^{n+}]}{[M^{n+}][L]} \tag{5.5}$$

The equilibrium constant K in this case is called the *formation constant* or *stability constant*, since it is measuring formation of a metal complex, and defining its thermodynamic stability. Experimentally, because we measure K values under non-ideal (in a thermodynamic sense) conditions, the term 'constant' here is not absolutely correct, as discussed above. Remember that it is defined only under the particular experimental conditions employed in reality, although it is fairly true to say that the value varies in only a limited way across the range of conditions that we are most likely to apply.

There is also a direct relationship between the stability constant and the free energy (ΔG^0, in kJ mol^{-1}) of a reaction, expressed in terms of the relationship (5.6):

$$\Delta G^0 = -2.303 \cdot R \cdot T \cdot \log_{10} K \tag{5.6}$$

where R is the gas constant and T the temperature in Kelvin. This means that the higher is K, the more negative is the free energy of the reaction. We feel this usually as a release of heat on complexation, because of the relationship between free energy and reaction enthalpy

(ΔH^0) and reaction entropy (ΔS^0), namely (5.7):

$$\Delta G^0 = \Delta H^0 - T \cdot \Delta S^0 \tag{5.7}$$

Examination of Equation (5.5) shows that a large value of K means a high concentration of ML^{n+} relative to M^{n+} and L; in other words, a large K means a strong preference for complex formation. The size of K with metal complexes is usually so large that we tend to report $\log_{10} K$ values, for ease of use; obviously, it's simply easier to refer to a (log) K of 7.5 rather than a K of 3.16×10^7.

In this discussion, we shall also meet another closely related type of stability constant, the *overall stability constant* (β), which represents the stability for a set of sequential complexation steps, rather than for an individual component step. It allows us to represent, for example, the stability constant for an overall reaction $M + nL$ forming ML_n, rather than just for a single ligand addition step such as $M + L$ forming ML. As for K, the larger is β, the more thermodynamically stable is the assembly. The overall stability constant is dealt with more fully in Section 5.1.3.

You may already be familiar with another form of equilibrium constant, the *acid dissociation constant* (K_a), which is so named because it describes the dissociation of the acid HX to its ions (Equation 5.8):

$$K_a = \frac{[H^+][X^-]}{[HX]} \tag{5.8}$$

This is, in effect, an *instability* constant, since it describes a break-up rather than a formation process; the equation for the constant is inverted relative to the form met for the stability constant of interest here. However, expressing it as $pK_a = -\log K_a$ makes it similar to a complex formation or stability constant expressed simply as $\log K$. We shall focus here on complex formation constants.

In the above discussion, we have tended to use a neutral ligand L in discussion, as this avoids dealing with charge variation in equations. This does not imply that the outcome is in any real way different for a neutral or anionic ligand; the same forms of equations apply. Whereas we can perhaps conceive more easily how an anionic ligand X^- may be attracted to and form a coordinate bond with a metal cation, recall that neutral ligands will display polarity as a result of different atoms in bonds. As a consequence, a H—N bond, for example, may be considered as a $^{\delta+}H—N^{\delta-}$ entity, and it is the negative end of the dipole (or, in effect, the lone pair) that attaches to the metal cation. As a consequence, one might anticipate that more polar neutral ligands will make better ligands. Most heteroatoms in molecules that act as efficient ligands to metal ions carry substantial partial negative charge.

5.1.2 Factors Influencing Stability of Metal Complexes

We can identify a number of factors that contribute to the size of stability constants, and it is appropriate to summarize and review effects. In doing so, we will also attempt to divide the effects into those more associated with the metal and those more associated with the ligand, since both contribute to the partnership and inevitably each brings something to the metal–ligand marriage.

Table 5.1 The influence of metal ion charge, size and (for d-block elements) crystal field effects on experimentally-determined stability constants of their complexes, illustrated with hydroxide and ammonia ligands.

Ionic charge effects:		$M^{z+}_{\ aq} + OH^-_{\ aq} \rightleftharpoons M(OH)^{(z-1)+}_{\ aq}$		
M(OH)	Li(OH)	$Mg(OH)^+$	$Y(OH)^{2+}$	$Th(OH)^{3+}$
log K	+0.3	+2.1	+7.0	+10
charge (z)	+1	+2	+3	+4
Metal ion size effects:		$M^{2+}_{\ aq} + OH^-_{\ aq} \rightleftharpoons M(OH)^+_{\ aq}$		
M(OH)	$Be(OH)^+$	$Mg(OH)^+$	$Ca(OH)^+$	$Ba(OH)^+$
log K	+7.0	+2.1	+1.5	+0.6
r (Å)	0.31	0.65	0.99	1.35
Crystal Field effects:		$M^{2+} + 6\,NH_3 \rightleftharpoons M(NH_3)_6^{\ 2+}$		
M	Co	Ni	Cu	Zn
log β	5.0	7.8	13.0	9.0

5.1.2.1 Factors Dependent on the Metal Ion

5.1.2.1.1 Size and Charge

Several factors based on pure electrostatic arguments contribute to a strong stability constant. First, since we are bringing together a positively-charged metal ion and anionic or polar neutral ligands carrying high electron density in their lone pairs, there is certainly going to be a purely electrostatic contribution. We can see this experimentally, as shown above in Table 5.1. As the charge on cations, all of similar size, varies from 1+ to 4+ across the series, the size of log K increases.

When the metal ion charge is fixed but the metal ion size is increased, the surface charge density decreases as the ionic radius increases. This means a less effective attractive force for the ligand applies, which leads to a fall in the size of log K (Table 5.1). The role of surface charge density is seen also if one plots stability constant against charge/surface ratio for a range of $M(OH)^{(n-1)+}$ complexes (Figure 5.1). Obviously, the effect of charge

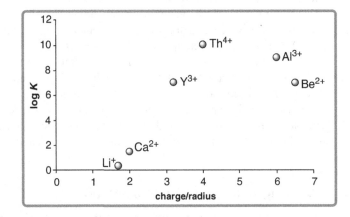

Figure 5.1
Variation of stability constant (K) with charge/radius ratio for some $M(OH)^{(n-1)+}$ complexes. This is a reasonable representation of how K varies will ion charge and size, but obviously the effect of charge has some dominance, given the drop in K for Al^{3+} and Be^{2+} ions despite their large charge/radius ratio.

appears more important than that of ionic radius, from the fall in $\log K$ values for very small but lower-charged ions (Al^{3+} and Be^{2+}). Of course, our simple model of the cations as hard spheres applied in this analysis is also imperfect; overall, nevertheless, it is possible to predict stability adequately if not finitely from the charge/radius ratio relationship. As a result, a large K is favoured by a large charge/radius ratio, or the smaller the ion and the larger its charge, the more stable will be its metal complexes.

Although our focus currently is on the metal, it should be recognized that the size (or molar volume) of the ligand plays a role in the electrostatic effect on stability of a complex, particularly for anionic ligands. This is sensible, since the ligand can be assigned a surface charge density (when anionic) in the same way that we have done for the cation, and obviously an anion with a high surface charge density should form stronger complexes from an electrostatic perspective. This is best illustrated by examining halide monatomic anions, where spherical surfaces have some meaning. With F^- (radius 133 pm) and Cl^- (radius 181 pm), Fe^{3+} forms complexes with $\log K$ of 6.0 and 1.3 respectively, reflecting the greater charged/surface ratio for the former. The concept of ion radius becomes diffuse when we move to molecular anions, however, whose shape may not be anywhere near spherical. However, at least for large reasonably symmetrically-shaped and thus pseudo-spherical anions like ClO_4^-, the very low stability of its complexes is fairly consistent with an electrostatic model, since this ion has a much greater radius than simple halide ions of the same charge.

5.1.2.1.2 Metal Class and Ligand Preference

We have examined the hard/soft acid/base 'like prefers like' concept as it applies to metal–ligand binding already in Chapter 3, so it is simply necessary to revise aspects here. Electropositive metals (lighter and/or more highly charged ones from the s, d and f block such as Mg^{2+}, Ti^{4+} and Eu^{3+}, belonging to *Class A*) tend to prefer lighter p-block donors (such as N, O and F donors). Less electropositive metals (heavier and/or lower-charged ones such as Ag^+ and Pt^{2+} belonging to *Class B*) prefer heavier p-block donors from the same families (such as P, S and I donors). A more significant M—L covalent contribution is asserted to apply in the latter case, along with other effects such as π back-bonding. Of course, in any situation where there are only two categories, there is a 'grey' area of metals and ligands who do not sit easily in either set. A summary is given in Table 5.2 below, in which a large number of the metal ions and simple ligands you are likely to meet appear.

Table 5.2 Examples for both ligands and metals of 'hard' and 'soft' character, with some less clearly defined intermediate cases also included.

Hard	Intermediate	Soft
Ligands		
F^-, O^{2-}, ^-OH, OH_2, OHR, $RCOO^-$, NH_3, NR_3, RCN, Cl^-, NO_3^-, CO_3^{2-}, SO_4^{2-}, PO_4^{3-}	Br^-, ^-SR, NO_2^-, N_3^-, SCN^-, H_5C_5N	PR_3, SR_2, SeR_2, AsR_3, CNR, CN^-, SCN^-, CO, I^-, H^-, R^-
Metal Ions		
Mo^{5+}, Ti^{4+}, V^{4+}, Sc^{3+}, Cr^{3+}, Fe^{3+}, Co^{3+}, Al^{3+}, Eu^{3+}, Cr^{2+}, Mn^{2+}, Ca^{2+}, Mg^{2+}, Be^{2+}, K^+, Na^+, Li^+, H^+	Fe^{2+}, Co^{2+}, Ni^{2+}, Cu^{2+}, Zn^{2+}, Pb^{2+}	Cu^+, Rh^+, Ag^+, Au^+, Pd^{2+}, Pt^{2+}, Hg^{2+}, Cd^{2+}

5.1.2.1.3   *Crystal Field Effects*

The above simple effects are not the sole metal-centric influences on K, as is evident from examining trends in experimental values with a wide range of ligands. This is particularly evident when one examines a series of transition metal 2+ ions forming complexes with ligands; data with ammonia appears in Table 5.1. Overall, between Mn^{2+} and Zn^{2+}, there is a general trend in K values essentially irrespective of ligand of the order:

$$Mn^{2+} < Fe^{2+} < Co^{2+} < Ni^{2+} < Cu^{2+} > Zn^{2+}.$$

The relative sizes of the cations vary in a not so different fashion, with radii following the trend:

$$Mn^{2+} > Fe^{2+} > Co^{2+} > Ni^{2+} < Cu^{2+} < Zn^{2+},$$

so the change is in part (but only in part) assignable to charge/radius ratio ideas. Across this series of metal ions, the number of protons in the nucleus is increasing, but shielding by the electron clouds is imperfect, so that as Z increases there is a progressively higher apparent nuclear charge 'seen' by the ligands, despite the common formal charge. This also contributes to a steady climb in the stability constants from left to right across the periodic block. However, there is an obvious discontinuity between Cu^{2+} and Zn^{2+}, with a drop that is consistently observed regardless of ligand type. The source of this effect is tied to the presence of an incomplete set of d electrons except for Zn^{2+}, and we shall return to it below.

5.1.2.1.4   *The Natural Order of Stabilities for Transition Metal Ions*

For transition metal ions with incomplete sets of d electrons, there is a contribution to stability from the crystal field stabilization energy (CFSE), whereas for d^{10} metal ions (such as Zn^{2+}), with a full set, there is no such stabilization energy. Crystal field stabilization energies of metal ions in complexes have been asserted to have a key influence on values of K for transition metals, reflected in the experimentally-determined stabilities for the series of metal ions from Mn^{2+} across to Zn^{2+}. This effect is overlaid on the general 'upward' trend from left to right across the row with increasing Z, associated with imperfect shielding of the nucleus, discussed earlier, and leads to what is usually called the *natural order of stabilities* (also called the Irving–Williams Series). This experimentally-based order states that the accessible water-stable M^{2+} octahedral first-row transition metal ions exhibit an order of stabilities with any given ligand which is essentially invariant regardless of the ligand employed, that is

$$Mn^{2+} < Fe^{2+} < Co^{2+} < Ni^{2+} < Cu^{2+} > Zn^{2+}.$$

This means the same trend in K values is seen *regardless of ligand* (unless some other special effect overrides it). This trend identifies copper(II) as the metal ion that forms the most stable complexes, irrespective of ligand. The series predicts a fall in K values in each direction from Cu^{2+}, irrespective of ligand. The trend is preserved for the series of metal ions with didentate chelates even when the type of donor atom or the size of the chelate ring is varied (Figure 5.2).

In this series, we are dealing, for high-spin complexes, with octahedral systems. Only for Cu^{2+} do we have an especially distorted octahedral shape. Electrons placed in the lower t_{2g} level are stabilized by $0.4\Delta_o$, whereas those placed in the upper e_g level are destabilized

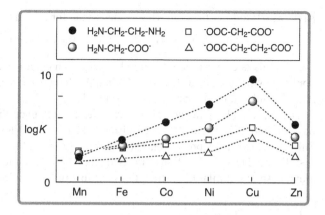

Figure 5.2
The *natural order of stabilities* in operation: consistent variation of measured stability constant with different chelate ligands for the series of transition metal(II) ions from manganese across to zinc.

by $0.6\Delta_o$. Overall, the calculated CFSE follows the order:

$$Mn^{2+} < Fe^{2+} < Co^{2+} < Ni^{2+} > Cu^{2+} > Zn^{2+},$$

with the ions at each end each having zero CFSE as a result of half-filled or filled d shells. This behaviour closely follows the observed (experimental) behaviour of K with varying M^{2+} (Figure 5.2) *except* for Cu^{2+}, where the apparently anomalous behaviour is related to the axially elongated structure met for this ion. This leads to an enhanced CFSE for Cu^{2+} as a result of a different d-orbital energy arrangement to that operating in the pure octahedral environment. Thus, the influence of the d-electron configuration, through the CFSE, is important in understanding the natural order of stabilities for light transition metal ions; it links the basic crystal field theory to experimental observations effectively.

5.1.2.2 Factors Dependent on the Ligand

5.1.2.2.1 Base Strength
One aspect of complex formation that became apparent at an early date was the relationship between the Brønsted base strength of a ligand and its ability to form strong complexes. This is sensible, in the sense that base strength is a measure of a capacity to bind a proton. Making a substitution of H^+ by M^{n+} seems reasonable, permitting basicity to define likely complex strength. From this perspective, the better base NH_3 should be a better ligand than PH_3 or H_2O, and F^- should be superior to other halide ions. This predicted behaviour holds quite well for s-block, lighter (first-row) d-block, and f-block metal ions, which have been defined as 'hard' or *Class A* metal ions. We can say with some certainty that the greater is the base strength of a ligand (that is, its affinity for H^+) the greater is its affinity for (and hence stability of complexes of) at least *Class A* metal ions.

Unfortunately, when one examines heavier metal ions and those in low oxidation states, the behaviour is not the same (hence their definition as 'soft' or Class B metals). Obviously, there are electrostatic contributions applying, but other influences are now important. A classical example is the reaction of the relatively large, low-charged Ag^+ ion in forming

AgX, which follows the order (log values of the formation constant in parentheses):

$$F^-(0.3) < Cl^-(3.3) < Br^-(4.5) < I^-(8.0).$$

Clearly, this does not follow the trend expected on electrostatic grounds, which should be opposite to that observed. The trend is thought to reflect increased covalent character in the Ag–X bond in moving from fluoride to iodide.

The steric bulk of ligands also introduces an influence that can act counter to a pure base strength effect. This has been discussed earlier for NR_3 compounds (Chapter 2.2.2). Substituted pyridines present another example of this influence; 2,6-dimethylpyridine is a poor ligand due to the location of the two methyl groups either side of the pyridine N-donor atom, despite a similar base strength to unsubstituted pyridine.

5.1.2.2.2   Chelate Effect

From a purely thermodynamic viewpoint, the equilibrium constant is reporting the heat released (the enthalpy change, ΔH^0) in the reaction and the amount of disorder (called the entropy change, ΔS^0) resulting from the reaction, related to the reaction free energy as defined in Equation (5.7). The greater the amount of energy evolved in a reaction the more stable are the reaction products; this heat change can sometimes be felt when holding a test-tube in which a reaction has been initiated, and is certainly experimentally measurable to high levels of accuracy. Further, the greater the amount of disorder resulting from a reaction the greater is the entropy change and the greater the stability of the products. This is a harder concept to grasp than some, but think of it in terms of particles involved in a reaction – if there are more particles present at the end of the reaction than at the start, or even if those present at the end are less structured or restricted in their locations, there is increased disorder and hence a positive entropy change. Again, this is experimentally measurable, but not as directly as a simple heat change associated with reaction enthalpy. In the coming together of two oppositely charged ions to form a complex, there is both a release of heat (increased enthalpy, ΔH^0) and a release of solvent molecules from the ordered and compressed layers around the ions (increased entropy, ΔS^0) on complexation. Moreover, the higher the charge on the metal ion, the greater is the effect.

When we employ molecules as ligands where they offer more than one donor group capable of binding to metal ions, there is the strong possibility that more than one donor group will coordinate to the one metal ion. We have entered the realm of polydentate ligands. Just because a ligand offers two donor groups does not mean that both can coordinate to the one metal ion. For example, the *para* and *ortho* diaminobenzene isomers can both act as didentate ligands, but the former must attach to two different metal ions because of the direction in which the rigidly attached donors point. Only the *ortho* form has the two donors directed in such a way that they can occupy two adjacent coordination sites around a metal ion. This isomer has achieved *chelation*. The same behaviour is displayed by the linked pyridine molecules 2,2′-bipyridine and 4,4′-bipyridine (Figure 5.3).

In general, chelation is beneficial for complex stability, and chelating ligands form stronger complexes than comparable monodentate ligand sets. For example, consider Equations (5.9) and (5.10):

$$Ni(OH_2)_6^{2+} + 6\,NH_3 \rightleftharpoons Ni(NH_3)_6^{2+} + 6\,H_2O$$

$$(log\ \beta_6 = 8.6) \qquad \beta_6 = \frac{[Ni(NH_3)_6^{2+}]}{[Ni(OH_2)_6^{2+}]\,[NH_3]^6} \qquad (5.9)$$

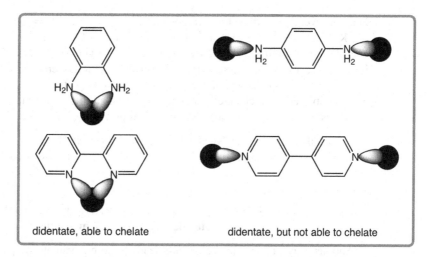

didentate, able to chelate   didentate, but not able to chelate

Figure 5.3
Lone pairs of *ortho*-diaminobenzene (top left) and *2,2'*-bipyridine (bottom left), and *para*-diaminobenzene (top right) and *4,4'*-bipyridine (bottom right) are arranged in space so that only the former pair can chelate to a single metal centre. The latter pair can only bind to two separate metal centres.

$$Ni(OH_2)_6{}^{2+} + 3\,NH_2CH_2CH_2NH_2 \rightleftharpoons Ni(NH_2CH_2CH_2NH_2)_3{}^{2+} + 6\,H_2O$$

$$(log\,\beta_3 = 18.3) \qquad \beta_3 = \frac{[Ni(NH_2CH_2CH_2NH_2)_3{}^{2+}]}{[Ni(OH_2)_6{}^{2+}]\,[NH_2CH_2CH_2NH_2]^3} \tag{5.10}$$

The three didentate chelate ethane-1,2-diamine ligands occupy six sites, as do the six monodentate ammonia molecules, and amine nitrogen donors are involved in each case; however, it is the chelation of the former that causes the enhanced stability. However, direct comparison of the overall stability constants reported above is inappropriate, since, as a result of the different terms in the equations for each β_n, the units are not equivalent, being M^{-3} for β_3 and M^{-6} for β_6. A better but still not perfect approach is to examine the experimental results for direct ligand exchange of two ammines by an ethane-1,2-diamine, such as reaction (5.11).

$$[Ni(NH_3)_2(OH_2)_4]^{2+} + H_2NCH_2CH_2NH_2 \leftrightarrows [Ni(H_2NCH_2CH_2NH_2)(OH_2)_4]^{2+}$$
$$+ 2\,NH_3\,\{log\,K = 2.4 \quad (\text{and } \Delta G^0 = -13.7\,\text{kJ mol}^{-1})\}. \tag{5.11}$$

It is obvious that there is a strong driving force for this ligand replacement reaction, with formation thermodynamically favourable, as seen by the negative ΔG^0 value. Introducing one $H_2NCH_2CH_2NH_2$ chelate in place of two ammine ligands leads to a complex that is ~300-fold more stable than the analogue with two ammonia molecules bound; this is enhanced substantially once all sites around the octahedron are occupied by chelated amine donors, as exemplified by the overall reaction (5.12).

$$[Ni(NH_3)_6]^{2+} + 3\,H_2NCH_2CH_2NH_2 \leftrightarrows [Ni(H_2NCH_2CH_2NH_2)_3]^{2+} + 6\,NH_3$$
$$log\,K = 9.3 \quad (\text{and } \Delta G^0 = -53\,\text{kJ mol}^{-1}). \tag{5.12}$$

For this overall reaction, the components of ΔG^0 are $\Delta H^0 = -16.8\,\text{kJ}\,\text{mol}^{-1}$ and (at 298 K) $T\Delta S^0 = +36.2\,\text{kJ}\,\text{mol}^{-1}$. Both contribute to the overall negative ΔG^0, but the contribution from the entropy-based term $T\Delta S^0$ is larger. The smaller ΔH^0 contribution is in part associated with increased crystal field stabilization, seen experimentally as a change in the maximum in the visible spectrum. We can interpret this reaction as dominated by a favourable entropy change associated with chelation (i.e. an increase in disorder and 'particles' or participating molecules), as exemplified in (5.13) and (5.14).

$$[Ni(OH_2)_6]^{2+} + 3\,(H_2NCH_2CH_2NH_2) \leftrightarrows [Ni(H_2NCH_2CH_2NH_2)_3]^{2+} + 6\,H_2O$$
$$\{(\textbf{4 molecules}) \qquad \rightarrow \qquad (\textbf{7 molecules})\}$$

$$(5.13)$$

versus

$$[Ni(OH_2)_6]^{2+} + 6\,NH_3 \leftrightarrows [Ni(NH_3)_6]^{2+} + 6\,H_2O$$
$$\{(\textbf{7 molecules}) \quad \rightarrow \quad (\textbf{7 molecules})\}$$

$$(5.14)$$

Another way of viewing the process leads to an equivalent interpretation. Adding the first amine of an ethane-1,2-diamine molecule or adding an ammonia molecule in place of a water molecule should occur with similar facility, as they are both strong basic N-donor ligands. However, the next substitution step in each molecular assembly is distinctly different, as there is no special advantage for a second ammonia entering, whereas the second amine group of the partly-bound ethane-1,2-diamine molecule is required to be closely located to a second substitution site as a result of the anchoring effect of the first substitution. Less translational energy is required and the coordination is more probable.

Some reactions appear to be driven almost entirely by a positive entropy effect. For the two related reactions (5.15) and (5.16) below, each leading to four Cd–N bonds forming, the ΔH^0 values are equivalent ($-57\,\text{kJ}\,\text{mol}^{-1}$ for the monodentate, $-56\,\text{kJ}\,\text{mol}^{-1}$ for the didentate chelate), and only the $-T\Delta S^0$ terms differ ($+20\,\text{kJ}\,\text{mol}^{-1}$ for the monodentate, $-4\,\text{kJ}\,\text{mol}^{-1}$ for the didentate chelate), favouring the latter.

$$\textit{Monodentate}: Cd^{2+}{}_{aq} + 4(H_2NCH_3) \leftrightarrows [Cd(H_2NCH_3)_4]^{2+} \quad (\log\beta = 6.5) \qquad (5.15)$$

$$\textit{chelate}: Cd^{2+}{}_{aq} + 2(H_2NCH_2CH_2NH_2) \leftrightarrows [Cd(H_2NCH_2CH_2NH_2)_2]^{2+} \quad (\log\beta = 10.5)$$

$$(5.16)$$

However, it is not correct to assume that a favourable entropy change always drives ligand substitution reactions, as there are examples where the overall entropy change opposes reaction, which is driven rather by a substantial negative enthalpy change. Factors that can contribute to the latter include CFSE variation, reduction in electrostatic repulsion terms on reaction, and solvation and hydrogen-bonding changes that favour reaction. However, entropy-based effects are both easier to visualize and comprehend and more often the dominant contributor, and so tend to attract greater attention.

With higher multidentates, a favourable entropy change can account for an equilibrium lying to the right (favouring the ligand with more donor groups), for example (where **NN** = didentate and **NNNN** = tetradentate ligands) in (5.17):

$$[Ni(\textbf{NN})_2(OH_2)_2]^{2+} + \textbf{NNNN} \leftrightarrows [Ni(\textbf{NNNN})(OH_2)_2]^{2+} + 2\,\textbf{NN}$$
$$\{(\textbf{2 molecules}) \qquad \rightarrow \qquad (\textbf{3 molecules})\}$$

$$(5.17)$$

The growing stability with number of potential donor groups can be seen by looking at the change in stability constants for occupying four coordination sites with four ammine molecules, two ethane-1,2-diamine (en) molecules,

Table 5.3 Comparison of formation constants and reaction energies from reaction of the nickel(II) ion $[Ni(OH_2)_6]^{2+}$ with monodentate and related didentate chelate ligands.

Monodentate	$\log \beta$	ΔG° (kJ mol^{-1})	Didentate	$\log \beta$	ΔG° (kJ mol^{-1})
with pyridine (py) or 2,2′-bipyridine (bipy)					
2 py	3.5	−20	1 bipy	6.9	−39
4 py	5.6	−32	2 bipy	13.6	−78
6 py	9.8	−56	3 bipy	19.3	−110
with ammonia(NH$_3$) or ethane-1,2-diamine (en)					
2 NH$_3$	5.0	−28	1 en	7.5	−43
4 NH$_3$	7.9	−44	2 en	13.9	−79
6 NH$_3$	8.6	−49	3 en	18.3	−104

or one *N,N′*-bis(2′-aminoethane)-ethane-1,2-diamine (trien) ligand. The overall stability constant ($\log \beta$ value) for the substitution of $[Ni(NH_3)_4(OH_2)_2]^{2+}$ to form $[Ni(H_2NCH_2CH_2NH_2)_2(OH_2)_2]^{2+}$ is 5.7 and for $[Ni(H_2NCH_2CH_2NH_2)_2(OH_2)_2]^{2+}$ to $[Ni(H_2NCH_2CH_2NHCH_2CH_2NHCH_2CH_2NH_2)(OH_2)_2]^{2+}$ is 4.0. Clearly, as a general rule, the higher the denticity (or number of bound donor atoms) of a ligand, the higher will be its stability constant with a particular metal ion.

Invariably, for occupancy of the same number of coordination sites around a metal ion, the outcome is greater stability (larger stability constants) for the chelate system over a set of monodentates [β(chelate) > β(monodentates)]. As the number of coordinated donor atoms in a ligand rises, almost invariably the size of the stability constant likewise grows. This effect is assigned in large part, as discussed above, to entropy considerations, associated with favourable entropy change (or disorder) in the system as a chelate ligand replaces a set of monodentate ligands. Examples of stability constants and reaction energies for monodentate and analogous chelate ligands appear in Table 5.3. Recall, however, that direct comparisons of data for monodentate/chelate substitution of the aqua metal ion are of limited value.

A more sophisticated aspect of chelation relates to the chelates 'readiness' for binding as a chelate. This is illustrated in Figure 5.4, where one didentate chelate is rigidly fixed into an orientation immediately appropriate for chelation (that is, it is '*pre-organized*'), whereas the other requires rotation of one component part (which costs energy) from its

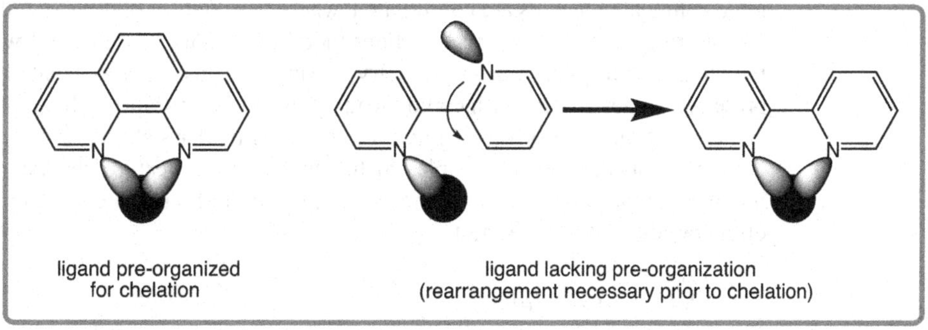

ligand pre-organized for chelation

ligand lacking pre-organization (rearrangement necessary prior to chelation)

Figure 5.4
How a pre-defined appropriate ligand shape (pre-organization) provides chelation at less energy cost than in circumstances where energy-demanding ligand re-arrangement must accompany chelation.

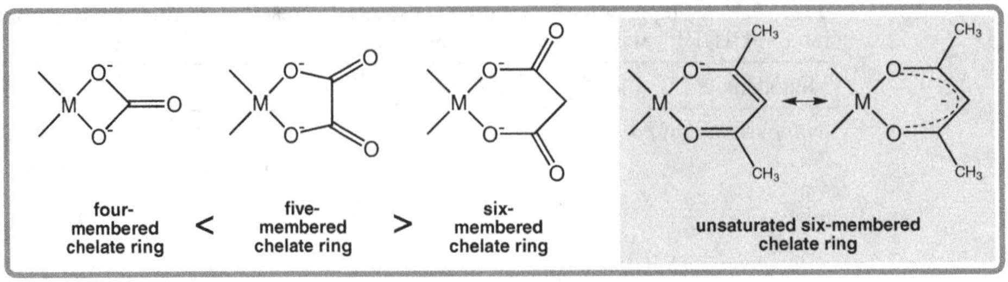

Figure 5.5
Influence of ring size on the stability of first-row d-block metal complexes; five-membered saturated rings are more stable. The exception is for chelation of unsaturated conjugated ligands, where six-membered rings (such as the acetylacetonate ligand illustrated) form complexes of enhanced stability.

preferred free ligand conformation prior to adopting a shape suitable for chelation. Usually, systems that are pre-organized for binding metal ions exhibit stronger stability constants than comparable systems where ligand rearrangement is required.

5.1.2.2.3   Chelate Ring Size

Notwithstanding the above discussion, the *size* of the chelate ring also influences the size of the stability constant, an aspect we have already touched on earlier, being at its largest for five-membered rings and conjugated six-membered rings. Since we are constraining the metal by binding to two linked donor atoms, it can be hardly surprising that there is a relationship between the formed chelate ring size and stability. For example, the smallest chelate ring practicable (three-membered) will reflect a tension between the preferred metal donor length and the preferred donor–donor length, since perfect compatibility is rarely reached. Compromise resulting, seen in terms of variation in intraligand angles and bond distances, has an influence on the stability of the assembly, as obviously a system under great strain is hardly likely to be of high stability. Simply, the size of the stability constant depends on the number of atoms or bonds in the ring. For saturated rings, five-membered rings where ligand donor 'bite' and preferred angles within the chelate ring are optimized are preferred for the lighter metal ions, with smaller or larger rings being of lower stability (Figure 5.5).

The exception to this observed preference for five-membered rings comes when unsaturated conjugated ligands are coordinated, where very stable complexes with six-membered chelate rings can exist with some light metal ions. This is associated with a shift from tetrahedral to trigonal planar geometry of ring carbons, with associated opening out of preferred angles around each carbon leading to a more appropriate ligand 'bite'. When the chelate ring size grows very large, there is no particular stability arising from chelation. As a consequence, most examples of the chelate effect feature chelates with five or six members in the ring; indeed, as one moves to lower or higher chelate ring size, the chelate effect rapidly becomes modest.

5.1.2.2.4   Steric Strain

Size matters. As ligands can vary so much more in size and shape than metal cations, there must be other consequences, including simply size effects in terms of 'fitting' around the central atom. These effects of ligand bulk, resulting from molecules being necessarily required to occupy different regions of space and thus required to avoid 'bumping' against each other when confined around a central metal ion, tend to be termed *steric* effects. As a general rule, the bulkier a molecule the weaker the complex formed when there is a set of ligands involved. Therefore one would expect NH_3 to be a superior ligand to $N(CH_3)_3$ despite both being strong bases, and this behaviour is experimentally observed in solution. In the gas phase, where there is no metal ion solvation and usually only low coordination number complexes form due in part to a low probability of metal–ligand encounters, discrimination in complex stability based more on base strength than steric bulk may apply. Thus $N(CH_3)_3$, which is a stronger base than NH_3, forms the stronger complexes in the gas phase, the opposite to behaviour in aqueous solution. However, gas phase coordination chemistry is not commonly met, and we will explore only solution chemistry. Nevertheless, even in solution, inherently sterically less demanding ligands can display stability constants that reflect basicity effects more; R–S–R ligands often form stronger complexes than R–S–H, for example.

Overall, we can assume that large, bulky groups that interact sterically (that is 'bump into' other ligands) when attempting to occupy coordination sites around a central metal ion usually leads to lower stability. The strain in such systems is seen in distortions of bond lengths and angles away from the ideal for the particular stereochemistry applying. It is possible to model these distortions effectively even with simple molecular mechanics that treats molecules as composed of atomic spheres joined by springs, based on Hooke's Law principles (see Chapter 8), although more sophisticated modelling evolving from atomic theory has developed in recent years. Ligand bulk is particularly significant with regard to its introduction in the immediate region around the donor atom, as congestion will become greater the closer bulky groups approach the central atom. Thus coordination of $N(CH_3))_3$ will lead to greater steric interaction with other ligands than will $^-OOC–C(CH_3)_3$, as the ligand bulk is displaced further out in space in the latter case. As a consequence, the carboxylate is termed a more 'sterically efficient' ligand than the amine.

5.1.2.3   *Sophisticated Effects*

Although we understand complexation reasonably well, and can make experimental measurements of stability constants using a number of different physical methods with high accuracy, the outcome of a suite of effects influencing the stability of metal complexes is that prediction is an approximate and not a perfect art. There are other effects that arise as a result of molecular shape that add complications. For example, large cyclic ligands that carry at least three donor atoms (macrocycles) have a central cavity or 'hole', and the fit of a metal ion into this hole is an important consideration. In fact fit (or misfit) of metal ions into ligands of pre-defined shape (or topology) is an important aspect of modern coordination chemistry, as we now tend to meet more and more sophisticated and designed ligand systems. However, it is more appropriate to leave this aspect for an advanced textbook, and we shall restrict ourselves here to an introduction to one type.

5.1.2.3.1   *The Macrocycle Effect*
We have already introduced the hole-fit concept for a macrocyclic ligand in Chapter 4 (see Figure 4.38), which effectively means that the most stable complexes form where the internal diameter of the ring cavity matches the size of the entering cation. The effect can be significant; for example, the natural antibiotic valinomycin is a macrocycle that binds potassium ion to form a complex $\sim 10^4$ times more stable than that formed with the smaller

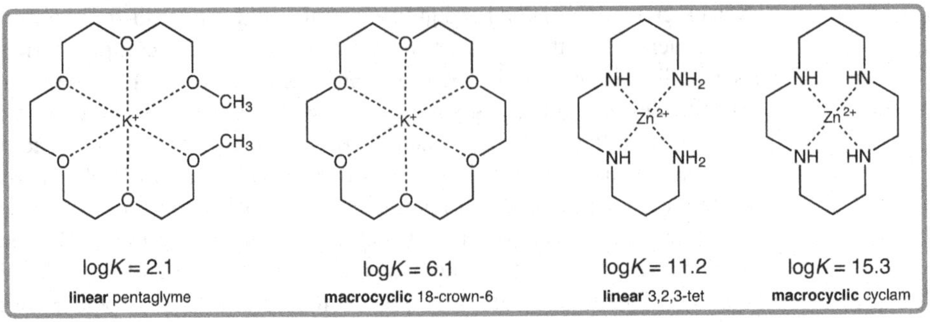

Figure 5.6
Comparison of the stability constants for (at left) a linear and a cyclic polyether binding to potassium ion and (at right) a linear and cyclic polyamine binding to zinc ion, illustrating the macrocyclic effect in action.

sodium ion, despite the chemical similarity of the cations. There is, however, an aspect of pre-organization that is important to macrocycle complexation also, and comparison of the stability constants for potassium ion binding to a linear polyether and a cyclic polyether with the same number of donors and linkage chain lengths (Figure 5.6) provide a clear illustration.

The flexible, long-chain linear molecule must undergo significant translational motion to 'stitch' itself onto the metal ion, which is not favoured. However, the cyclic molecule has the donors pre-organized in more appropriate positions for binding to the metal ion, and its coordination is thus favoured. The higher stability achieved is mainly entropy-driven, although both enthalpy and entropy can contribute, as measured for the polyamines in Figure 5.6, for which changes in both ΔH^0 and ΔS^0 occur. This type of enhanced stability for complexation of macrocycles over acyclic analogues is shown by a wide range of macrocycles of different size, donor type and number, and is well established; a typical example involving polyamines appears in Figure 5.6.

5.1.3 Overall Stability Constants

There is another effect introduced in the experimental data above also – the use of a set of ligands rather than a single ligand in binding the metal in many examples, and it is this that we shall examine more fully next. Because metal ions typically provide more than one coordination site, they can bind more than one donor group, or in most cases (except where one polydentate ligand fully satisfies the coordination sphere demands of a metal ion) more than one ligand. Thus we need to extend the equations for stability constants developed above for attachment of one ligand to account for attachment of a set of n ligands. This process occurs sequentially, not all at once. This is because ligand replacement is the result of molecular encounters, with the complex unit required to make contact with an incoming ligand with sufficient velocity and with an appropriate direction of approach so as to permit a ligand exchange to occur. The probability of all of a set of L replacing all of a set of coordinated water molecules at once (in a concerted reaction) is far too low to be considered a viable pathway. Hence, after ML, a series of other complexes ML_2, ML_3, ..., ML_x form in a sequence of steps. Although we refer to the stability constants for each of these processes as stepwise stability constants, there is no sudden 'step' from all of

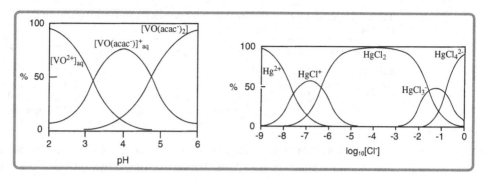

Figure 5.7
Speciation diagrams resulting from metal–ligand titrations, with the concentration profiles of the various ML_n species throughout the process of successive addition of the ligand identified. At left is the pH-dependent formation of vanadyl complexes with the acetylacetonate anion; at right is the chloride ion concentration-dependent formation of chloromercury(II) species.

one species to all of another occurring in solution. Rather, and depending on the reaction conditions and relative sizes of the stepwise stability constants for each species, more than one complex will exist in solution at any one time, although one may be dominant. We can see this experimentally from analysis of accurate metal–ligand titrations, where formation curves resulting from data analysis track the concentrations of sequentially formed species. Two examples appear in Figure 5.7.

We can represent the sequential substitution steps for formation of ML_x by a series of equilibria (5.18):

$$
\begin{aligned}
M^{n+} + L &\leftrightarrows ML^{n+} & (K_1) \\
ML^{n+} + L &\leftrightarrows ML_2{}^{n+} & (K_2) \\
ML_2{}^{n+} + L &\leftrightarrows ML_3{}^{n+} & (K_3) \\
&\ \ \vdots & \vdots \\
&\ \ \vdots & \vdots \\
ML_{(x-1)}{}^{n+} + L &\leftrightarrows ML_x{}^{n+} & (K_x)
\end{aligned}
\tag{5.18}
$$

For *each* step, a stepwise stability constant can be written of the form (5.19):

$$
K_x = \frac{[ML_x{}^{n+}]}{[ML_{(x-1)}{}^{n+}][L]}
\tag{5.19}
$$

The *overall* reaction occurring through combination of the above steps can be written as (5.20):

$$
M^{n+} + xL \leftrightarrows ML_x{}^{n+} \quad (\beta_x)
\tag{5.20}
$$

for which we can define an *overall* stability constant (5.21)

$$
\beta_x = \frac{[ML_x{}^{n+}]}{[M^{n+}][L]^x}
\tag{5.21}
$$

This overall stability constant simply defines the formation of the overall or 'final' complex, where all of one ligand (here, water) can be considered replaced by another; it does not infer anything about the mechanism of the process. Actual examples have appeared earlier in Equations (5.9) and (5.10).

Table 5.4 Successive stability constants for several metal ion complexes with simple monodentate ligands.

Reaction (where m = 0 → 6)	$\log K_1$	$\log K_2$	$\log K_3$	$\log K_4$	$\log K_5$	$\log K_6$
$Ni(NH_3)_m{}^{2+} + NH_3 \leftrightarrows Ni(NH_3)_{(m+1)}{}^{2+}$	2.7	2.1	1.6	1.1	0.6	−0.1
$AlF_m{}^{(3-m)+} + F^- \leftrightarrows AlF_{(m+1)}{}^{(2-m)+}$	6.1	5.0	3.9	2.7	1.6	0.5
$Cu(NH_3)_m{}^{2+} + NH_3 \leftrightarrows Cu(NH_3)_{(m+1)}{}^{2+}$	4.0	3.2	2.7	2.0	< 0	≪ 0
$CdBr_m{}^{(2-m)+} + Br^- \leftrightarrows CdBr_{(m+1)}{}^{(1-m)+}$	2.3	0.8	−0.2	+0.1	not formed	not formed

It is a straightforward task to show that the overall stability constant can be represented in terms of the component stepwise stability constants by Equations (5.22) and (5.23):

$$\beta_x = K_1 \cdot K_2 \cdot K_3 \cdots K_x \tag{5.22}$$

$$\log \beta_x = \log K_1 + \log K_2 + \log K_3 + \cdots + \log K_x \tag{5.23}$$

From examinations of experimental results, is has been observed that, for sequential ligand substitution without change in coordination number, there is a constant size order for the sequential stability constants, shown in Equation 5.24.

$$K_1 > K_2 > K_3 > \cdots > K_x \tag{5.24}$$

Thus, as more and more of one type of ligand are introduced, the gain in stability in each step keeps diminishing. This occurs irrespective of whether neutral or anionic ligands are involved, as illustrated in Table 5.4.

This trend is explained to a modest level of satisfaction by a simple *statistical* argument. Consider a partly-substituted metal complex containing both L and OH_2 ligands. If one reaches in and plucks out 'unseen' and at random any ligand from the coordination sphere, the probability of removing L (rather than OH_2) from ML_m is greater than the probability of removing L from $ML_{(m-1)}$, simply because there is one more to choose from in the former case. At the same time, if you reach in and pluck out a ligand at random and replace it by a new L ligand (irrespective of whether you first pluck off an L or an OH_2), in some cases you will advance the number of L ligands (when you replace an OH_2 by an L), and in others you will make no change by inadvertently replacing one L by another L. From a statistical standpoint, the probability of adding L to $ML_{(m-2)}$ is greater than the probability of adding L to $ML_{(m-1)}$, because there is one extra OH_2 site to choose from in the former complex. Let's represent this in a 'real' case (5.25) below.

$$[M(OH_2)_6] \underset{\text{-NH}_3\ (+OH_2)}{\overset{\text{+NH}_3\ (\text{-}OH_2)}{\rightleftharpoons}} [M(NH_3)(OH_2)_5] \underset{\text{-NH}_3\ (+OH_2)}{\overset{\text{+NH}_3\ (\text{-}OH_2)}{\rightleftharpoons}} [M(NH_3)_2(OH_2)_4] \tag{5.25}$$

Six water substitution sites are available in $[M(OH_2)_6]$, with five water substitution sites available in $[M(NH_3)(OH_2)_5]$, so it is more probable for addition of ammonia to occur in $[M(OH_2)_6]$ than in $[M(NH_3)(OH_2)_5]$. For the reverse step of removal of ammonia, $[M(NH_3)_2(OH_2)_4]$ is more likely to lose one, as two are available versus only one in $[M(NH_3)(OH_2)_5]$. So there is a greater driving force for adding an NH_3 in the first step, and a greater driving force for removing an NH_3 in the second step. This means the concentration $[M(NH_3)(OH_2)_5]$ versus $[M(OH_2)_6]$ will be relatively greater than that of $[M(NH_3)_2(OH_2)_4]$ versus $[M(NH_3)(OH_2)_5]$, and hence K_1 will be larger than K_2. This statistical argument can be made for each sequential substitution. Indeed, it is possible to use the above modelling to calculate the ratio of successive stability constants, and these statistical predictions match at least modestly to experimental values. Clearly, it is not the sole contributor to sequential stability (for example, the relative sizes of the two competing ligands might be expected to have some role to play when building up a set of ligands around a single central metal), but it plays a significant role for at least simple ligands.

Stability constants for Ni^{2+}/NH_3 and Al^{3+}/F^- follow the predicted trend smoothly, regardless of one system using a neutral ligand and the other an anionic ligand. For Cu^{2+}/NH_3, although the appropriate trend is followed, the far larger than usual drop from $\log K_4$ to $\log K_5$ (Table 5.2) is evidence for the operation of the Jahn–Teller effect, which imposes elongation along one axis that does not favour strong complexation in the fifth and sixth axial positions.

There are also conditions that can lead to an exception to this general rule of size order of stability constants. For example, experimental $\log K_m$ determined for $[Cd(OH_2)_6]^{2+}$ reacting with Br^- follow the usual trend from K_1 to K_3, but then there is a rise to K_4 and there are no further values measurable (Table 5.4); clearly something different is happening here. As it happens, we know from physical and structural studies that the complex resulting from introduction of four Br^- ions is a tetrahedral species $[CdBr_4]^{2-}$, whereas the immediately prior species with three Br^- ions coordinated is an octahedral species $[CdBr_3(OH_2)_3]^-$. Thus, it is clear that in the fourth substitution step, there is a change from six-coordinate octahedral to four-coordinate tetrahedral coordination. Yet sequential addition of Br^- still occurs, so why should it matter so much? The answer to this lies in another quite different effect, associated with reaction entropy, which can be envisaged as reflecting the change in the level of disorder in the system upon reaction. Let's see this from the participating molecule aspect, introduced earlier. The change in the K_4 step can be written as (5.26):

$$[CdBr_3(OH_2)_3]^- + Br^- \leftrightarrows [CdBr_4]^{2-} + 3\,H_2O$$
$$\text{6−coordinate} \qquad\qquad \text{4−coordinate} \qquad\qquad (5.26)$$
$$\{\textbf{(2 molecules)} \qquad \rightarrow \qquad \textbf{(4 molecules)}\}$$

versus the usual situation met where there is no change in coordination number, as for K_3 (5.27):

$$[CdBr_2(OH_2)_4] + Br^- \leftrightarrows [CdBr_3(OH_2)_3]^- + H_2O$$
$$\text{6−coordinate} \qquad\qquad \text{6−coordinate} \qquad\qquad (5.27)$$
$$\{\textbf{(2 molecules)} \qquad \rightarrow \qquad \textbf{(2 molecules)}\}$$

The difference is the sudden release of three water molecules when the coordination number falls from six to four. This means there is an increase in 'disorder' in this first case (or a favourable entropy change) and not in the other, influencing K_4 but not K_3 and other prior steps. While this entropy-based effect has been illustrated for simple ligands through a reaction where change in coordination number is the key step, in effect any mechanism that leads to an unprecedented change in disorder (or a 'burst' of released molecules) will produce an enhanced stability constant.

5.1.4 Undergoing Change – Kinetic Stability

Reactions have a start and an end; thermodynamics is concerned with the end, whereas kinetics is concerned with the processes leading from the start to that end. Thermodynamic stability is concerned with the extent of formation of a species under certain specified

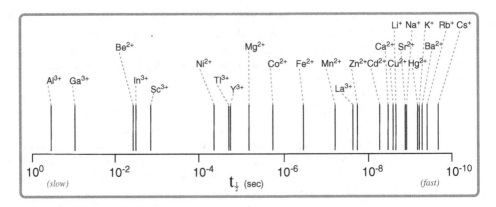

Figure 5.8
Half-lives for water exchange of aquated metal ions.

conditions when the system has reached equilibrium. The rate of formation of a species *leading to equilibrium* is a measure of what we can call *kinetic stability*. Reactions occur with a vast spread of half-lives (the time by which half the precursor molecules have reacted) or reaction rates. Complexes of metal ions which undergo reactions *rapidly* are termed *labile*. Complexes of metal ions which undergo reactions *slowly* are termed *inert*. Henry Taube suggested a suitable definition was for a reaction with a half-life of less than a minute to equate with a labile complex. Thermodynamic stability does not imply kinetic inertness, nor does thermodynamic instability imply lability. For example, in acidic solution Co(III) amine complexes are thermodynamically unstable, but are inert towards dissociation, and as a result most can be stored in solution for extensive periods of time (even decades).

A key basic reaction in aqueous solution is *water exchange*, which is the process whereby a metal ion changes its coordination sphere of water molecules for other solvent water molecules. This process defines aqua metal ions as dynamic species that undergo a series of exchange reactions continuously (5.28):

$$[M(OH_2)_x]^{n+} + x\ H_2\underline{O} \rightleftharpoons [M(O\underline{H}_2)_x]^{n+} + x\ H_2O \tag{5.28}$$

The half-life for water exchange spans a broad range, and follows the pattern shown in Figure 5.8.

A wide variation in rates is seen. Factors influencing how fast ligands are exchanged are metal ion size, metal ion charge and (to some extent for transition elements) electronic configuration. For example, as the ionic radius increases from Mg^{2+} to Ca^{2+}, the exchange rate increases from $\sim 10^5$ to $\sim 10^8$ s^{-1}. The left-hand side of Figure 5.8 equates with inert compounds, and ions here have a large charge/radius ratio. The right-hand side equates with labile compounds. There is obviously a 'grey' area, as will always be the case where we partition into two classes. However, it is clear that complexes need not be static species, and coordinated water molecules undergo exchange with their solvent environment, measurable where practicable by employing isotopically distinctive $^{18}OH_2$ or $^{17}OH_2$ as solvent and following its introduction into the metal coordination sphere in place of normal $^{16}OH_2$ (such as by using ^{17}O NMR spectroscopy). While we can measure these processes, understanding how they occur is another thing altogether.

Because we can initiate and observe reactions, we know that we start with certain reagents and end with others. We can usually separate and quantify the products of a reaction. However, we cannot really understand the reaction unless we can answer the question of how the reactants have been converted into the products. This process is the *mechanism* of the reaction, and it is difficult to tease out because the stages along the way involve only transient non-isolable species – at best, we can infer their nature from a range of experiments and by analogy to known stable species. At the core of reaction mechanisms is the concept of the transition or activated state, the assembly present at the peak of the activation barrier prior to relaxation to form the products. In coordination chemistry, the predicted activated state species are considered to be based on known geometries, but ones that are not inherently stable for the particular system under investigation; there is good supporting evidence based on reactivity and products that support this approach.

5.2 Complexation – Will It Last?

You've probably heard the doomsayers, in commenting on a new partnership, predict that 'it will never last'. Of course, given the fragility of life, this is hardly a prophetic statement, but more of a comment on longevity. At the molecular level, we are also dealing with finite rather than infinite time periods, and the longevity of a molecular partnership is not constant for all assemblies. Coordination complexes are not uniform in their behaviour – in fact, each complex has unique physical properties, which include unique thermodynamic and kinetic stabilities. The fate of a complex relates to its environment; in solution in particular, it will need to deal with the presence of other compounds, which include the solvent itself. How it conducts itself, and whether it retains its integrity, depends on both its thermodynamic and kinetic stability. Let's look at a human-scale example: we recognize a bank (usually, at least) as a stable place that coordinates money, but if it parts with its money in too undisciplined a way it can become, over time, insolvent and cease to exist. The money has not disappeared, merely moved elsewhere. Coordination compounds follow the same path; they may appear to have a certain form and stability, yet may eventually under external influences undergo further change that leads to their demise – all a bit like life, really.

5.2.1 Thermodynamic and Kinetic Stability

If a complex has a large thermodynamic stability that means the complex is a greatly favoured product, resulting from a particular reaction, and possibly even the only significant product, although one shouldn't immediately assume that this is the case. This behaviour says very little about the manner by which and pace at which it undergoes its formation and any following reactions, however. Having a high stability constant may mean that a particular product forms essentially exclusively, but its accommodation of subsequent changes in its environment relates in part to the rate at which it can undergo reactions. There are, nevertheless, relationships between thermodynamic and kinetic parameters. Consider a simple equilibrium involving M, L and ML. Let's define reactions that involve formation of ML (the 'forward reaction') and the reverse decomposition of ML (the 'back reaction') with rate constants for the forward (k_f) and back (k_b) reactions as follows

(5.29 and 5.30):

$$M + L \xrightarrow{kf} ML \tag{5.29}$$

$$ML \xrightarrow{kb} M + L \tag{5.30}$$

We can then consider combining Equations 5.29 and 5.30 into one equation (5.31), namely:

$$M + L \overset{k_f}{\underset{k_b}{\rightleftharpoons}} ML \tag{5.31}$$

This expression, as written, also corresponds to the one we use to define the equilibrium constant, K. It can be shown that there is a simple relationship between these terms (5.32), namely:

$$K = \frac{k_f}{k_b} \tag{5.32}$$

This holds provided that no stable intermediate is involved in the reaction. If the rate at which ML dissociates into M and L (defined by the rate constant, k_b) is slower than the rate at which M and L assemble into ML (defined by the rate constant k_f), then there will always at equilibrium be a larger amount of ML present than of M and L, which translates into a large thermodynamic stability constant K, which equates with the ratio of these two kinetic terms. In effect, for such simple steps, there is a thermodynamic–kinetic link. At this level, however, we shall not dwell on this too deeply.

5.2.2   Kinetic Rate Constants

Reactions can occur in a number of ways, but we measure the processes under way in a common manner, by observing the change in the amount of reactants and/or products over time. This can be done by employing one or more of a wide variety of physical methods; the key requirement is that the technique selected involves a measurable change, and that the scale of the change with respect to the amounts of compounds is either essentially linear or of known variability in the range of interest.

There are different types of behaviour for reactions; for example, in zeroth-order reactions, the change in the amount of a reactant or product is essentially linear over time; in first-order reactions, it is exponential. The first-order reaction process is extremely common in coordination chemistry, and this is the type that we shall restrict ourselves to here. In its simplest form, the concentration of a reacting species varies over time as (5.33):

$$[S]_t = [S]_o \cdot e^{-k \cdot t} \tag{5.33}$$

where $[S]_t$ is the concentration at a particular time t during the reaction, $[S]_o$ is the original concentration at time zero and k is the rate constant, which is the characteristic measure of the rate of the reaction. This equation may also be expressed as (5.34):

$$\ln([S]_t) - \ln([S]_0) = -k \cdot t \tag{5.34}$$

so that a plot of $\ln([S]_t)$ versus time (t) will be linear with a slope equal to the negative rate constant (k). This rate constant, usually expressed in units of s^{-1}, is a constant only under the experimental conditions operating (as it varies with temperature, solvent and added electrolyte). Typically, rate constants double for about every 5 °C rise in temperature.

Measurement of variation in rate constants with temperature allow determination of the activation parameters (activation enthalpy, $\Delta H^{\neq}$, and activation entropy, $\Delta S^{\neq}$) applying in the reaction, which assist in elucidating the mechanism.

In some reactions, the rate constant is seen to vary with concentration of one of the reactants (say molecule X). If a plot of the measured rate constant (k_{obs}) determined under particular defined conditions versus [X] present is made, where reaction with various concentrations of X has been examined, and is found to increase linearly with increasing [X], then the behaviour can be represented as (5.35)

$$k_{obs} = k_X \cdot [X] \tag{5.35}$$

and k_X is then called a second-order rate constant and will have units of $M^{-1} s^{-1}$, the experimentally observed rate constant (k_{obs}, s^{-1}) being dependent on the molar concentration of X. This type of behaviour means that X is involved in the reaction in some way; it may appear as part of the product, although it may alternatively participate only in the transition state and not appear in the product. These are brief aspects of reactions that form a background to our discussion of the mechanisms of reactions below.

5.2.3 Lability and Inertness in Octahedral Complexes

We have already introduced the concept of rate of change for complexes, expressed in terms of the two extremes of *labile* (fast) and *inert* (slow), where charge and size are key factors. Experimental observations of reactivity allow us to define, at least approximately, which metal ions fall into which category. For transition metals, although size/charge effects are important, they do not fully explain experimental observations. It was Henry Taube who showed that the d-electron configuration played an important role, and that high-spin complexes are generally labile. The following groupings were identified from experimental observations of octahedral complexes:

labile – all complexes where the metal ion has electrons in $e_g{}^*$ orbitals [e.g. Co^{2+} (d^7: $t_{2g}{}^5 e_g{}^2$); high spin Fe^{3+} (d^5: $t_{2g}{}^3 e_g{}^2$)], and all complexes with less than three d electrons [e.g. Ti^{3+} (d^1)];

inert – octahedral d^3 complexes [e.g. Cr^{3+} (d^3: $t_{2g}{}^3 e_g{}^0$)], and low spin d^4, d^5 and d^6 complexes [e.g. Co^{3+} (d^6: $t_{2g}{}^6 e_g{}^0$)].

An analysis in terms of the relative energies of the precursor and reaction intermediate predicts lability for octahedral complexes with populated $e_g{}^*$ levels, in line with the experimental observations.

There are, of course, 'grey' areas. Even within a traditionally inert system, reactivity may vary markedly with the type of ligand undergoing reaction. For $[Co^{III}(NH_3)_5X]^{n+}$ reacting in water to replace the X-group by a water molecule, the rate constant varies by an order of $\sim 10^{10}$ from NH_3 ($k = 5.8 \times 10^{-12} s^{-1}$ at room temperature, a half-life of 3800 years) to $OClO_3{}^-$ ($k = 1.0 \times 10^{-2} s^{-1}$, a half-life of 7 seconds). Such variations reflect the nature of the ligand more than the metal, since the ligand lability or capability as a good leaving group tends to extend across a wide range of metal ions.

5.3 Reactions

We can categorize reactions in a number of ways. So far, we have done so in terms of rate (how fast or slow). This approach, however, tells us nothing about what is happening in a reaction or how a particular reaction occurs. We shall now define reactions in terms of what happens. We shall explore examples of the following reactions in Chapter 6, when describing examples of chemical synthesis in coordination chemistry.

Inorganic reactions involving metal complexes fall into three major categories:

1. reactions involving the inner *coordination sphere* of the complex;
2. reactions involving the *oxidation state* of the central metal; and
3. reactions involving the *ligands* themselves.

It is possible also to include metal ion exchange as another category, but since this involves a change in the inner coordination sphere of each metal participating in this process, it is appropriate to consider this as a subset of (1) above. Indeed, within each category there may be several further convenient subdivisions applicable. In particular, the sub-categories may be defined for *reactions involving the coordination sphere* as:

(a) increase in coordination number
   (*addition:* $ML_x + nL \rightarrow ML_{(x+n)}$);
(b) decrease in coordination number
   (*elimination:* $ML_xY_n \rightarrow ML_x + nY$);
(c) partial or complete ligand replacement
   (*substitution:* $ML_x + nZ \rightarrow ML_{(x-n)}Z_n + nL$);
(d) change in shape for a fixed coordination number
   (*stereoisomerization:* e.g. tetrahedral $\rightarrow$ square planar);
(e) change of relative positions of atoms for a fixed coordination number
   (*isomerization:* e.g. *cis*-$ML_xZ_n \rightarrow$ *trans*-ML_xZ_n);
(f) change of central metal for a fixed ligand set
   (*metal ion exchange:* e.g. $ML + M' \rightarrow M'L + M$).

These cover a vast array of reactions, and we shall restrict ourselves here to a limited selection of some important reactions, in order to further define their character. We shall do this in terms of not only the nature of the reaction, but also the manner in which it occurs, or the *mechanism* of the reaction.

When we initiate and observe a reaction, we can of course clearly define the chosen reactants, and we are able also, through their separation and characterization, to identify the stable products. What we cannot define so readily is how the process of change from reactants to products has occurred. The process by which change occurs is, as stated above, the reaction *mechanism*. Reaction theory proposes that reactants pass through a *transition state*, which, as its name implies, is a short-lived intermediate or activated state of higher energy. Energy is required to reach this transition state, with stable products of lower energy then forming from this transition state. The change in energy of reactants in reaching the transition state is the *activation energy* for the reaction. The change in energy between reactants and products is the *reaction energy*. The former relates to the pathway of transition to the products, and is associated with the kinetics, the latter relates to the relative stabilities of reactants and products, and is a thermodynamic term. The key to a reaction mechanism lies in understanding the transition state. This is something we cannot usually detect directly and cannot isolate, so our identification of it must rely on inference from secondary observations. For example, the variation of rate constants with concentrations of various reagents, the measured activation parameters, the range and various amounts of products, and incorporation of isotopically labelled components of the reaction mixture in products are devices that can assist the chemist in predicting the way a reaction proceeds.

5.3.1 A New Partner – Substitution

We shall examine in some detail here one of the most important, simplest and best understood classes of reactions in coordination chemistry, namely *substitution*. This involves one (or several) ligand(s) departing the coordination sphere to be replaced by one (or several) others, without a change in coordination number and basic shape between the reactant complex and its product. It is usually a clearly observable reaction, at least where the nature of the ligands involved in the substitution are distinctively different, and can often be followed by simple approaches such as observing the change of colour as reaction proceeds.

In considering how a substitution can occur, there are really a very limited number of sensible options, which are:

(a) the departing ligand can leave first (leaving a temporary 'vacancy' in the coordination sphere), and then the incoming ligand enters in its place;
(b) the entering ligand can add first (causing a short-lived increase in the number of ligands around the metal ion), and then the departing ligand leaves; or
(c) one ligand can leave while the other enters, in a concerted manner (where a 'swapping' of these ligands occurs with neither favoured in the transition state).

In all cases, regeneration of the preferred coordination number and geometry in forming the stable product occurs.

In the above options, (a) involves ligand loss (or *dissociation*) as the key initiation step, whereas (b) involves ligand gain (or *association*) as the key initiation step. The third option (c) involves concerted replacement (or *interchange*) as the key step. As a consequence, they tend to be referred to as *dissociative* (abbreviated as D), *associative* (A) and *interchange* (I) mechanisms in turn.

The concepts rely also on our understanding of preferred coordination numbers and geometries. For example, for many metal ions six-coordination is common, but some stable examples of both five- and seven-coordination are known. Therefore, it isn't too much of a jump to consider a short-lived five- or seven-coordinate species existing for a species stable only in six-coordination – in other words, a transition state of lower or higher coordination number. As an example, six-coordinate octahedral is the overwhelmingly dominant coordination mode for stable cobalt(III) complexes. Yet, in recent years, rare examples of isolable but usually very reactive compounds with five-coordination and seven-coordination have been prepared. It doesn't take too much of an act of faith to assert that such geometries form as short-lived transition state species in substitution reactions of this whole family of complexes.

5.3.1.1 *Octahedral Substitution Mechanisms*

Since octahedral complexes dominate coordination chemistry of particularly lighter elements, the mechanisms of these reactions are both important and well studied. There are two limiting mechanisms, *dissociative* and *associative*; they are distinguished by the order

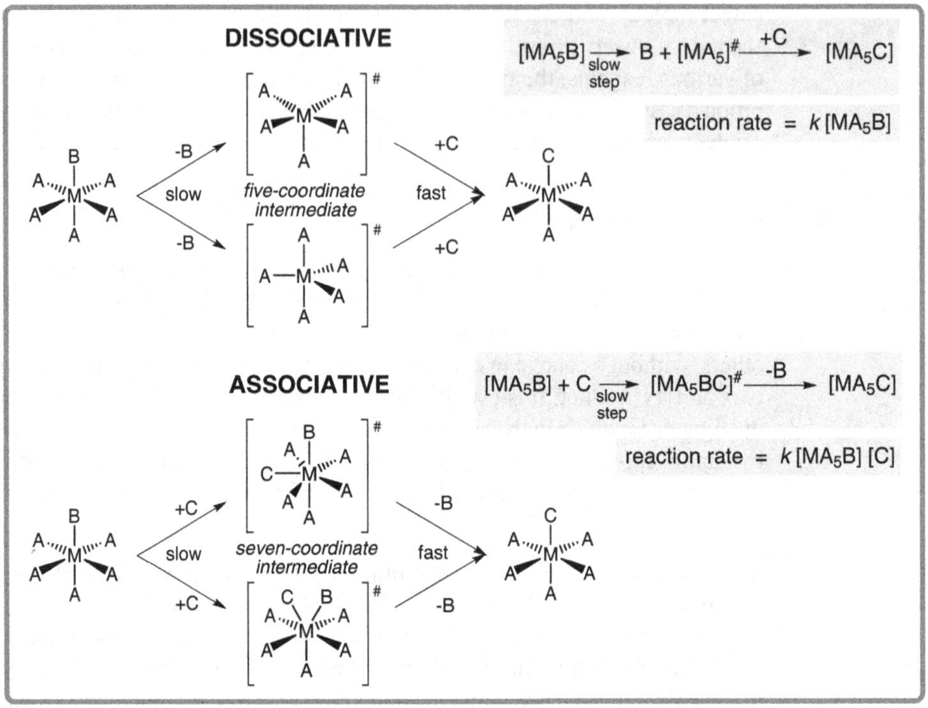

Figure 5.9
The operations involved in the dissociative and associative reaction mechanisms for octahedral complexes. Two different shapes for intermediates are based on known shapes for these coordination numbers.

in which the process of substitution of the ligand to be replaced (the leaving group) by a new ligand (the entering group) occurs, as outlined above, and in Figure 5.9. The outcome is in each case identical, of course – one ligand has been replaced by another. The mechanism that operates cannot be observed directly, but must be inferred from various experiments.

The above discussion could suggest that a successful forward step always occurs. If energy acquired in an encounter or collision process between reacting molecules is insufficient to allow the reactant to reach the energy level of the activated state, then no reaction can occur; this is a key aspect of activation theory. Moreover, the transition state, when reached, does offer a 'forward or back' option in both mechanisms. With reference to Figure 5.9, should B re-enter the coordination sphere (in the D mechanism) or C depart before B (in the A mechanism), then the initial reactant is regenerated. Thus there is a potential reversibility in the reactions, although this has been neglected here for the sake of simplification. The observation of change in a reaction tells us that there is a driving force towards the products.

5.3.1.1.1   Dissociative Mechanism

Slow and therefore rate-determining loss of one ligand to produce a *five-coordinate intermediate* is the key to this process. This intermediate is a short-lived transition state of higher energy than the reactants; with energy (the activation energy) required to reach this intermediate state, this is the defining slow (or rate determining) step in the overall reaction

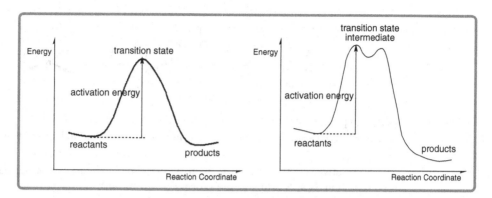

Figure 5.10
Defining the barrier to reaction (activation energy) in terms of a reaction coordinate, which represents the progress of a reaction. Shown is the simple process where no intermediate species with any significant lifetime can be considered to form (left) and a process where there is an intermediate of limited stability and lifetime formed (right).

process, which can be expressed as (5.36):

$$[MA_5B] \xrightarrow[\text{slow}]{-B} [MA_5]^{\#} \xrightarrow[\text{fast}]{+C} [MA_5C] \tag{5.36}$$

Once reached, the transition state intermediate can 'fall back' to reform the reactants (so nothing is achieved), or else can 'fall forward' to yield the products (and a reaction is observed to occur). The process occurring is illustrated in terms of a reaction coordinate diagram in Figure 5.10.

The *dissociative*, or **D**, mechanism is also sometimes called an S_N1 reaction, a form of nomenclature used in organic reaction mechanisms. In the representation S_N1, **S** refers to the process (substitution), subscript **N** to the character of the leaving and entering group (a nucleophile, equated in this case to a Lewis base with a lone pair of electrons), and **1** to the number of molecules involved in the rate-determining step (unimolecular). For the D mechanism, the reaction rate can be expressed by Equation 5.37:

$$\text{rate} = k[MA_5B] \tag{5.37}$$

It is first-order in reactant only, since the first step of ligand loss from the precursor complex, which involves no other molecule, is rate-determining.

5.3.1.1.2 Associative Mechanism
Slow rate-determining gain of one ligand to produce a *seven-coordinate intermediate* is the key to this process. We can represent this reaction as a sequence of a slow and therefore rate-determining formation of a seven-coordinate intermediate, followed by the fast loss of a ligand which is only significant if the ligand lost differs from the one added (5.38).

$$[MA_5B] \xrightarrow[\text{slow}]{+C} [MA_5BC]^{\#} \xrightarrow[\text{fast}]{-B} [MA_5C] \tag{5.38}$$

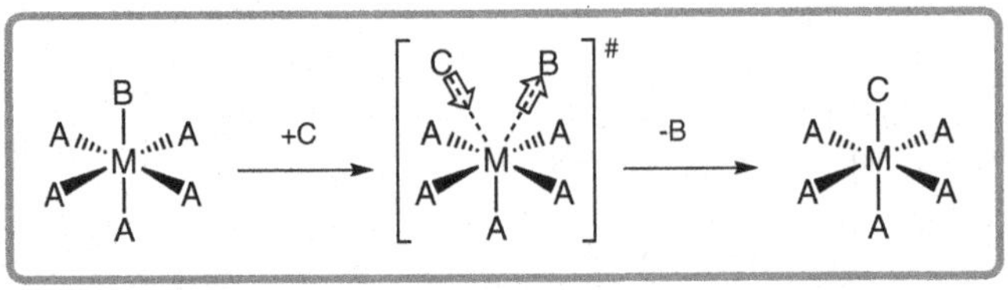

Figure 5.11
A schematic representation of the interchange mechanism for octahedral substitution.

The associative, or **A**, mechanism is also sometimes called an S_N2 mechanism. In the representation S_N2, **S** refers to the process (substitution), subscript **N** to the character of the entering group (a nucleophile), and **2** to the number of molecules involved in the rate-determining step (bimolecular, as the precursor complex and the entering group join together in forming the activated state). For the A mechanism, the reaction rate can be expressed by Equation 5.39

$$rate = k[MA_5B][C] \qquad (5.39)$$

Since the rate-determining step involves two species, it is bimolecular and two concentration terms appear in the rate expression.

For the dissociative (D) mechanism *bond-breaking* is important in the rate-determining step, and this should not be influenced too much by varying the entering group. Thus we expect the rate constant to be largely independent of the entering group. For the associative (A) mechanism *bond-making* is important in the rate-determining step, and clearly this process will be influenced by the character of the group entering to form the new bond. As a result, we expect k to be dependent on the entering group. As a consequence, examining how the rate constant changes with change in entering group may provide evidence for a particular mechanism. Unfortunately, 'pure' A or D mechanisms are uncommon; as a result, the distinction is not clear-cut.

5.3.1.1.3   Interchange Mechanism
The interchange (I) mechanism is offered as a 'half-way house' between the limiting A and D mechanism, involving concomitant entry and departure of the two participating ligands (Figure 5.11).

As a result of the process proposed, you could view the transition state as a five-coordinate species with two tightly bound outer-sphere exchanging ligands, or else as a seven-coordinate species with two very long bonds to the exchanging ligands. The descriptions amount to the same thing, in effect.

5.3.1.1.4   Probing the Transition State
The problem with the transition or activated state is its extremely short-lived nature, which means we cannot 'see' it easily by any physical method. However, this does not prevent experiments being devised that probe the transition state through inspecting the products. There are several long-standing experiments that do this neatly. One is to make use of the opportunity for stereochemical change in passing through particular transition states, experiments most accessible using inert complexes for which both the precursors and products can be easily isolated and characterized. The sole example we shall use to illustrate this aspect is the reaction of *cis-* or *trans-*[CoCl(en)$_2$(OH)]$^+$ to form [Co(en)$_2$(OH)(OH$_2$)]$^{2+}$,

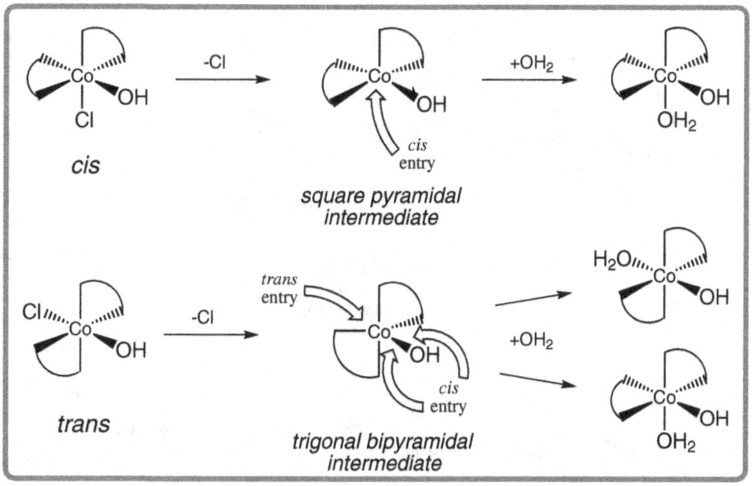

Figure 5.12
Stereochemical control of the substitution reaction in *cis*- and *trans*-[CoCl(en)$_2$(OH)]$^+$ via different transition state geometries. The square pyramidal intermediate offers only one choice to the entering group and 100% retention of geometry, the trigonal bipyramidal intermediate offers three entry points around the triangular central plane, two leading to *cis* entry and one to *trans* entry, for a 2:1 ratio.

which is a simple hydrolysis substitution reaction proceeding (as for the vast majority of cobalt(III) complexes) via a dissociative mechanism. In the D mechanism in Figure 5.9, two different five-coordinate intermediates were suggested – trigonal bipyramidal and square-based pyramidal. This reaction offers us the opportunity to both support this mechanistic assertion and identify the likely geometry of the transition state. It does this by examining the stereochemistry of the products. For *cis*-[CoCl(en)$_2$(OH)]$^+$ reacting, there is 100% conversion to *cis*-[Co(en)$_2$(OH)(OH$_2$)]$^{2+}$, whereas for *trans*-[CoCl(en)$_2$(OH)]$^+$ reacting there is ~70% *cis* and ~30% *trans* isomers of the product formed. Given the very close similarity in the precursor complexes, a common gross dissociative mechanism is presumed – so why the change in isomer ratio? The answer lies in the form of the transition state.

For a square-based pyramid, full retention of stereochemistry for the *cis* isomer is predicted (Figure 5.12) as only one site of attachment is available for the incoming water group to regenerate the original stereochemistry, whereas the trigonal bipyramidal shape for the *trans* isomer allows choice, because water entry can occur from any of three different positions around the triangular plane, some of which lead to different geometric isomers (in fact, a theoretical ratio of 2:1 *cis:trans* is predicted, not too distant from the experimental results). Thus, simple experiments with geometric isomers, involving separating and identifying product isomers, allows the reaction mechanism to be probed in some detail.

5.3.1.1.5 *Base Hydrolysis*
Although most reactions we meet occur in aqueous solution, we have not yet looked at the effect of one obvious variable in aqueous solution, which is pH change. Reactions may occur in neutral, acidic or basic solution. What has been observed is that some reactions are accelerated by protons and others by hydroxide ions. Such observations infer a role for

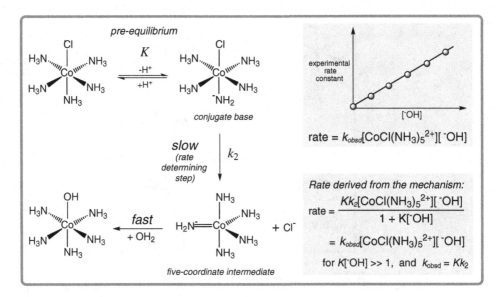

Figure 5.13
Mechanism for base hydrolysis of cobalt(III) complexes. A pre-equilibrium ammine deprotonation step is followed by slow loss of the leaving group (Cl⁻ here) to form a short-lived intermediate that reacts readily with solvent water to form the hydrolysed product. A first-order dependence on [⁻OH] is observed experimentally (top inset), with the mechanism yielding the equivalent expression (lower inset).

these ions in the mechanism. We shall examine the origin of such effects in one case only here, with hydroxide ions.

Reactions in aqueous solution in the presence of added base (hydroxide ion) may lead to substitution of one donor group by hydroxide, a process termed *base hydrolysis*. As some ligands are not readily replaced by hydroxide ion, the leaving group is not selected arbitrarily. Some ligands are more susceptible to reactions than others. A simple example is the complex ion $[CoCl(NH_3)_5]^{2+}$, where the NH_3 groups are only very slowly replaced, whereas the Cl^- ion is much more readily removed, leading to a specific reaction (5.40).

$$[CoCl(NH_3)_5]^{2+} + {}^-OH \rightarrow [Co(NH_3)_5OH]^{2+} + Cl^- \tag{5.40}$$

The reaction is usually catalysed by base, seen experimentally by the reaction getting faster as the concentration of base increases. For this behaviour, Equation (5.41) fits the experimental behaviour.

$$\text{rate} = k_{\text{obsd}}[CoCl(NH_3)_5^{2+}][{}^-OH] \tag{5.41}$$

Upon initial inspection, this expression looks like that given earlier for an A mechanism, but is this the case? There is strong support for a separate, distinct mechanism in this case, involving *conjugate base* formation. From a range of studies, the now accepted mechanism involves a pre-equilibrium deprotonation and a following D mechanism, as illustrated in Figure 5.13.

The process involves formation of the conjugate base of the complex by proton loss from an ammonia ligand that remains intact in the complex, this amide ion being reprotonated to regenerate the ammonia in the fast final step where the intermediate reacts with addition of a water molecule as its two component parts, the proton adding to the deprotonated ligand and the hydroxide ion to the vacancy in the coordination sphere (Figure 5.13). The rate law resulting from the mechanism, shown in Figure 5.13, reduces to fit the experimentally observed expression under expected conditions. The intermediate is a five-coordinate species (as in a standard D mechanism), stabilized further in this case by the special $^-NH_2$ bonding. This can be visualized as having double-bond character through π-type interaction with the metal ion of the additional lone pair created upon deprotonation, a process that is unavailable to an ammonia molecule, which has only the one lone pair. The activation energy for forming this intermediate is lowered compared with the usual D mechanism, and therefore catalysis has occurred. This is related to its being more stable, though still not sufficiently so as to be isolable. This mechanism obviously relies on the capacity of an amine ligand to deprotonate, and the general observation that N—H bonds are very rapidly replaced by N—D bonds in alkaline D_2O solution supports facile deprotonation/reprotonation reactions for coordinated ammonia ligands. Moreover, if all ammonia ligands are replaced by the aromatic amine pyridine, which has no N—H bond and is thus unable to react by the same mechanism, the reaction is very much slower, despite the bulkier ligands possibly promoting a dissociative process.

5.3.1.1.6 Ion Pair Formation

Another reaction that poses similar problems in interpretation is where an anionic group other than hydroxide replaces another group in a cationic complex. For this type of reactions, as the cation and anion are oppositely charged, they can undergo strong electrostatic attraction to form what is called an *ion pair* in an equilibrium reaction. This process precedes the actual substitution, and therefore is again called a *pre-equilibrium* (5.42).

$$[MA_5B]^{n+} + X^{m-} \underset{pre\text{-}equilibrium}{\overset{K_{IP}}{\rightleftharpoons}} \{[MA_5B]^{n+}.X^{m-}\}^{\#} \xrightarrow{k'} [MA_5X]^{p+} + B^{q-} \quad (5.42)$$

A complication is that either the free cation or the ion pair can proceed forward to form the final products. The unpaired cation will react differently to the ion-paired cation (which is, apart from the ion-pair formation, otherwise identical). If the reaction is accelerated due to ion-pair formation, we can derive Equation (5.43):

$$\text{rate} = k' K_{IP}[MA_5B][X] = k[MA_5B][X] \quad \text{(where } k = k' K_{IP}) \quad (5.43)$$

which cannot be distinguished from the direct A mechanism, which has the same apparent rate law. Fortunately, ion-pair strength for complexes in aqueous solution is usually low, so it is not usually a mechanistic issue in water, but may be important in other solvents.

5.3.1.1.7 Substitution by Solvent

Other types of reactions may not present themselves as mechanistically obvious due to complicating factors. For a reaction where a solvent molecule is replacing another ligand, namely Equation 5.44:

$$[MA_5X] + \text{solvent} \rightarrow [MA_5(\text{solvent})] + X \quad (5.44)$$

we cannot distinguish between a dissociative mechanism (5.45), where

$$\text{rate} = k[MA_5X] \quad (5.45)$$

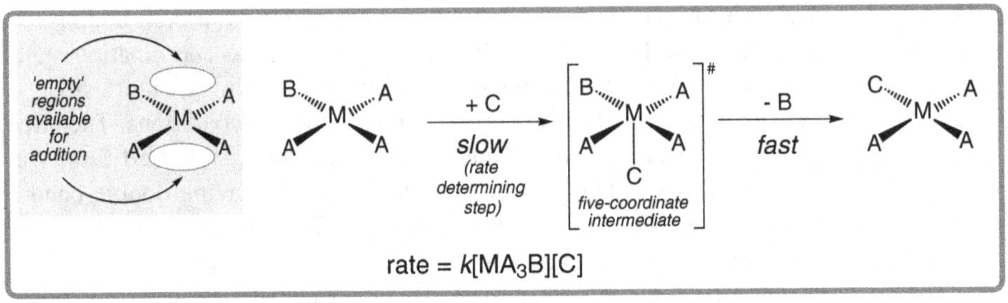

$$\text{rate} = k[MA_3B][C]$$

Figure 5.14
Associative mechanism for substitution reactions in square planar complexes. Sites above or below the plane are inherently readily accessible to incoming ligands.

and an associative mechanism (5.46), where

$$\text{rate} = k[MA_5X][\text{solvent}] = k'[MA_5X] \quad (\text{where } k' = k[\text{solvent}]) \tag{5.46}$$

because the concentration of solvent is high, and cannot be varied. Changing solvent concentration by dilution through addition of another solvent changes the resultant solvent properties, since no solvent is 'innocent' and without effects, invalidating the process. This means that an associative mechanism would appear experimentally as a simple first-order expression. Thus the D and A mechanisms present the same apparent kinetic behaviour.

These examples are presented to show that defining a reaction mechanism is not a simple field of study, since we are dealing with transient species that cannot be isolated or often even sensed directly by any method. Nevertheless, the field has progressed due to a suite of sophisticated experiments to the point where basic concepts, as discussed above, are fairly well understood.

5.3.1.2 The Square Planar Substitution Mechanism

Since the square planar geometry has a coordination number of only four, and there are no ligands bound above and below the plane of the metal and its surrounding ligands, substitution where the entering ligand attaches first in an axial position prior to departure of the leaving group seems very reasonable. Although 'empty' regions exist above and below the metal and ML_4 plane, metal orbitals are directed in these directions and may be accessible to entering nucleophiles. This process, with bond-making dominant, is an associative mechanism (Figure 5.14). There is good experimental evidence to support it as the dominant mechanism for square planar complexes. A dissociative mechanism would generate a three-coordinate intermediate, a rare geometry for coordination complexes, which suggests it may be disfavoured compared with a five-coordinate intermediate as formed in the associative mechanism.

Evidence favouring the A mechanism can be exemplified for the well-known square planar d^8 Pt(II) complexes, with three pieces of evidence relevant:

1. The rates of *aquation* (the process where a ligand, in this example chloride ion, is replaced by water) of $[PtCl_4]^{2-}$, $[PtCl_3(NH_3)]^{1-}$, $[PtCl_2(NH_3)_2]^0$ and $[PtCl(NH_3)_3]^{1+}$ are all similar, that is the rate is independent of complex charge. This is consistent with rate-determining attack by neutral OH_2 in an A mechanism. For a D mechanism, rate-determining release of the anion Cl^- should be affected significantly by the original

complex charge, as one would expect that it should be easier for an anion to leave an anionic complex than a neutral or cationic one, and this is clearly *not* the case experimentally.

2. Reactions show a dependence on the entering group concentration. For the halide substitution reaction (5.47), the rate = k[PtCl(NH$_3$)$_3$][Br$^-$], which is behaviour consistent with an A mechanism.

$$[PtCl(NH_3)_3]^+ + Br^- \rightarrow [PtBr(NH_3)_3]^+ + Cl^- \qquad (5.47)$$

While this could be interpreted as a dissociative mechanism with an ion-pair pre-equilibrium, discussed above for octahedral complexes, it is notable that the similar substitution reaction of octahedral complexes that usually react by dissociative processes shows no dependence on entering group concentration.

3. The effect of changing the leaving group on the rate is not as large as usually observed with dissociative mechanisms. For the series of complexes [Pt(dien)X]$^+$ reacting with pyridine (py) to form [Pt(dien)py]$^{2+}$ and to release X$^-$, the observed rate constant varies only by $\sim 10^3$-fold across a wide range of monodentate ligands; softer leaving groups lead to slower reaction, suggesting the differences have their source in inherent bond strength differences, and not a mechanistic change.

Discussion of square planar complexes in terms of mechanistic concepts developed in more detail for octahedral complexes serves to illustrate that the concepts devised initially for octahedral geometry can be transferred in large part to other coordination numbers and stereochemistries. In doing so, of course, it is necessary to consider the most likely way a reaction may occur, as exemplified above for square planar geometry.

5.3.2   A New Body – Stereochemical Change

As outlined at the commencement of Section 5.3, there are other types of reactions apart from substitution reactions. One of these of key importance involves what we can describe as a change in the stereochemistry. This may involve a change in gross shape of the molecule, such as transition from square planar to tetrahedral, or, more often observed, a change in the relative position or three-dimensional arrangement of ligands in a particularly-shaped complex.

5.3.2.1   *Change in Overall Shape of the Complex*

A complex prepared and isolated usually exists in a thermodynamically stable form. As such, there seems likely to be little reason for the complex to undergo a change in gross shape or stereochemistry, since this would involve energy and not usually lead to a more stable complex. For a complex that exists in a particular geometry to undergo a change in its overall shape without any change in the ligand set, there must be some change in its environment. This may involve a change of temperature or else dissolution in a different solvent that may favour a different shape. An example of this type of behaviour is transition from square planar to tetrahedral shape, for which the activation barrier can be low. This has been exemplified earlier (Figure 4.12). Another example involves the reaction below

(5.48):

$$(5.48)$$

This system displays a relatively small activation barrier of $\sim$45 kJ mol^{-1}, with a rate constant of $\sim$10^5 s^{-1} at 298 K.

Molecules that exhibit stereochemical non-rigidity are said to display *fluxional character*. All molecules undergo vibrations about an equilibrium position that does not alter their average spatial location; for a limited number, however, rearrangement that changes the configuration can occur. One of the commonest ligands, ammonia, is a simple example of the concept (Equation 5.49), since as a pyramidal molecule it can invert.

$$(5.49)$$

planar
intermediate

This has a low energy barrier ($\sim$25 kJ mol^{-1}) and occurs rapidly ($\sim$2 $\times$ 10^{10} s^{-1}). However, for pyramidal PR$_3$ compounds, the activation barrier has climbed to over 100 kJ mol^{-1}, so inversion is very slow, sufficiently so to even allow enantiomer separation in the case of chiral phosphines. Coordination of the lone pair freezes the configuration. Examples of fluxional complexes tend to be found mostly, but not exclusively, amidst organometallic compounds. Spectroscopic methods, particularly NMR spectroscopy, assist in defining the fluxional behaviour.

5.3.2.2 Change in the Mode of Ligand Coordination – Linkage Isomerization

Whereas many ligands offer a single lone pair on only one atom or at least a single donor, and are thus are able to coordinate at one site in only one manner, a range of *ambivalent* ligands were introduced in Chapter 4 that offer two or more potential donors carrying lone pairs of electrons – put simply, they have a choice of donors by which they may be attached to a particular site. While isolated species are usually the thermodynamically stable form, it may be possible to prepare a complex where the ligand is bound in its less stable form. For example, a route to the less stable $[Co(NH_3)_5(SCN)]^{2+}$ (S-coordinated form of thiocyanate) has been devised. When dissolved in solution, this complex is seen to spontaneously undergo ligand donor group exchange, forming the thermodynamically more stable $[Co(NH_3)_5(NCS)]^{2+}$ (N-coordinated form). This process is termed linkage isomerism, the name referring to a change in the way the ligand is linked to the metal centre.

The process can be represented in a general form as Equation 5.50:

$$[L_nM(X–Y)]^{m+} \rightarrow [L_nM(Y–X)]^{m+} \qquad (5.50)$$

where there is no change to the coordination sphere except for that involving the binding mode of the ambivalent ligand. Because the two alternate donors are usually distinctly different, a distinctive change in the colour of the complex occurs in this reaction. Another example of an ambivalent donor is the nitrite ion (NO_2^-), which can coordinate via either the N or an O atom. Coordination modes for *nitrito* (O-bound; nitrito-*O*) and *nitro* (N-bound; nitrito-*N*) linkage isomers are shown in Equation (5.51), along with the intermediate that may be involved in the process of linkage isomerization.

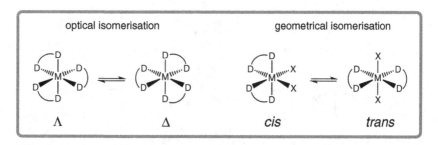

$$\text{(5.51)}$$

nitrito isomer   symmetrical intermediate   **nitro** isomer

Because isotopic labelling studies have shown that the reaction above occurs without the coordinated anion departing the coordination sphere, the mechanism of isomerization for the nitrito–nitro reaction has been proposed to involve a symmetrical intermediate where both oxygen atoms and the nitrogen atom are disposed equidistant from the metal ion in the transition state. Relaxation from this activated state occurs by capture of an oxygen atom (which means no effective reaction occurs, apart from possibly oxygen 'scrambling') or capture of the nitrogen alone (leading to the more stable isomer).

5.3.2.3   Change in the Relative Position of Ligands – Geometrical Isomerization

Ligand rearrangement of the type mentioned in the previous section involves a change in the way a particular ligand is bound. Another process focussed on ligand arrangement involves changing the way a set of ligands are assembled around a central metal. We have already established in Chapter 4 that ligands can in some cases organize themselves in a number of ways to form *geometric isomers*. Apart from separating and observing geometric isomers, it became apparent early on that interconversion between geometric isomers was possible. Many of these interconversions are accompanied by a colour change, while analysis shows clearly that the set of ligands originally present have been maintained in the reaction. Two key conclusions were made: the way ligands are arranged around a central metal ion influences physical properties such as colour; and there must be a sufficiently low energy barrier to reaching the activated state to permit the observed interconversions.

In effect, there are two types of isomerization reactions (Figure 5.15), and we shall examine each in turn. However, it is appropriate initially to identify the two and the differences. In one form, called optical isomerization, change in optical properties alone

Figure 5.15
Optical and geometrical isomerization represent the two classes of reaction where change of relative position of ligands is involved.

(such as the direction of rotation of plane polarized light interacting with a solution of the complex) occurs. Because no other physical change is observed, this type of reaction is not readily 'seen', as it requires the use of devices that measure specific optical properties. In the other and more familiar form, geometrical isomerization, change in most physical properties (such as colour) occurs, so this type is readily 'seen', often even by the naked eye. The reason for the stark difference between observable behaviour lies in the way the set of donors in the coordination sphere is affected. In optical isomerization, one optical isomer is converted into the other; this occurs without any change in the positional arrangement of the set of donors themselves, and thus has no influence on most physical properties. In geometrical isomerization, although the set of donors remains unchanged, the actual physical arrangement of donors in space around the central metal changes. This is accompanied by a change in molecular symmetry, with subsequent effects on physical properties.

The mechanism by which one geometric isomer converts to another can employ the type of intermediates used in substitution reactions, discussed earlier. The clear difference here is that no new introduced ligand can replace one already bound; any additional molecule present in the coordination sphere in the activated state must be non-competitive as a ligand, allowing the original coordination sphere to remain intact. While both A and D mechanisms can account for isomerization reactions represented in Figure 5.15, another prospect is for the molecule to undergo a bond angle distortion and twisting (without any bond-breaking or participation of any entering group) that takes it to a more symmetrical trigonal pyramidal geometry in the activated state from which it can revert to the original isomer or continue adjusting to form the other isomer (Figure 5.16).

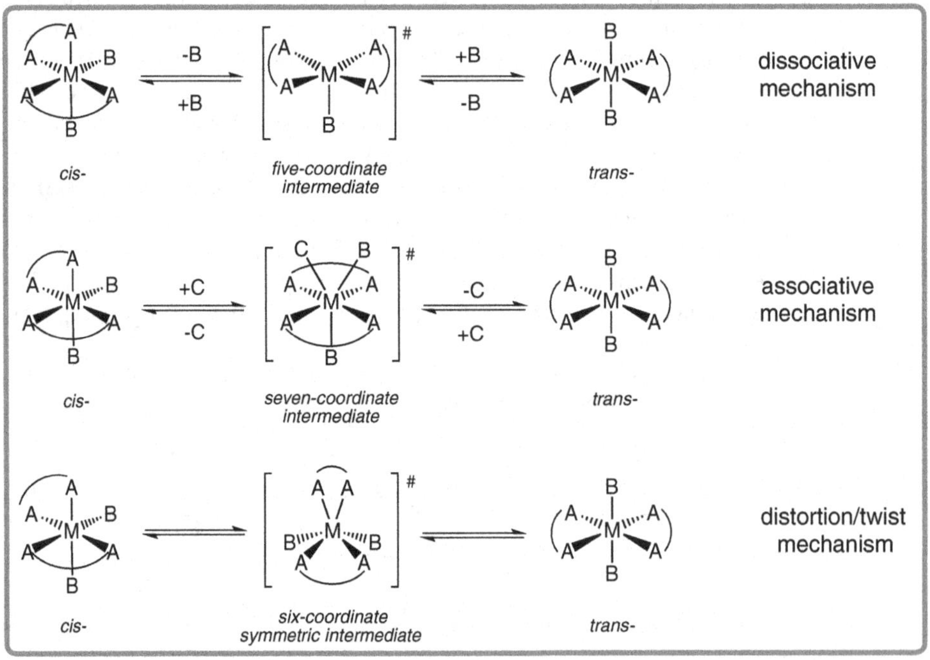

Figure 5.16
Mechanisms for geometrical isomerization: dissociative, associative and distortion/twist mechanisms.

While geometric isomerization without any ligand substitution processes interfering is well known, it is not unusual to observe substitution reactions *also* occurring along with geometrical isomerization. An example is given in Equation (5.52).

$$trans\text{-}[CoCl_2(en)_2]^+ + OH_2 \rightarrow \textit{cis}\text{-}[CoCl(en)_2(OH_2)]^{2+} + Cl^- \tag{5.52}$$

Introduction of a different ligand alters the relative stabilities of *cis-* and *trans-* isomers, so there may be a driving force for isomerization, leading to this occurring concomitant with substitution. In circumstances where two geometric isomers are energetically similar, both may exist in solution once thermodynamic equilibrium is reached, with the amount of each form present defined by the relative stability of the isomers.

5.3.2.4 *Change in the Relative Position of Ligands – Optical Isomerization*

Optical isomerism is a hidden but constant component of coordination chemistry. It is a constant because molecules that have a non-superimposable mirror image are very common, whereas it is hidden since the physical properties of optical isomers are identical. It is necessary to separate the Δ and Λ optical isomers by resolution before their hidden optical properties can be exposed, and then it requires special instrumentation to record those properties. If this is the case, one might ask why the interest in optical isomerism anyway? One answer lies in biological systems, where optical isomers do behave differently as a result of the way they match with a specific optically active (chiral) environment. This is merely an extension of the way optical isomers are separated in chemistry; while the Δ and Λ form of a complex behave identically as isolated species, when partnered with another chiral entity Δ', the $\{\Delta \cdot \Delta'\}$ and $\{\Lambda \cdot \Delta'\}$ assemblies are no longer equivalent in their properties, allowing their separation. Presented with a naturally chiral protein, the Δ and Λ forms of a complex may likewise interact differently. Natural coordination complexes may exist as one particular optical isomer, as exemplified in Chapter 8. Another observation is that one of the spectroscopic methods for reporting optical activity, circular dichroism, is important in providing information on the splitting of energy levels in complexes under reduced symmetry, and also is very sensitive to the surrounding environment of a complex, providing a method for probing outer-sphere interactions of complexes. Apart from these aspects, it is simply important to understand the full breadth of the chemistry of complexes, of which chirality is an important component.

It has been known for decades that an optically pure complex (100% of either the Δ or the Λ form) can undergo a process called racemization, whereby it returns to an optically inactive racemic (50:50 Δ:Λ) mixture of its two optical isomers. For this to occur, it is assumed that it must proceed through a symmetrical transition state, from which there is no particular preference regarding whether it reverts to the original or the opposite optical isomer, thus providing a process that will lead to a racemate. For most complexes, it is possible to devise an appropriate transition state by invoking traditional D or A mechanisms. In addition, the twist mechanism developed above for geometrical isomerism can be usefully applied to optical isomerism; unlike the other two mechanisms, this involves no bond-breaking or bond-making but merely bond angle distortion in moving to and from the activated state. These mechanisms are illustrated in Figure 5.17 for one of the classical examples of optical activity in coordination complexes, octahedral compounds with three didentate chelate ligands.

All three mechanisms involve a transition intermediate of higher energy and lower stability than the two optical isomers, and so there is an activation barrier to overcome.

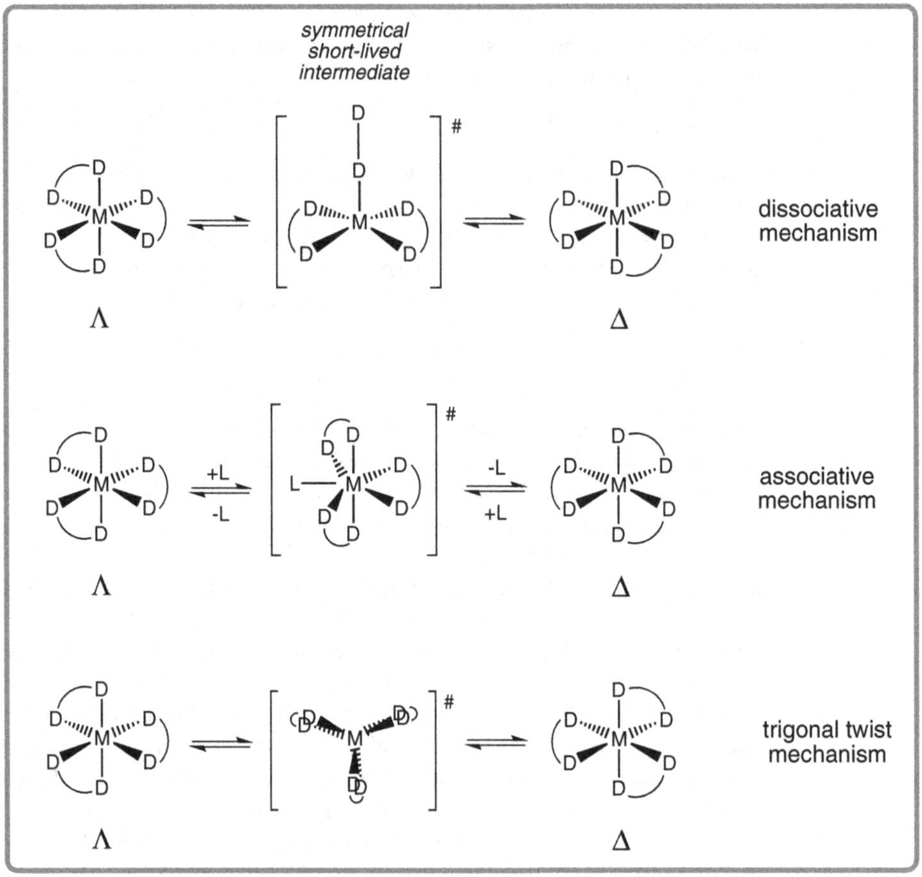

Figure 5.17
Mechanisms for isomerization of an octahedral complex with three didentate chelates, proceeding through a symmetrical transition state.

For some complexes, the isomerization is very slow (as for the inert cobalt(III) complex $[Co(en)_3]^{3+}$), whereas for others the process is sufficiently fast (as for the nickel(II) analogue $[Ni(en)_3]^{3+}$) that the complex cannot be easily resolved into its optical forms through conventional crystallization methods. The isomerization reaction can be readily followed by observing the loss of optical rotation at a selected wavelength over time; this is usually a simple first order exponential decay process.

5.3.3 A New Face – Oxidation–Reduction

Oxidation–reduction (electron transfer) reactions are important in chemistry and biology. When a chemical oxidation of *A* by *B* occurs, *B* itself is reduced – an *electron transfer* process has occurred. For such chemical processes, there is always a partnership between an oxidant (which is reduced in carrying out its task) and a reductant (which is oxidized in the reaction); thus we frequently talk of oxidation–reduction, or (for ease of use) redox, reactions for what are essentially electron transfer processes. Of course, an electron can be

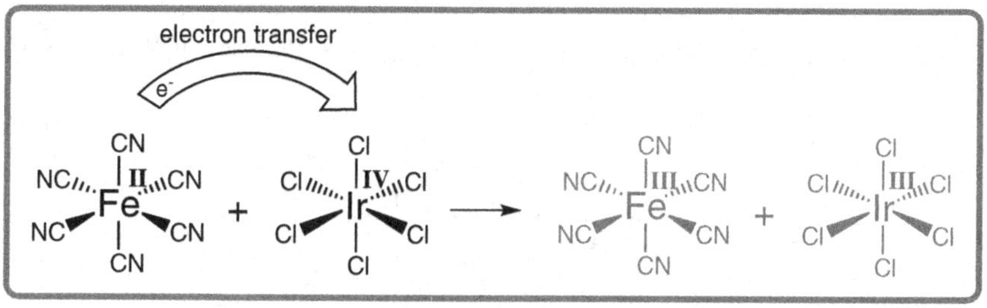

Figure 5.18
A reaction involving electron transfer alone. The Fe(II) centre is oxidized to Fe(III) and the Ir(IV) centre is reduced to Ir(III), without any change in the ligand set for either complex.

supplied or accepted electrochemically via an electrode rather than through using a chemical reducing or oxidizing agent, and this is another way to initiate change in the oxidation state of complexes. Here we shall concentrate mainly on chemically based systems, where an oxidant and reductant of appropriate potentials need to be combined. A reaction will be favourable if $E^0 > 0$, with E^0 being the difference between the standard potentials for the two half-reactions (one for the oxidation part, one for the reduction) that are combined to form the overall reaction.

There are two processes that are important in inorganic redox reactions. The key process is, of course, *electron transfer*. However, some reactions also involve *atom transfer*, whereby a component of one reacting molecule is transferred to another during the reaction. It is not essential for both to occur, and many reactions are purely electron transfer reactions. A classical example of a pure electron transfer alone is the reaction shown in Figure 5.18.

This reaction of the two octahedral complexes occurs without any change in the coordination spheres, or ligand sets, of either metal. However, if you inspect the two metal centres, you will note that the iron complex (the reductant) is oxidized from Fe(II) to Fe(III) and at the same time the iridium complex (the oxidant) is reduced from Ir(IV) to Ir(III) – an electron has been transferred from one metal to the other. This reaction can be conveniently followed, since the colour of each species changes as it is converted from one oxidation state to another.

Atom transfer alone can also occur, although it is a more difficult concept to come to grips with. Think of it as an atom moving with its normal complement of valence electrons, and no others, from one central atom to another. *Oxidative addition reactions* in organometallic chemistry can be considered as a form of atom transfer. A typical example of this type is given in Equation 5.53.

$$[IrCl(CO)(PR_3)_2] + Cl_2 \rightarrow [IrCl_3(CO)(PR_3)_2] \tag{5.53}$$

Some reactions occur with both electron transfer and atom transfer. A simple example of this class is the reaction of two octahedral complexes shown in Figure 5.19.

In the above example, an electron is transferred from the Cr(II) to the Co(III) and a chloride ion is transferred from the cobalt to the chromium ion as well. The Cr(III) product is inert, allowing it to be isolated and identified, thereby defining the presence of the chloride ion in its coordination sphere. Colour changes in this reaction allow the reaction to be readily monitored spectrophotometrically.

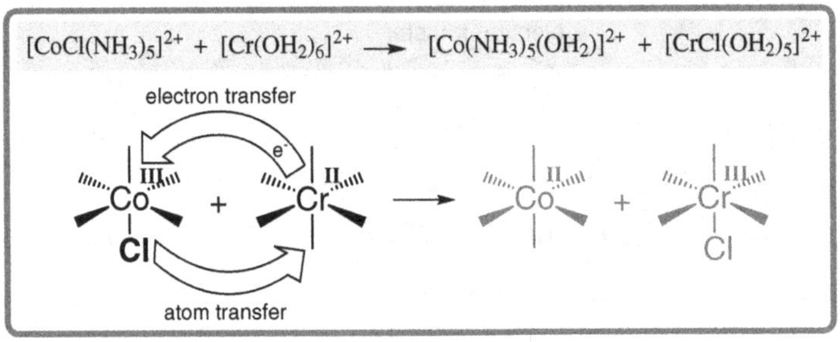

$$[CoCl(NH_3)_5]^{2+} + [Cr(OH_2)_6]^{2+} \longrightarrow [Co(NH_3)_5(OH_2)]^{2+} + [CrCl(OH_2)_5]^{2+}$$

Figure 5.19
An example of electron transfer with associated atom transfer.

Clearly, the different processes of electron transfer alone and electron transfer along with atom transfer, illustrated above, need not, and most likely do not, occur by a common mechanism – in fact, there are two general mechanisms for electron transfer involving metal complexes. These are called the *outer sphere mechanism* and the *inner sphere mechanism*, and we shall examine each in turn to learn about their characteristics. Both, of course, involve electron movement from one metal to another, a process involving 'tunnelling'. This is a quantum-mechanical process associated with the wave nature of electrons that permits passage through an energy barrier that would otherwise be too high to allow the process to proceed; an understanding of this process is, fortunately, not essential to obtaining an understanding of the mechanisms at a basic level.

5.3.3.1 The Outer Sphere Electron Transfer Mechanism

The key to the outer sphere mechanism is that electron transfer from reductant to oxidant occurs with the coordination shells (or spheres) of each reactant staying intact throughout. Since the coordination (or inner) sphere, that is the set of bound ligands, is not changed during the reactions, it appears that the key to the process lies beyond these, in the outer sphere around the reactants. We have seen an example earlier involving two different metal centres. Another classical example of pure electron transfer alone, involving two oxidation states of the one metal ion, is Equation (5.54).

$$[\mathbf{Os}(2,2'\text{-bipy})_3]^{3+} + [\underline{\mathbf{Os}}(2,2'\text{-bipy})_3]^{2+} \rightarrow [\mathbf{Os}(2,2'\text{-bipy})_3]^{2+}$$
$$+ [\underline{\mathbf{Os}}(2,2'\text{-bipy})_3]^{3+} \qquad (5.54)$$

This reaction of the octahedral complex with 2,2'-bipyridyl chelates occurs rapidly ($k = 5 \times 10^4$ M^{-1} s^{-1}), but there appears to be nothing happening at first glance. However, if you inspect the two differentiated Os complex ions (differentiated above by one being underlined), you will note that one is Os(III) on the left and Os(II) on the right, and vice versa – an electron has been transferred from one complex ion to another, although there is no colour change in the reaction as the products equate with the reactants exactly. The only way this reaction can be conveniently followed, and indeed shown to occur, is by using an optically active form of the complex in one of the two oxidation states in the initial mixture, since its optical properties change as it is converted from its original oxidation state to the other, even though the overall percentage of Os(II) and Os(III) complex remains constant.

Another example involves the $Ru^{3+}_{aq}/Ru^{2+}_{aq}$ system, where the *self-exchange* process may be distinguished through having different oxygen isotopes present in the two aquated complex ions, where we can write the reaction as in Equation (5.55).

$$[Ru^{II}(^{16}OH_2)_6]^{2+} + [Ru^{III}(^{18}OH_2)_6]^{3+} \rightarrow [Ru^{III}(^{16}OH_2)_6]^{3+} + [Ru^{II}(^{18}OH_2)_6]^{2+}$$

$$(5.55)$$

There is no scrambling of the isotopically-labelled water between metal ions; all water molecules remains completely with the one metal centre throughout the reaction, because the rate at which water is exchanged with bulk water is very slow compared with the rate at which the electron transfer reaction occurs. The reaction is observed to be first order with respect to both reactants, consistent with a bimolecular encounter process.

There is a sequence of elementary steps that must occur for an outer sphere reaction to proceed. Firstly, the oxidant and reductant complexes must come together sufficiently close and in an arrangement in space appropriate for electron transfer to proceed, through collision and rotational orientation processes. Because electron transfer over long distances is a high energy process, close approach of reactants to short separation distances is important; the upper limit to this distance between metal centres for an effective reaction is uncertain, but appears to be $\sim$1000 pm, although examples of exchange over longer distances are known. To achieve optimal contact, the complexes usually shed some of their outer sphere layer of tightly bound solvent molecules so as to make a close and specifically oriented approach appropriate for orbitals involved in the electron transfer to interact sufficiently, sometimes termed 'forming an encounter assembly' or 'precursor assembly'.

Next, this activated assembly (sometimes called an 'ion pair') undergoes extremely rapid electron transfer, with subsequent adjustment of the successor complexes to their new oxidation states (sometimes termed 'relaxation'). Since metal–ligand bond lengths vary for many metal ions in different oxidation states, the size of the ions may change as a result of electron transfer, which becomes a key part of the adjustment process. For example, Co(III)–N distances are typically 195 pm, whereas Co(II)–N distances are near 210 pm. For cobalt complexes of N-donor ligands, there is consequently a significant change in complex ion size as a result of electron transfer associated with the shrinking or expanding first coordination sphere tied to these bond length variations. For some ions, however, the bond distance change with metal oxidation state alteration is small; ruthenium is such an ion, with Ru(III)–N distances of 210 pm and Ru(II)–N distances of 214 pm. Low-spin iron cyanide complexes are another system displaying small changes in bond distances, as exemplified by Fe(III)–CN distances of 192 pm and Fe(II)–CN distances of 195 pm.

A significant change in ion size will perturb the solvent molecules in the immediate outer sphere of the complex ions, and solvent reorganization processes that occur will add to the free energy of activation. This is shown by comparing reactions involving Ru(III) and Ru(II), where substantially smaller changes occur than for Co(III) and Co(II). Because there is little expansion or contraction of the complex ions, electron transfer reactions between Ru(III)/Ru(II) compounds will involve less energy-demanding solvation sheath rearrangement effects than for Co(III)/Co(II), and reactions should be faster; this is observed experimentally. For simple systems, rates can vary significantly, even allowing for different ligand sets; for example, $[Co(NH_3)_6]^{3+/2+}$ (10^{-6} M^{-1} s^{-1} in water at 298 K), $[Ru(OH_2)_6]^{3+/2+}$ (45 M^{-1} s^{-1}) and $[Fe(CN)_6]^{3-/4-}$ (700 M^{-1} s^{-1}). This is consistent with an observation that electron exchange is more rapid wherever there is no movement of atoms (no stretching or compression of metal-ligand bonds) in reactants and products (called the Frank–Condon principle).

Finally, following electron transfer, there is no need for the complex ions, often of the same charge, to remain in close contact, and they relax and move apart. This process of the

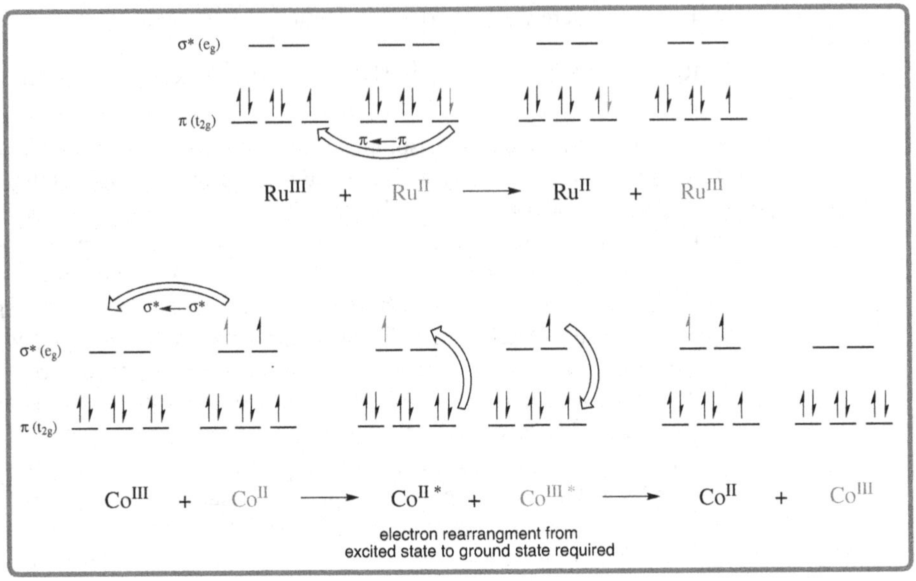

Figure 5.20
Electron transfer processes octahedral $Ru^{III/II}$ (top) and $Co^{III/II}$ (bottom) complexes.

successor assembly breaking up to form separated ions is a form of dissociation, but in this case of whole complex ions.

If we return to ligand field theory, recall that the d electrons for an octahedral complex lie in the t_{2g} and $e_g{}^*$ levels, also designated as π and σ^* respectively to equate with the character of bonding in which they participate in a complex exhibiting both σ-donor and π-donor/acceptor bonding character. For electron transfer, a metal d electron needs to move from a location in a π or σ^* orbital on one metal ion to a π or σ^* orbital on the other metal ion. Generally, it will be more favourable for an electron to move between orbitals of the same symmetry (or from like to like orbitals); that is $\pi \rightarrow \pi$ or $\sigma^* \rightarrow \sigma^*$ transitions are energetically favoured. Further, the character of π and σ^* orbitals differ, and so the electron transfer process will be affected by the nature of the donor and acceptor orbitals. Models predict that the $d\pi$ orbitals are more 'exposed' than $d\sigma^*$ orbitals, thus more able to interact with orbitals on a different metal complex, and as a result it is anticipated that $\pi \rightarrow \pi$ electron transfer should occur faster than $\sigma^* \rightarrow \sigma^*$ electron transfer. Let's examine this for a Ru(III)/Ru(II) and a Co(III)/Co(II) couple.

For Ru(III)/Ru(II), we have a process operating as defined in Figure 5.20, with a vacant position in the π orbital of a Ru(III) ion able to accommodate an electron transferred from the π orbital of a Ru(II) ion – a favourable $\pi \rightarrow \pi$ process. Moreover, the bond lengths for similar donor groups in Ru(III) and Ru(II) complexes are very similar (typically within 5 pm), so there is little solvent rearrangement required. As a result of efficient electron transfer between π orbitals along with small bond length changes following electron transfer, the electron transfer reaction rate is predicted to be fast in this example. For Co(III)/Co(II), we have a process operating (Figure 5.20) with vacant positions only in the σ^* orbitals of a low-spin Co(III) ion able to accommodate an electron transferred from the σ^* orbital of a Co(II) ion; a less favourable $\sigma^* \rightarrow \sigma^*$ process operates.

Moreover, the resultant electron arrangement of this electron transfer, shown in the lower centre of Figure 5.20, is an unfavourable excited state for both the Co(II) and Co(III) ions; both need to undergo electron rearrangement (a change in spin multiplicity) to restore their normal ground state arrangements. In addition, the bond lengths for similar donor groups in Co(III) and Co(II) complexes differ significantly (typically by 20 pm or more), so there is substantial solvent rearrangement required. As a result of a $\sigma^* \rightarrow \sigma^*$ electron transfer process and significant bond length changes following electron transfer, the electron transfer reaction rate is predicted to be slow in this example. Experimental support of the above arguments comes from the observation that outer sphere electron transfer reactions of Ru systems are commonly at least 10^6-fold faster than for equivalent Co systems.

The reactions of cobalt(III) in mixed-metal outer sphere electron transfer reactions are usually slow also, as a result of the influence of effects discussed above. The reaction (5.56) of the hexaamminecobalt(III) ion with the hexaaquachromium(II) complex

$$[Co^{III}(NH_3)_6]^{3+} + [Cr^{II}(OH_2)_6]^{2+} \xrightarrow{slow} [Co^{II}(NH_3)_6]^{2+} + [Cr^{III}(OH_2)_6]^{3+}$$
$$\downarrow H^+_{aq} \tag{5.56}$$
$$Co^{2+}_{aq} + 6\, NH_4^+$$

occurs slowly, with a rate constant of $\sim 10^{-3}$ M^{-1} s^{-1}. Decomposition of the initial Co(II) ammine complex product in the aqueous acid reaction conditions to release ammonium ion and form the Co^{2+}_{aq} ion is characteristic of the outcome for Co(III) systems upon forming a labile Co(II) product.

However, the similar reaction (5.57) is substantially faster (a rate constant of $\sim 10^6$ M^{-1} s^{-1}).

$$[Co^{III}(NH_3)_5Cl]^{2+} + [Cr^{II}(OH_2)_6]^{2+} \longrightarrow [Co^{II}(NH_3)_5(OH_2)]^{2+} + [Cr^{III}(OH_2)_5Cl]^{2+}$$
$$\downarrow H^+_{aq} \tag{5.57}$$
$$Co^{2+}_{aq} + 5\, NH_4^+$$

This is despite the fact that the sole change in this reaction compared with that above it, is the replacement of one ammonia ligand by one chloride ligand in the precursor Co(III) complex. Such a substantial change of $\sim 10^9$-fold for such an apparently minor change is unprecedented if the same mechanism is operating. The obvious conclusion is that a *different* mechanism occurs in this example. It was observations like this that pointed to an alternative mechanism, the so-called inner sphere mechanism.

5.3.3.2   *The Inner Sphere Electron Transfer Mechanism*

The key to the inner sphere mechanism is that electron transfer from reductant to oxidant occurs across a *bridging group*, which is a ligand shared between the reductant and the oxidant, and in the coordination sphere of both during the activation step. This bridging group has one obvious requirement – it must be able to bind to two metals simultaneously, which means that it must have two lone pairs on one or several donors orientated in such a way as to accommodate this shared arrangement. Of course there are many examples of stable compounds where ligands bridge between two metal centres, in dimer and higher oligomer complexes, so invoking such a short-lived intermediate in a reaction profile does not introduce a large leap away from conventional chemistry. The chloride ion is an example of a single-atom ligand capable of bridging, being replete with lone pairs of electrons; many examples of M—Cl—M coordination exist. Another type commonly met is an ambivalent ligand with two different donor atoms available, such as thiocyanate (SCN⁻), which offers both a N and a S donor, and is able to form M—SCN—M linkages.

The basic process can be exemplified by the reaction of the cobalt(III) complex $[Co(NH_3)_5(NCS)]^{2+}$ (the oxidant) with the chromium(II) complex $[Cr(OH_2)_6]^{2+}$ (the reductant), where thiocyanate acts as the bridging ligand. The overall reaction in aqueous acidic solution is shown in Equation (5.58).

$$[Co^{III}(NH_3)_5(\underline{N}CS)]^{2+} + [Cr^{II}(OH_2)_6]^{2+}(+5H^+) \rightarrow [Cr^{III}(OH_2)_5(\underline{S}CN)]^{2+}$$
$$+ Co^{2+}_{aq} + 5(NH_4^+) \quad (5.58)$$

The original inert Co(III) complex with N-bound thiocyanate has been reduced to a labile Co(II) complex, which has dissociated into free protonated ligands and the Co(II) aqua ion. Not only is the labile Cr(II) centre oxidized to an inert Cr(III) complex, but an S-bonded thiocyanate group has been introduced into its coordination sphere. The only source of this ligand is the original Co(III) complex. While this anion may have been captured from solution following dissociation of the reduced Co(II) species, the observation that it is bonded entirely by the S-donor group suggests that it may have been transferred in a 'handshake' operation where it is shared between the cobalt and chromium centres; an intermediate of the type $\{[(H_3N)_5Co-NCS-Cr(OH_2)_5]^{4+}\}$ is presumed to form, where one water group in the labile Cr(II) complex has been displaced and replaced by the thiocyanate S atom, as the N atom is already attached to the Co centre. The overall mechanism that has been proposed is shown in Figure 5.21, for a general bridging ligand represented as X—Y.

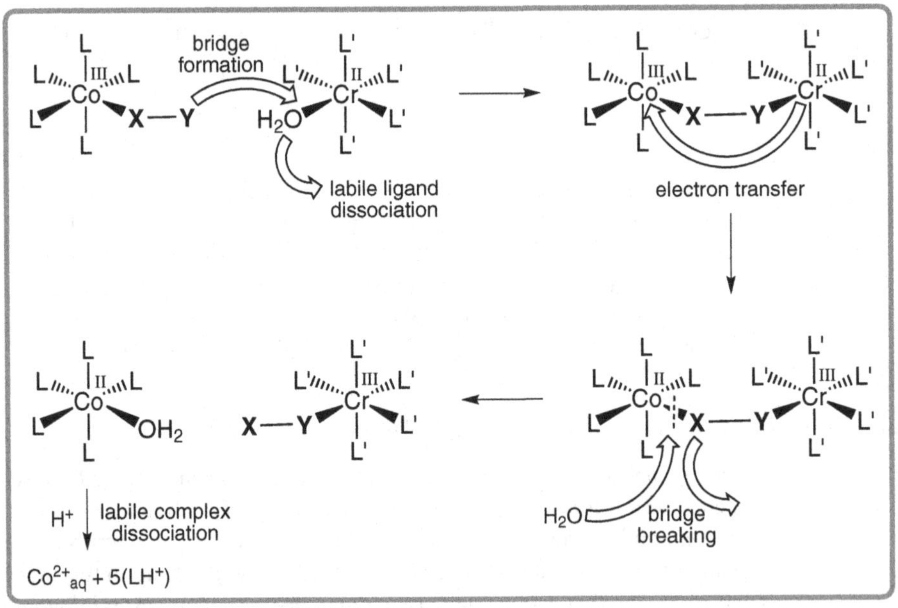

Figure 5.21
Electron transfer between Co(III) and Cr(II) complexes by the *inner sphere* mechanism, involving bridge formation in the intermediate. Transfer of the bridging group from the oxidant to the reductant is not required, but is characteristic of the mechanism where it does occur.

Key experiments by Nobel laureate Henry Taube and co-workers, involving complexes where chloride ion acts as a bridging ligand, cemented the mechanism. One reaction probed was the classical reaction mentioned earlier (Equation 5.59):

$$[Co^{III}Cl(NH_3)_5]^{2+} + [Cr^{II}(OH_2)_6]^{2+}(+ 5H^+) \rightarrow [Cr^{III}Cl(OH_2)_5]^{2+} + Co^{2+}{}_{aq} + 5(NH_4{}^+)$$
(5.59)

which is similar to the thiocyanate-bridged reaction discussed above, but with chloride ion as the potential bridging group. The study made use of chloride ion introduced as an enriched isotope, rather than as the natural isotopic mixture; by following the fate of the isotope in the reaction, mechanistic information could readily be gleaned. Using isotopic (or 'labelled') chloride ion (^{36}Cl), it was found that the original cobalt-bound halide transferred completely from precursor to the chromium product. If the labelled chloride was on the cobalt initially, it was transferred fully to the chromium product, even in the presence of added natural-isotope chloride ion in solution; conversely, using unlabelled cobalt-bound chloride and labelled free chloride in solution, labelled chloride is not introduced into the chromium product. These experiments effectively require a 'tight' interaction between oxidant and reductant, best viewed as one where an intermediate that shares the transferring ligand exists, that is $\{[(H_3N)_5Co-Cl-Cr(OH_2)_5]^{4+}\}$, which means an *inner sphere* mechanism.

The process of bridge formation can sometimes lead to a marked acceleration of the electron transfer process compared with outer sphere reactions of related compounds. In general, the key requirements for an inner sphere mechanism to operate can be summarized as follows:

(a) one reactant must possess at least one ligand capable of binding simultaneously to two metal ions in a bridging arrangement (this reactant is often the oxidant); and
(b) at least one ligand of one reactant must be capable of being replaced by a bridging ligand in a facile substitution process (this replaced ligand in many examples is found on the reductant).

The latter requirement means a relatively labile site must be available, and often involves a coordinated water group. Note that atom transfer is *not* a requirement in this mechanism. However, it can occur, and its occurrence is usually good evidence for the mechanism. For example, when atom transfer occurs with an ambidentate ligand (like SCN^-), the donor preferred and bound by the precursor metal ion is often not the one attached to the stable product, as described earlier; this helps define the mechanism. Nevertheless, the bridging ligand *may* remain with its parent metal ion.

The type of bridging ligand can affect the observed electron transfer rate markedly. Its role is to bring the metal ions together, and mediate the transfer through itself. For the reaction discussed above but extended to employ a range of halide ions (Equation 5.60), the rate increases with increasing size of the halide ion in the ratio for F : Cl : Br : I of 2 : 6 : 14 : 30.

$$[Co^{III}(NH_3)_5X]^{2+} + [Cr^{II}(OH_2)_6]^{2+} \rightarrow [Co^{II}(NH_3)_5(OH_2)]^{2+} + [Cr^{III}(OH_2)_5X]^{2+}$$
(5.60)

This is assigned to increasing polarizability with increasing halide ion size. The higher-charged Co(III) attracts more halide electron density towards its side of the bridging unit, depleting the side near the Cr(II) from where the electron departs and thus facilitating attraction of the transferring electron to the bridging ligand.

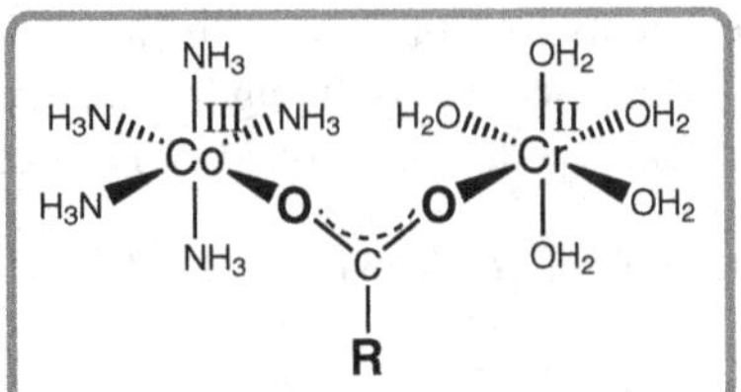

Figure 5.22
The proposed intermediate in the reaction of $[Cr(OH_2)_6]^{2+}$ and $[Co(NH_3)_5(OOCR)]^{2+}$; the size of the R-group impacts on the reaction rate.

The effect of altering the bridging group can be seen more starkly in the reaction of $Cr^{2+}{}_{aq}$ with the carboxylate complex $[(NH_3)_5Co(OOCR)]^{2+}$, where it arises from a different cause. Attack by the Cr(II) ion at the carboxylate carbonyl of the Co(III) complex leads to the proposed *intermediate* in Figure 5.22, and rate constants clearly vary with the size of the R group, as a result of increasing steric bulk of the R-group limiting facile bridge formation; that is, there is a clear steric effect on the electron transfer rate.

Another obvious expectation of the inner sphere mechanism is that the electron is transferred via the ligand; think of it like a ball rolling across a bridge. If this is the case, then for some time, albeit very short, the electron will reside on the bridging ligand, leading to a ligand radical. This has also been probed, using as a bridging ligand one that includes an aromatic heterocyclic group. Such groups help stabilize ligand radicals, and the presence of the radical intermediate has then been detected by electron spin resonance spectroscopy, which is a highly sensitive method for detecting radicals. Thus, overall, experimental evidence for the inner sphere mechanism is strong.

The two mechanisms of electron transfer, or oxidation–reduction reactions, for metal complexes have some common aspects, with which we shall finish this section. These are the keys to electron transfer reactions. To occur efficiently, the molecular orbital in which the donated electron originates, and that to which it moves, should be of the same type. The most efficient *outer sphere* electron transfer requires transfer between π orbitals. Effective *inner sphere* electron transfer occurs with transfer between either both π or both σ^* orbitals. The chemical activation process prior to electron transfer will be much greater if the above does not apply, meaning that under such circumstances reactions may not occur, or occur only very slowly. Further, structural deformation and/or electron configuration changes may be necessary (costing energy, and thus raising the activation barrier), and reactions under these conditions will, in general, be slower than those requiring little solvent reorganization and/or no electron configuration change. Overall, only where the activation barrier leading to the intermediate is sufficiently low will reactions proceed at measurable rates.

5.3.3.3 Electrochemical and Radiolytic Electron Transfer Reactions

It is useful to note that there are other methods for supplying electrons to complexes that will lead to a change in oxidation state. Whereas oxidation–reduction reactions are partnerships between two compounds, an alternative is to offer a direct source of or sink for electrons; this is achieved by electrodes in an electrochemical cell. At the appropriate potential, a complex may be reduced by transfer of an electron from an electrode to the complex; oxidation can occur by transfer of an electron to an electrode. What is perhaps apparent is that, for a solid electrode and a solution of a complex, it is only near the surface of the electrode that the process can occur efficiently, so both the rate of transport to an electrode surface, as well as the rate at which an electron is transferred between solid surface and dissolved complex ion, are of relevance. In general, we can write a simple expression for the process (represented here for reduction) (Equation 5.61):

$$[ML_m]^{n+} + e^- \rightarrow [ML_m]^{(n-1)+} \tag{5.61}$$

The oxidation and reduction potentials of complexes are usually probed by the experimental methods of *voltammetry*, which is a sensing rather than a complete conversion technique; the full reduction or oxidation of a complex is achieved by *coulometry*, using large surface area electrodes and stirring to enhance mass transport and thus speed up the process. These techniques and their application are beyond the scope of this textbook.

Yet another, and perhaps more exotic, method of supplying an electron in solution is to use radiolysis, where a very high energy source is directed into an aqueous solution to generate a range of products from reaction with the abundant water molecules, with the aquated electron (e^-_{aq}) and the hydroxyl radical ($^{.}OH$) being dominant species, and of particular importance. The former is a powerful one-electron reductant, the latter a powerful one-electron oxidant. The technique requires the use of high energy devices like a van der Graaf generator or synchrotron, so is not widely available. However, the powerful radical species formed can initiate definitive one-electron reduction or oxidation of complexes, either at the metal centre (direct reduction or oxidation) or at the ligand (with metal reduction or oxidation via a following intramolecular electron transfer possible). The former process is a form of outer sphere reduction, with e^-_{aq} as the reductant here. The latter process is followed by electron transfer from the ligand radical to the metal ion, which is, in effect, 'half' of the reaction in the electron transfer through a bridge in a chemical inner sphere process. When the high energy source is pulsed (the technique is called *pulse radiolysis*), a reaction in solution can be initiated rapidly and the outcome followed kinetically. Electron transfer reactions between two metal complexes can be initiated by pulse radiolysis also under certain circumstances. For example, if a relatively high concentration of Zn(II) and a low concentration of a Cr(III) complex are both present in solution, creation of e^-_{aq} causes it almost exclusively to react rapidly with the Zn(II), because it is in such very large excess, to form the extremely rare and unstable Zn(I) ion. Then two reaction can occur with the highly reducing Zn(I), in competition: *intermolecular electron transfer* between Zn(I) and the Cr(III) complex; and *disproportionation* of Zn(I) to form equal amounts of Zn(II) and Zn(0). By examining colour change associated with the chromium centre, the intermolecular reaction can be examined. However, this technique is not one you are likely to meet often.

5.3.4 A New Suit – Ligand-centred Reactions

Changing ligands is a common feature of reactions that we have discussed in the section above. However, it is possible to chemically alter a ligand while it remains attached to the metal centre. Coordination need not prohibit chemistry going on with the bound ligand; in fact, it may promote reaction as a result of electronic and positional influences resulting from coordination. Because reactions of coordinated ligands is a topic intimately tied up with synthesis, since new stable complex molecules are formed and from which new ligands can be isolated, it is addressed in detail in Chapter 6.

Concept Keys

The thermodynamic stability of a complex can be expressed in terms of a stability constant, which reports the ratio of complexed ligand to free metal and ligand in an equilibrium situation.

The stability of metal complexes is influenced by an array of effects associated with metal or ligand: the size of and charge on the metal ion, hard–soft character of metal and ligand, crystal field effects, base strength of the ligand, ligand chelation where it is possible, chelate ring size where applicable, and ligand shape influences.

For complexation of a set of ligands, each successive addition has a stability constant (K_n) associated with it, with each successive stability constant progressively smaller for a fixed stereochemistry. An overall stability constant ($\beta_n = K_1 \cdot K_2 \cdot \ldots \cdot K_n$) expresses the stability of the completed assembly.

Whereas thermodynamic stability is concerned with complex stability at equilibrium, kinetic stability is concerned with the rate of formation of a complex leading to equilibrium. Complexes undergoing reactions rapidly are labile, those reacting slowly are inert.

Reactions of complexes may involve changes to the coordination sphere, the metal oxidation state and/or the ligands themselves. Partial or complete ligand replacement (substitution) is the most common reaction met in metal complexes.

Octahedral substitution reactions may occur through a dissociative mechanism featuring a five-coordinate transition state (ligand loss rate-determining), an associative mechanism featuring a seven-coordinate intermediate (ligand addition rate-determining), or else by an interchange mechanism where ligand exchange happens in a concerted manner. Similar concepts can be applied to other coordination numbers and shapes.

Mechanisms for optical and geometrical isomerization reactions similar to those employed for substitution reactions can be envisaged. Additionally possible is a twist mechanism, involving distortion of the polyhedral framework in the activated state but in which no ligands depart or join the coordination sphere.

Oxidation–reduction (or electron transfer) reactions involving two metal complexes may occur by one of two mechanisms: outer sphere (no direct involvement of the coordination sphere) or inner sphere (where one ligand on one complex forms a bridge to the other metal in the transition state).

Electron addition or extraction from a complex to cause reduction or oxidation may also be achieved electrochemically (at an electrode in an electrochemical cell) or radiolytically (using an aquated electron or hydroxyl radical).

Further Reading

Atwood, J.D. (1997) *Inorganic and Organometallic Reaction Mechanisms*, 2nd edn, Wiley-VCH Verlag GmbH, Weinheim, Germany. Classical examples are covered clearly for the student in good depth; reactions, procedures for their examination, and mechanisms by which they react are explained – with the bonus of coordination complexes and organometallic systems presented in a single text.

Kragten, J. (1978) *Atlas of Metal–Ligand Equilibria in Aqueous Solution*, Ellis Horwood, Chichester, UK. A dated but still useful resource book for stability constants, should you seek numerical data.

Martell, A.E. and Motekaitis, R.J. (1992) *Determination and Use of Stability Constants*, 2nd edn, John Wiley & Sons, Inc., New York, USA. Provides a description of the potentiometric method commonly used to determine stability constants along with how complex speciation is determined, with examples and software.

Wilkins, R.G. (1991) *Kinetics and Mechanisms of Reactions of Transition Metal Complexes*, Wiley-VCH Verlag GmbH, Weinheim, Germany. A somewhat older text, but one of the classics in the field, that gives a fine student-accessible coverage of the basic concepts, which have remained fairly constant over time.

6 Synthesis

6.1 Molecular Creation – Ways to Make Complexes

Synthesis, the process of making compounds, lies at the heart of chemistry, with the preparation of new and possibly useful compounds occupying a great number of chemists. Every new compound presents challenges not only in its preparation and isolation, but also there are the challenges of characterization and defining its reactivity and possible uses. Whereas synthesis directed towards formation of a specific product is a mature field in organic chemistry, the challenges to the synthetic chemist when a metal ion is present are more profound, so coordination chemistry remains at a relatively more developmental stage. Nevertheless, a number of clear strategies exist, which we shall explore. Knowledge of thermodynamic and kinetic stabilities of complexes, developed in Chapter 5, is invaluable as an aid in devising methods for the synthesis of coordination compounds and/or understanding synthetic reactions. In this chapter we will draw on categories of reactions developed in Section 5.3 for discussion of synthetic approaches. Prior to this detailed look at synthesis, it would seem appropriate to provide a brief overview of the complex chemistry of metals.

6.2 Core Metal Chemistry – Periodic Table Influences

The periodic 'membership' of metallic elements (see Figure 1.3), and location in the Periodic Table impacts on their chemical behaviour and reactivity. Prior to an examination of synthetic strategies, it is useful to introduce the various groups of metallic elements that act as the central atom in most coordination complexes, to identify aspects of their chemistry that impact on complexes that are accessible and the subsequent development of synthetic procedures. In general, the periodic block origins of a metal will play a part in the type of complexes that can be easily and conveniently synthesized.

6.2.1 s Block: Alkali and Alkaline Earth Metals

The metals of the s block of the Periodic Table have properties that can be interpreted in terms of the trend in their ionic radii down each periodic column. There is a very strong tendency towards formation of M^+ (for ns^1 or alkali) and M^{2+} (for ns^2 or alkaline earth) metal ions, and other oxidation states are not important. With relatively high surface charge density, the alkali and alkaline earth metal ions are 'hard' Lewis acids and have a preference for small 'hard' Lewis bases. They particularly like O-donors, but can also accommodate N-donors, especially when present as part of a molecule offering a mixed O,N-donor set. The number of metal–donor bonding interactions varies a great deal, depending in one

Introduction to Coordination Chemistry   Geoffrey A. Lawrance
© 2010 John Wiley & Sons, Ltd

respect on the shape of the ligand and orientation of potential donors. For example, Mg^{2+} forms a complex with the polyaminoacid anion $EDTA^{4-}$ where all of the six heteroatom groups in the molecule are bound as a N_2O_4-donor set, in additional to an extra water molecule. Some of the strongest complexes they form are with large cyclic or polycyclic polyether molecules (see Figure 4.42). In these situations, the size of the ligand cavity relative to the size of the metal cation is an important factor in defining the 'strength' of the complex formed (concepts developed in Figure 4.38 earlier). Solvated compounds aside, few complexes with other simple ligands that offer just one donor atom occur, apart from some reported for beryllium. However, some metal cation–organic anion compounds (such as $Li_n(CH_3)_n$) can form. The Na^+ and Mg^{2+} cations in particular have important biological roles.

6.2.2 p Block: Main Group Metals

Main Group chemistry is a vast area of research, best addressed through specialist textbooks on the topic. The main group is considered most important as the source of the nonmetal donor heteroatoms such as N, O, S and P that metal ions use in forming coordination complexes. It also houses the metalloids, which display mixed nonmetal–metal character, and engage in coordination chemistry. However, the p block is a source of a number of important metals as well, including the most common metal in the Earth's crust, aluminium. The metallic elements of the p block form cations that are generally good Lewis acids in their common oxidation states, as these oxidation states are often high as a result of the number of valence electrons in the elemental form being at least three. As in the s block, metals of the p block tend to form stable ions through complete loss of their valence shell of electrons to leave the inert gas core electrons only:

$$\textbf{Al [Ne]}3s^2 3p^1 \rightarrow \textbf{Al}^{3+} \textbf{[Ne]}$$
$$\textbf{Sn [Kr]}5s^2 5p^2 \rightarrow \textbf{Sn}^{4+} \textbf{[Kr]}$$

However, elements such as tin and lead form both M^{2+} and M^{4+} compounds, and most of their stable coordination compounds involve the M(II) oxidation state; likewise, indium forms In^{3+} and In^+. We can interpret this simply by seeing it as loss of at least sufficient electrons to leave a full subshell:

$$\textbf{Sn [Kr]}5s^2 5p^2 \rightarrow \textbf{Sn}^{2+} \textbf{[Kr]}5s^2$$
$$\textbf{In [Kr]}5s^2 5p^1 \rightarrow \textbf{In}^+ \textbf{[Kr]}5s^2$$

However, it is only the heavier metals of the p block that tend to display this trait. Thus aluminium exists in its ionic form almost exclusively as Al^{3+}, and, as a consequence of being a light element with a high charge, has a high surface charge density and is a 'hard' Lewis acid. It has, like the s-block elements, a preference for O-donor ligands, but binds efficiently to chelating ligands that also carry other types of donors. It forms a $[Al(OH_2)_6]^{3+}$ complex, although this readily undergoes proton loss (hydrolysis) from some coordinated water groups to form bound HO^- as the pH rises, due to the influence of the highly-charged metal ion on the acidity of the bound water groups. Chelates of particularly O-donor or mixed O,N-donor ligands form relatively strong complexes with Al(III).

6.2.3   d Block: Transition Metals

The d-block transition metals, which form a group of elements ten-wide and four-deep in the Periodic Table associated with filling of the five d orbitals, represent the classical metals of coordination chemistry and the ones on which there is significant and continuing focus. In particular, the lighter and usually more abundant or accessible elements of the first row of the d block are the centre of most attention. Whereas stable oxidation states of p-block elements correspond dominantly to empty or filled valence shells, the d-block elements characteristically exhibit stable oxidation states where the nd shell remains partly filled; it is this behaviour that plays an overarching role in the chemical and physical properties of this family of elements, as covered in earlier chapters.

The ability of complexes of d-block metal ions to readily undergo oxidation or reduction involving one or several electrons is a key feature of their chemistry. Because many transition elements have this capacity to exist in a range of oxidation states, they offer different chemistry even for the one element as a result of the differing d electrons present in the different oxidation states; each oxidation state needs to be considered separately. Lighter elements of the d block tend to prefer O-, N- or halide ion donor groups, whereas heavier elements lean towards coordination by ligands featuring heavier p-block elements such as S- and P-donors. The diversity of oxidation states, shapes and donor groups met in transition metal chemistry makes it both fascinating and frustrating – it is not simple chemistry.

The d-block elements display perhaps the greatest variation of the various groups, with the first series row (with a 4s3d valence shell) in particular distinct from the second (5s4d) and third (6s5d) series rows. The fully synthetic fourth (7s6d) row has little chemistry on which to report yet. Radii of heavier elements and ions are larger than those of the first series, although lanthanide contraction determines that radii of the 5d series differ little from radii of the 4d series, despite increased atomic number. *Higher oxidation states* are much more stable for 4d and 5d than for 3d compounds; some oxidation states have no analogues in the 3d series. The +II oxidation state is of relatively little importance in 4d, 5d series (except for Ru), but is of major importance in the 3d series. The +III oxidation state is dominant in the 3d series, but relatively unimportant in 4d, 5d (exceptions are for Rh, Ir, Ru and Re). Rather, +IV, +V and +VI are more often met with 4d and 5d metals. The 4d, 5d rows are much more prone to *metal–metal bonding* than the 3d row, and multiple metal–metal bonds are common (Figure 6.1). For 3d, M—M bonds are only common in metal carbonyls. *Polynuclear (and cluster) compounds* are more common for 4d, 5d than for 3d species. *Magnetic properties* differ, with heavier elements tending to form dominantly low-spin compounds. *Higher coordination numbers* (>6) are also more common for 4d, 5d compounds.

If we examine a few of the columns within the d block selectively, the chemistry behind these generalizations becomes a little clearer. It is in group 9 (cobalt, rhodium, iridium) that the greatest similarities exist, so it is instructive to explore it first. Cobalt is a classical 3d element, with Co(II) and Co(III) the sole significant oxidation states (although Co(IV), Co(I) and Co(0) are known but rare), with a preference for N-, O- and halide donor atoms. Complexes dominantly feature octahedral stereochemistry, particularly for Co(III); four- and five-coordinate Co(III) are very rare and usually found only with bulky ligands, although reactivity of such complexes is high, and known Co(III) metalloenzymes which participate in fast reactions have five-coordination. Four-coordinate Co(II) compounds are known with halide and soft donor groups. Whereas Co(IV) is extremely rare, it is more common for Rh and Ir, with particularly stable Ir(IV) compounds such as $[IrCl_6]^{2-}$ existing, albeit as a

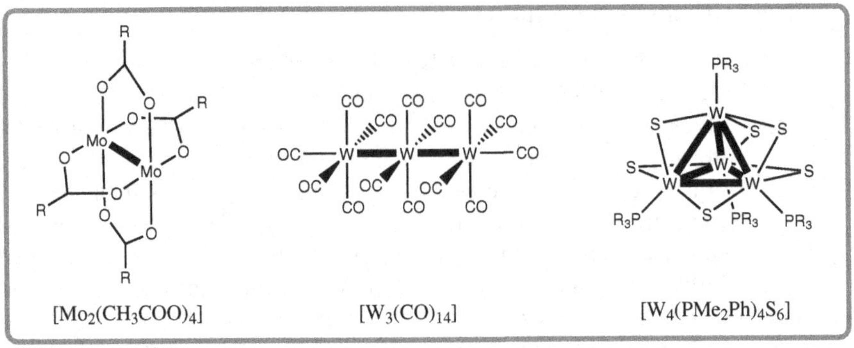

[Mo₂(CH₃COO)₄]   [W₃(CO)₁₄]   [W₄(PMe₂Ph)₄S₆]

Figure 6.1
Examples of metal–metal bonded complexes of the second and third row d-block elements molybdenum and tungsten. The Mo—Mo bond is quite short, and this is an example of a compound with a multiple metal–metal bond; other examples display single metal–metal bonds.

powerful oxidant capable of oxidizing hydroxide ion in aqueous solution to oxygen. The M(III) oxidation state is common for all of Co, Rh and Ir, but the inertness of complexes increases down the column; typical reactivity trends for the complexes Co:Rh:Ir are ~1000 : 50 : 1. This reactivity trend is characteristic of the d block in general. It is for the M(II) state that differences are starkest; Co(II) is common and stable, whereas Rh(II) and Ir(II) form few stable monomeric complexes. However, in yet lower oxidation states we find that all three metals form similar and fairly common compounds. Despite some common behaviour, there are clear differences across their chemistries: for example, polyhalo complexes such as $[IrCl_6]^{3-}$ and $[RhCl_5(OH_2)]^{2-}$ are stable, whereas no more than three halide ions, as in $[CoCl_3(NH_3)_3]$, can be present for stability in Co(III); also, hydride (H^-) complexes of Rh and Ir are common, with even the simple species $[RhH(NH_3)_5]^{2+}$ formed, whereas Co does not form such species.

Group 7 (manganese, technetium, rhenium) is one with clearer differences between the chemistry of the lightest member and the two heavier members. Manganese is known in oxidation states all the way from Mn(I) to Mn(VII), with a changeover from preferred six- to four-coordinate complexes occurring around Mn(V). The Mn(II) oxidation state is common, with higher oxidation states progressively less common, although some very important compounds exist at higher oxidation states, namely the simple oxide $Mn^{IV}O_2$ and the powerful and popular oxidant $Mn^{VII}O_4^-$. Technetium and rhenium have no analogue of Mn^{2+}_{aq} and form very few M(II) species; indeed, they have little cationic chemistry in any oxidation state. Unlike Mn, they have an extensive chemistry in the M(IV) and M(V) oxidation states; the latter is the least common for Mn. The formation of clusters and M—M bonds is much more common for Tc and Re than with Mn, and is a feature of the (II) and (IV) oxidation states for Tc and Re. Much of Tc, Re chemistry resembles more that of adjacent neighbours Mo, W than they do Mn, despite their different valence electron sets. Diversity is a key expectation of d-block chemistry.

6.2.4 f Block: Inner Transition Metals (Lanthanoids and Actinoids)

However, it is not entirely time to despair, since at least one family of metals show a remarkable consistency in their chemistry. This occurs in the oft-ignored f block of the

Periodic Table. The first row of that block is the *lanthanoids*. The lanthanoids (also called the lanthanides), were once called the 'rare earths', but are not particularly rare elements. They exist as the first row of the 14-wide f-block transition elements, where filling of the seven f orbitals is the key to their chemistry. All but promethium (for which the most stable isotope has a half-life of only 2.6 yrs) occur naturally. The scarcest naturally occurring element (thulium) is as common as bismuth and more common than arsenic, cadmium, mercury and selenium. They are named after lanthanum (La), electronic configuration $[Xe]6s^2 5d^1$, a *d*-block element. Immediately after this element the 4f orbitals lie slightly lower in energy than the 5d, and so fill first (with 14 electrons) up to Lu ($[Xe]4f^{14}6s^2 5d^1$) before returning to filling of the d shell. This means the f block breaks up the d block of the Periodic Table after the first column, but it is more traditional (and convenient, in the context of the shape of most printed pages) to represent the f block essentially as a subscript to the main Periodic Table in what is sometimes referred to as the short form of the Table. Because f orbitals occupy quite different spatial positions, the shielding of one f electron by others from the effect of the nuclear charge is weak. Thus, with increasing atomic number (and nuclear charge), the effective nuclear charge experienced by each f electron increases, causing a shrinkage in the radii of the atoms or ions from La (1.06 Å) across to Lu (0.85 Å), called the *lanthanide contraction*.

As splitting of the degenerate f-orbital set in crystal fields is small, so crystal field stabilization issues are only of minor importance. Preferences between different coordination numbers and geometries are controlled dominantly by metal ion size and ligand steric effects. *Coordination numbers* greater than six are usual (Figure 6.2), with seven, eight and nine all important. Examples with coordination numbers up to twelve exist. Coordination numbers of greater than nine are usually restricted to the f block, and rarely found in the d block. Some low coordination numbers (<6) are known but are rare, and only exist when stabilized with particular ligands, such as aryloxy (^-OR) or amido ($^-NR_2$) ligands. Dominantly, the lanthanoids exist in only one oxidation state, M(III), through loss of nominally the $6s^2 5d^1$ outer electrons. Certain metals form M(II) or M(IV) ions; their occurrence is related to the formation of especially stable *empty*, *filled* or *half-filled* f shells, but these are readily and/or rapidly oxidized or reduced to the M(III) ion. Because the 4f electrons are essentially inner electrons, due to effective shielding, their spectroscopic properties are little affected by the surrounding ligands. Their coordination chemistry tends to be very similar for the whole family of elements, and as hard Lewis acids they have a common

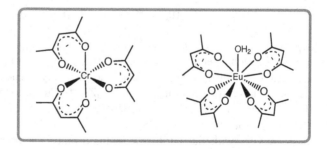

Figure 6.2
High coordination numbers are common in the f block. The didentate chelate ligand 2,4-dioxopentan-3-ido (acac$^-$) forms six-coordinate octahedral complexes with first-row d-block metal ions (like Cr(III), shown at left), but eight or nine coordinate complexes with f-block lanthanide ions (like Eu(III), shown at right).

preference for O-donors, though they can accommodate other donors, even including forming some metal–carbon bonded species. All form $[M(OH_2)_n]^{3+}$ (where $n > 6$), such as $[Nd(OH_2)_9]^{3+}$, although these readily hydrolyse; this tendency increases from La to Lu (as the ionic radius decreases). Chelating ligands give the most stable complexes, in line with expectations developed for d-block metals in Chapter 5; basically, they follow the normal rules of complexation developed in detail for d-block elements.

The row below the lanthanoids is the *actinoids* (also called the actinides), which are mainly synthetic elements. Those that are found on Earth naturally are isotopically long-lived thorium and uranium, but all are radioactive. After their d-block parent actinium, in principle the f orbitals are then filled for the following elements. However, energies of 5f and 6d are so close that elements immediately following Ac (and their ions) may have electrons in both 5f and 6d orbitals, at least until 4 or 5 electrons have been entered, when 5f alone seems to be more stable. This means that the early actinoid elements tend to show more d-block character (variable oxidation states and associated chemistry). Consequently, resemblance of the series to the parent is less marked than with the lanthanoids, at least until americium. Only after americium (about half-way along the series) are the elements similar and lanthanoid-like in chemistry, with only the M(III) oxidation state stable. Earlier elements, such as uranium, display oxidation states of up to M(VI); in fact, U(VI) is the most common oxidation state for that element. High coordination numbers (up to fourteen) are characteristic; for example, $[Th(NO_3)_6]^{2-}$ is twelve-coordinate, as each nitrate ion acts as an O,O-chelate ligand. Their solution chemistry is often complicated; hydrolysis in water (even to oxo species) is common. Elements above fermium are short-lived, and isolable only in trace amounts; others can be prepared in gram or even kilogram amounts. Being radioactive and in most cases rare synthetic elements, they are not met by most scientists let alone in everyday life, although they sometimes find an application.

6.2.5   Beyond Natural Elements

The Periodic Table has been extended from U ($Z = 92$) to currently $Z = 112$ since 1940 by synthetic methods. Some synthetic elements have extremely long half-lives (e.g. $^{248}_{96}Cm$, $t_{1/2}$ 3.5×10^5 yr), others moderate half-lives (e.g. $^{249}_{97}Bk$, $t_{1/2}$ 300 d) or short half-lives (e.g. $^{261}_{104}Rf$, $t_{1/2}$ 65 s). Their syntheses have involved fusion and bombardment reactions, for example:

$$^{249}_{97}Bk + {}^{18}_{8}O \rightarrow {}^{260}_{103}Lr + {}^{4}_{2}He + 3\,{}^{1}_{0}n$$

Separation of a new element is a key problem. Separations involve methods such as volatilization, electrodeposition, ion-exchange, solvent extraction and precipitation/adsorption. Separation relies on the unique chemistry of each element; although not heavy elements, but useful as an illustration, $^{64}_{30}Zn/^{64}_{29}Cu$ are separated by dissolution in dilute HNO_3 followed by selective electrodeposition of Cu (a very simple task, as the $Cu^{II/0}$ and $Zn^{II/0}$ redox potentials differ by ~ 1 V).

The totally synthetic fourth row of the d block has now been created fully, with all member elements prepared, albeit in tiny amounts, and characterized. Lifetimes of these new radioactive elements are not long, so their coordination chemistry has not been explored in any detail. However, it is very likely they will behave like their third row analogues. They are more chemical curiosities than applicable species at this time; however, they do stand as monuments to the human inventive spirit and technological capacity.

What may be apparent from this very brief overview is that the Periodic Table location of metals plays a strong role in determining their coordination chemistry – to the point that each has truly unique coordination chemistry. However, certain 'global' traits exist to guide the synthetic chemist. The above notes may serve to support the following specific discussion of synthetic methods.

6.3 Reactions Involving the Coordination Shell

Often, simple commercially-available hydrated salts of metal ions (chlorides, sulfates or nitrates in particular) are the starting point of much synthetic chemistry involving the coordination shell. Oxides are occasionally used, but tend to be highly insoluble and their use can involve dissolution with acid to form simple salts as a first step in any case. Further, anhydrous simple salts (dominantly halides) may be employed, particularly where reaction is to proceed in nonaqueous solvent; however, some of these are polymeric and as a result are often converted first to monomeric compounds that include coordinated solvent, which are more soluble and reactive. For example, polymeric $TiCl_3$ is reacted in tetrahydrofuran (THF) to form the monomer $[TiCl_3(THF)_3]$, and polymeric $MoCl_4$ is reacted in acetonitrile (MeCN) to form the monomer $[MoCl_4(MeCN)_2]$. Ligand substitution reactions using precursors where a didentate chelate ligand is replaced by stronger chelates is another approach; typical complexes for such reactions are $[M(CO_3)_3]^{x-}$ and neutral $[M(acac)_n]$, featuring chelated carbonate and acac$^-$ 2,4-dioxopentan-3-ido) ligands respectively. Much organometallic chemistry relies on metal carbonyl compounds, such as $[Cr(CO)_6]$, as starting points for the substitution chemistry. However, ligand substitution can occur with an almost infinite number of starting materials, providing thermodynamics and kinetics for the proposed reaction are appropriate.

6.3.1 Ligand Substitution Reactions in Aqueous Solution

The most common synthetic method used in coordination chemistry involves ligand substitution in an aqueous environment. It is a relatively cheap approach, employs the commonest and safest solvent, and has the advantage that, as many reagents and complexes are ionic, the reactants often have good solubility in water. Moreover, most metal salts are supplied commercially as hydrated forms (such as $Cu(SO_4) \cdot 5H_2O$), since they are prepared and isolated from aqueous solution during manufacture. As a consequence, this class of reaction often involves aqua ligand replacement, although substitution of other ligands is well known.

As an example, consider reaction (6.1). Here, the $Cu^{2+}{}_{aq}$ ion has water ligands in its coordination sphere substituted by a stronger ligand, in this case ammonia, to form $[Cu(NH_3)_4]^{2+}$. Multiple ligand substitution is assisted by the use of an excess of the incoming ligand to drive equilibria towards the fully-substituted compound.

$$Cu^{2+}{}_{aq} + aq. \ NH_3 \ (excess) \longrightarrow [Cu(NH_3)_4]^{2+}{}_{aq} \tag{6.1}$$

light blue

dark blue-purple

$\downarrow (X^-)_{aq}$

crystallization (of $[Cu(NH_3)_4]X_2$ salt)

There is no change in oxidation state in this substitution chemistry, and usually (though not always) the coordination number is also preserved. What is seen visually is a very rapid colour change, associated with the replacement of coordinated water groups by coordinated ammonia groups. Colour change is a common characteristic signal of change in the coordination sphere for metal complexes that absorb light in the visible region. The rapidity of the reaction is an indication that we are dealing with a *labile* metal ion, one that exchanges its ligands rapidly. We might more properly represent Cu^{2+}_{aq} as the distorted octahedral complex ion $[Cu(OH_2)_6]^{2+}$, but due to significant Jahn–Teller elongation of the axial bonds the axial water groups exchange much faster than equatorial groups, indicative of poor ligand binding in these sites. Consequently, stability constants for the $[Cu(NH_3)_5(OH_2)]^{2+}$ and $[Cu(NH_3)_6]^{2+}$ species are very low ($K_4 \gg K_5 > K_6$), so in practice the substitution in aqueous solution effectively halts after four ammonia ligands are added around the copper ion, making the tetraammine complex the dominant species in strong aqueous ammonia solution. On crystallization, it is this species that is usually isolated. Knowledge of spectroscopic behaviour and stability constants supports an understanding of the outcome.

Often, substitution is a simple process of one type of ligand being replaced fully by another type of ligand, such as for reaction of $[V(OH_2)_6](SO_4)$ in (6.2).

$$V^{2+}_{aq} \ + \ 3 \ en \ \longrightarrow \ [V(en)_3]^{2+} \tag{6.2}$$

$$(en = H_2N\text{-}CH_2\text{-}CH_2\text{-}NH_2)$$

However, the outcome of such reactions does depend on the entering ligand, as the above reaction performed with the aromatic monodentate amine pyridine instead of 1,2-ethanediamine leads to only four pyridines binding, with $[V(py)_4(SO_4)]$ as the product.

Substitution with change in coordination number may also occur in some cases. A clear example of this occurs for Ni^{2+}_{aq} reacting with pyridine, which undergoes the process (6.3).

$$Ni^{2+}_{aq} \ + \ \text{aq. pyridine (excess)} \ \longrightarrow \ [Ni(py)_4]^{2+} \tag{6.3}$$

$$\text{light green} \qquad\qquad\qquad \text{brown}$$

$$(X^-) \downarrow$$

$$\text{crystallization (of } [Ni(py)_4]X_2 \text{ salt)}$$

The strong-field ligand pyridine supports the spin-paired d^8 diamagnetic square planar geometry for the product, whereas the relatively weaker ligand water supports the spin-unpaired paramagnetic octahedral geometry in the precursor. Polyamines and other strong-field ligands generally tend to yield square planar complexes with this metal ion.

Complex formation is but the first stage of synthesis, as it is usually true that we require the product in a solid form, isolated as a salt or neutral compound. There are a number of approaches to the isolation of a solid:

- cooling, usually in an ice-bath, which may lower solubility sufficiently to permit crystallization of the product from solution, with addition of a seed crystal of the product from an earlier synthesis an option for assisting the process;
- concentration to a reduced volume, usually using a rotary evaporator that operates at reduced pressure and elevated temperature, followed by cooling, may be necessary if the solubility of the product is too high to permit precipitation from a dilute reaction mixture;

- slow addition to the reaction mixture, with stirring, of a water-miscible nonaqueous solvent (such as ethanol) in which the product is only sparingly soluble, until the commencement of precipitation;
- addition of an excess of a different simple salt of an anion (often in the Na^+ or K^+ form) that provides a high concentration of a different anion that forms a less soluble complex salt, allowing its ready crystallization;
- evaporation to dryness, appropriate where the product is sufficiently stable and unreacted species (such as excess ammonia, for example) are removed by evaporation, with possibly following recrystallization from a nonaqueous solvent in which the product is sparingly soluble;
- sublimation or distillation of the product following removal of solvent, which is not usually applicable with ionic products, but finds limited use with some neutral complexes.

As an example of the value of anion exchange reactions, the product from reaction of $Ni(SO_4)$ with pyridine in water, the complex cation $[Ni(py)_4](SO_4)$, is highly soluble. However, addition of excess sodium nitrite leads to ready precipitation of the much less soluble $[Ni(py)_4](NO_2)_2$ complex; change of the counter ion alone has occurred (6.4).

$$[Ni(py)_4](SO_4) + Na(NO_2) \longrightarrow [Ni(py)_4](NO_2)_2$$

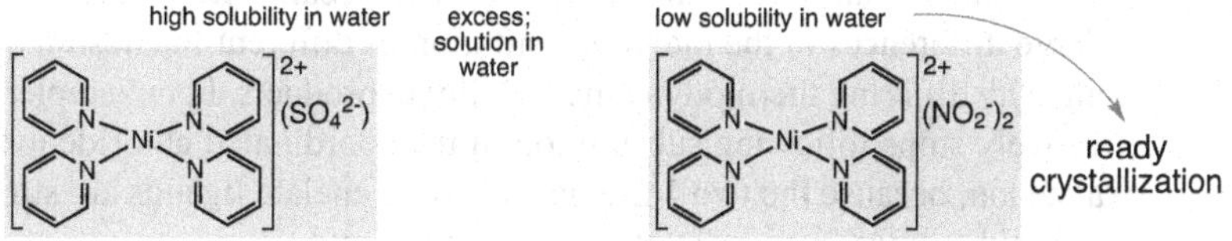

(6.4)

Where the product of a reaction is neutral, as occurs when red anionic $[PtCl_4]^{2-}$ has two chloride anions replaced by two neutral ammonia ligands to form yellow neutral $[PtCl_2(NH_3)_2]$, the solubility of the product may be inherently lower than for an ionic product, leading to its ready and selective precipitation from aqueous solution (6.5).

$$K_2[PtCl_4] + 2\,NH_3 \longrightarrow [PtCl_2(NH_3)_2] + 2\,KCl$$

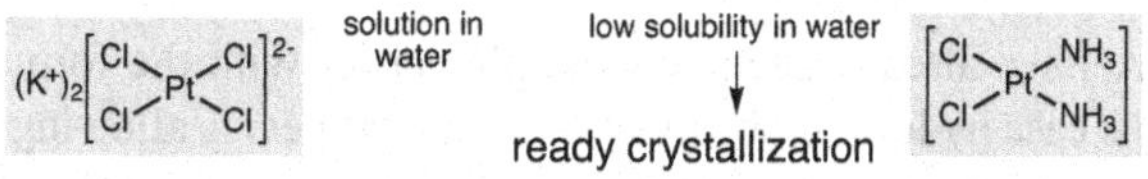

(6.5)

Reaction (6.5) is also an example where substitution of a ligand other than coordinated water is occurring, reminding us that ligand substitution does not inherently limit what the leaving and entering group can be.

Indeed, there is no requirement that only one type of ligand can be replaced in the one reaction; for example, purple $[CoCl_3(NH_3)_3]$ reacted with 1,2-ethanediamine (en) forms on heating yellow $[Co(en)_3]^{3+}$, where both chloride and ammonia ligands are substituted, the reaction (6.6) driven by the high stability of the chelate complex formed:

$$[CoCl_3(NH_3)_3] + 3\,(en) \longrightarrow [Co(en)_3]Cl_3 + 3\,NH_3$$

(6.6)

Whereas reactions of metal ions like Cu(II), Co(II), Mn(II) and Zn(II) are fast (these being *labile* complexes), metals ions like Ni(II) react somewhat slower, and Pt(II) much slower. Metals like Co(III), Rh(III), Cr(III) and Pt(II) are *inert*, and to make their reactions occur in a reasonable timeframe, it is often necessary to heat the reaction mixture. This is helpful because reactions typically double in rate for approximately every five degrees rise in temperature. An example is the inert red-purple $[RhCl_6]^{3-}$ ion, which reacts when boiled in aqueous solution with the chelating ligand oxalate ($C_2O_4{}^{2-}$) in approximately two hours to form yellow $[Rh(C_2O_4)_3]^{3-}$, whereas reaction at room temperature does not occur over reasonable time periods (6.7).

$$K_3[RhCl_6] + 3\,Na_2(ox^{2-}) \xrightarrow[\text{hours}]{\text{heat;}} K_3[Rh(ox)_3] + 6\,NaCl$$

(6.7)

Most examples such as the ones above have involved total ligand exchange to introduce just a single new type of ligand, driven usually by use of excess incoming ligand. This does not mean that partial substitution cannot occur. Often, partial substitution is driven by large differences in the rate of substitution of different ligands on the reacting complex along with differing thermodynamic stability in products. For example, in (6.8) the reaction effectively stops following substitution of the coordinated chloride ions, even with excess oxalate ion, because the two 1,2-ethanediamine chelate ligands are strongly bound and not readily substituted even by another chelating ligand.

$$cis\text{-}[CoCl_2(en)_2]Cl + Na_2(ox^{2-}) \longrightarrow [Co(en)_2(ox)]Cl + 2\,Nacl$$

(6.8)

We have already seen an example earlier with the formation of $[PtCl_2(NH_3)_2]$, driven in this case by low solubility of this neutral species allowing it to crystallize out of the reaction mixture rather than continue reaction to form ionic $[Pt(NH_3)_4]^{2+}$. Another way to achieve partial substitution is through, use of a stoichiometric amount of a reagent. For example, reaction (6.9) may occur, where only one of two available coordinated chloride ions are substituted because of the availability of only one molar equivalent of added cyanide anion.

$$cis\text{-}[CoCl_2(en)_2]Cl + Na(CN) \longrightarrow [CoCl(CN)(en)_2]Cl + NaCl$$

(6.9)

However, partial substitution is compromised in many cases by the formation of lesser amounts of species with both greater and lower levels of substitution than the target complex. This is a particularly common outcome with labile systems; for example, addition of two molar equivalents of ammonia to $Cu^{2+}{}_{aq}$ will not lead to only $[Cu(NH_3)_2]^{2+}{}_{aq}$, but also

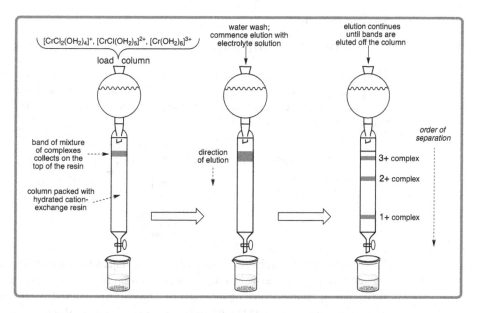

Figure 6.3
Simple chromatographic separation of differently-charged chromium(III) complexes by ion chromatography using a column packed with an aqueous slurry of cation-exchange resin beads to form a bed of resin.

to some $[Cu(NH_3)]^{2+}_{aq}$, $[Cu(NH_3)_3]^{2+}_{aq}$ and possibly even $[Cu(NH_3)_4]^{2+}$, the outcome depending on the relative stability constants of the various species.

Where a mixture of products result from a reaction, it may be possible to separate these by selective crystallization. However, this is not always successful and is difficult where small amounts of a particular product are present. An option in these circumstances is to employ chromatography to separate the various complexes present in the reaction mixture. For ionic and neutral complexes, it is possible to separate mixtures in aqueous solution reliably and readily using cation or anion exchange chromatography. Separation of cations, for example, can be carried out successfully using either an acid-stable resin such as Dowex® 50W–X2 or a neutral pH resin such as SP-Sephadex® C–25, typically employing as eluting reagents acids (0.5–5 M) for the former and neutral salts (0.1–0.5 M) for the latter. These resins separate first by charge, with separation also influenced by molecular mass and shape. The order of elution of complexes of comparable molecular masses from a column will be zero-charged $\gg$ 1+ complex $>$ 2+ complex $>$ 3+ complex. This can be illustrated for a heated aqueous solution of commercial chromic chloride, which contains a mixture of complex ions of different charge (Figure 6.3).

One observation from a large collection of experimental results is that ligands not undergoing substitution themselves ('spectator' ligands) can influence substitution at sites directly opposite them (*trans*) and, to a lesser extent, at adjacent sites (*cis*). The best-known examples lie with Pt(II) chemistry, where some groups exhibit a strong *trans* effect, causing groups opposite them to be more readily substituted than those in *cis* dispositions. Groups opposite a chloride ion, for example, are substituted more readily than those opposite an ammine group. Extensive studies have produced an order of *trans* effects for various ligands bound to Pt(II), being:

$$CO \sim CN^- > PH_3 > NO_2^- > I^- > Br^- > Cl^- > NH_3 > {}^-OH > OH_2$$

The *trans* influence means that reaction of $[PtBr_4]^{2-}$ with two molecules of ammonia will lead preferentially to the *cis*-$[PtBr_2(NH_3)_2]$ isomer being formed, as the second substitution step occurs selectively opposite a bromide ion, which has a stronger *trans* influence than ammonia (6.10).

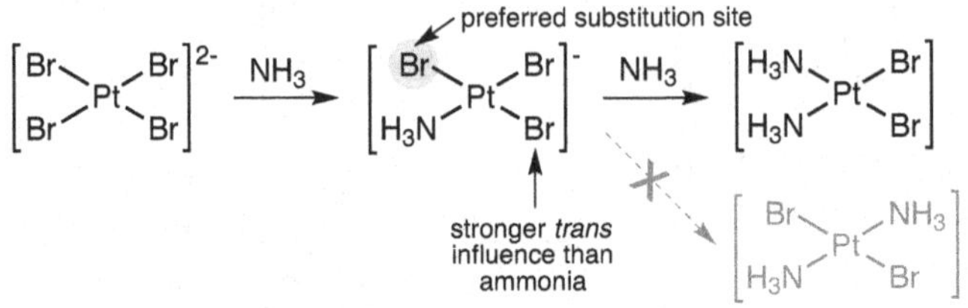

(6.10)

Taking account of the influence of different *trans* groups allows for construction of different isomers through stepwise substitution processes from different precursors. The complexes are displaying what is called *regiospecificity* – directing reactions to a specific site or region of the molecule. The influence of 'spectator' ligands on substitution processes relates to their capacity to withdraw electron density from or donate it to the metal centre, which influences their opposite partner; we see these influences in the variation in metal–donor distance for a particular group as the group opposite is varied.

Although all of the examples above relate to the synthesis of mononuclear complexes, it should be appreciated that polymetallic compounds are accessible, and sometimes form preferentially, depending on the ligands involved. A simple example involves the reaction of copper(II) carbonate with a carboxylic acid, which can produce a dinuclear compound where the carboxylate acts as a bridging didentate ligand (6.11). The Cu–Cu separation above is too great to be considered as a bonding distance, although analogues with heavier metals do display metal–metal bonds.

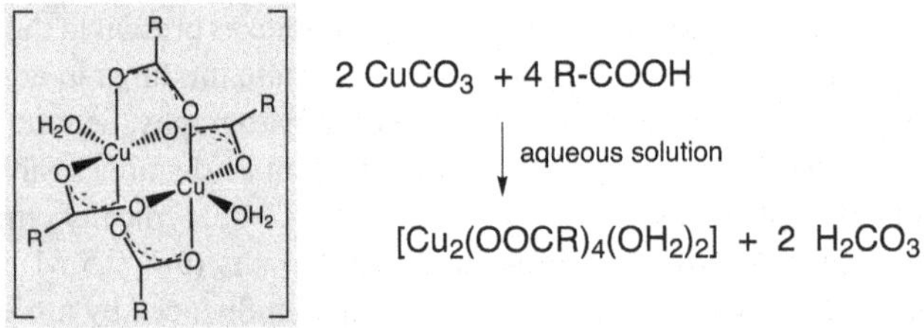

$$2\ CuCO_3 + 4\ R\text{-}COOH$$

aqueous solution

$$[Cu_2(OOCR)_4(OH_2)_2] + 2\ H_2CO_3$$

(6.11)

Limited examples of other dinuclear and higher polynuclear complexes will appear herein. However, the focus here will remain on mononuclear systems.

6.3.2 Substitution Reactions in Nonaqueous Solvents

Water may not be a suitable solvent in all cases due to reactant insolubility (of the precursor complex and/or the ligand, but usually the latter), extreme inertness that demands use of a higher boiling point solvent for reaction, or high stability of undesired hydroxo or oxo species that prevent or interfere with formation of the desired products. This is usually solved by using another solvent, either a conventional molecular organic solvent or a low melting point ionic liquid.

The aqueous chemistry of particularly Al(III), Fe(III) and Cr(III) involves formation of strong M—O bonds, and in basic aqueous solution, hydroxide species (or unreactive oligomers) usually precipitate preferentially and rapidly because many added ligands are strong bases that cause a rise in solution pH. The behaviour of Fe(III) in a weakly basic aqueous solution is a good example, and can be represented simplistically by (6.12).

$$
\begin{bmatrix} \text{H}_2\text{O} & \text{OH}_2 \\ & \text{Fe} \\ \text{H}_2\text{O} & \text{OH}_2 \\ & \text{OH}_2 \end{bmatrix}^{3+} \xrightarrow[-3\text{H}^+]{\text{OH}^-} \begin{bmatrix} \text{H}_2\text{O} & \text{OH}_2 \\ & \text{Fe} \\ \text{H}_2\text{O} & \text{OH} \\ & \text{OH} \end{bmatrix} \xrightarrow[-3\text{H}_2\text{O}]{\text{rapid dehydration and precipitation}} \text{Fe(OH)}_3 \downarrow \qquad (6.12)
$$

deprotonation to produce coordinated hydroxides

Hydroxide is an effective ligand, and where the hydroxo complexes are more stable than those of the added basic ligand, it is the role of the added ligand as a base, to promote formation of hydroxo complexes, that dominates its potential role as a ligand. This can be avoided simply by working in the absence of water, using instead an anhydrous aprotic solvent. For example, using a hydrated salt of chromium(III) in water (6.13) or an anhydrous salt in anhydrous diethyl ether (6.14) with 1,2-ethanediamine as introduced ligand has distinctly different outcomes, due to the absence of water in the latter reaction.

$$
\begin{bmatrix} \text{H}_2\text{O} & \text{OH}_2 \\ & \text{Cr} \\ \text{H}_2\text{O} & \text{OH}_2 \\ & \text{OH}_2 \end{bmatrix}^{3+} \quad \text{Cr}^{3+}_{\text{aq}} + 3(\text{en}) \xrightarrow{\text{water}} \text{Cr(OH)}_3 \downarrow \begin{array}{c}\text{insoluble}\\\text{chromium(III)}\\\text{hydroxide}\end{array} + 3(\text{enH}^+) \qquad (6.13)
$$

violet   *green*

$$
\begin{array}{c}\text{anhydrous}\\\text{chromium(III)}\\\text{chloride}\end{array} \quad \text{CrCl}_3 + 3(\text{en}) \xrightarrow{\text{ether}} [\text{Cr(en)}_3]\text{Cl}_3 \qquad \begin{bmatrix} \text{H}_2\text{N} & \text{NH}_2 \\ & \text{Cr} \\ \text{H}_2\text{N} & \text{NH}_2 \\ & \text{NH}_2 \end{bmatrix}^{3+} (\text{Cl}^-)_3 \qquad (6.14)
$$

purple   *yellow*

In the first case, the basic diamine effectively extracts protons from coordinated water molecules, leaving an insoluble metal hydroxide, rather than initiating any ligand substitution. In the aprotic solvent where no water is present, thus preventing formation of any hydroxo species, the diamine achieves coordination. It is also possible to prepare solvated salts other than hydrated ones for use. For example, $[\text{Ni(OH}_2)_6](\text{ClO}_4)_2$ may be replaced by $[\text{Ni(DMF)}_6](\text{ClO}_4)_2$ in dimethylformamide (DMF) as solvent, providing a way of ensuring that the initial complex coordination sphere is nonaqueous in form as well as the solvent.

Where solubility alone is the issue, simply changing solvent to permit all species to be dissolved allows the chemistry to proceed essentially as it would in aqueous solution were species soluble. Typical molecular organic solvents used in place of water include other protic solvents such as alcohols (e.g. ethanol), and aprotic solvents such as ketones (e.g. acetone), amides (e.g. dimethylformamide), nitriles (e.g. acetonitrile) and sulfoxides (e.g. dimethylsulfoxide). Recently, solvents termed ionic liquids, which are purely ionic material that are liquid at or near room temperature, have been employed for synthesis; typically, they consist of a large organic cation and an inorganic anion (e.g. *N,N'*-butyl(methyl)-imidazolium nitrate) and their ionic nature supports dissolution of, particularly, ionic complexes.

In some case, where hydrolysis reactions are not of concern, mixtures or organic solvents and water can be employed. For example, it is possible to mix an aqueous solution of the metal salt with a miscible organic solvent containing the organic ligand, with the two reactants sufficiently soluble in the mixed solvent to remain in solution and react. Further, it is possible sometimes to react an insoluble compound with a dissolved one with sufficient stirring and heating, since 'insoluble' usually does not mean absolute insolubility, with sufficient dissolving to permit reaction, and its 'removal' from solution by complexation allowing further compound to dissolve and react.

Some complexes may not form, or else are not stable, in water because they are not thermodynamically stable in that solvent. Their formation in the total absence of water can be achieved, however, as exemplified by the sequence of reactions below where the neutral tetrahedral complex can be isolated readily and is stable in the absence of water, although dissolution in water leads to very rapid formation of the octahedral Co^{2+}_{aq} cation and dissociation of the initial ligands (6.15).

$$\text{(6.15)}$$

Neutral complexes of the type [M(acac)$_2$] and [M(acac)$_3$], where acac$^-$ is the 2,4-dioxopentan-3-ido anion, usually show good solubility in organic solvents and serve as useful synthons, since they undergo ligand substitution reactions by other chelates fairly readily. Moreover, they tend to stabilize otherwise relatively unstable oxidations states; for example, the Mn(III) complex [Mn(acac)$_3$] (6.16) is very stable, whereas the hydrated ion Mn^{3+}_{aq} readily undergoes a diproportionation reaction (6.17) in water (to Mn(II) and Mn(IV)). The former complex is itself readily prepared from reaction of $MnCl_2 \cdot 4H_2O$ and 2,4-dioxopentane (acacH) in basic solution with a strong oxidant (6.16).

$$\text{(6.16)}$$

$$\text{(6.17)}$$

Subsequent reaction of [Mn(acac)$_3$] in an organic solvent with potential polydentate ligands offers a route to a wide range of complexes.

6.3.3 Substitution Reactions without using a Solvent

In some reactions the ligand to be introduced is a liquid, and where there are no issues relating to a need for stoichiometric amounts of reagents it is possible to use the ligand in excess as effectively its own solvent in the reaction. This is particularly appropriate where the ligand can be removed readily by evaporation or distillation, or where the complex product precipitates from the reaction mixture and can be separated directly by filtration. The best example of this type of reaction is with ammonia, which has a boiling point of only −33 °C. For example, (6.18) is a reaction that proceeds readily, and avoids the hydrolysis problems that lead to preferential formation of $Cr(OH)_3$ in aqueous ammonia.

$$CrCl_3 + 6(NH_3) \xrightarrow[\text{ammonia}]{\text{liquid}} [Cr(NH_3)_6]Cl_3 \quad \text{(6.18)}$$

While this reaction works well with most anhydrous metal ion salts, it is not necessarily convenient because of the handling difficulties involving liquid ammonia, and some salts form isolable ammine complexes sufficiently well using aqueous ammonia, as is the case with Ni(II) and Cu(II), for example.

Higher boiling point and also chelating amines can be reacted effectively with anhydrous metal salts in the same manner as described for ammonia. These reactions do generate heat, as expected in an acid/base (metal cation/ligand) reaction, so careful mixing of reagents is required. In principle, this technique also offers a route to a wide range of $[M(solvent)_6]^{n+}$ salts through reaction of an anhydrous salt with traditional molecular solvents, provided the solvent is an effective ligand. Nitriles, amides, sulfoxides and alcohols may be introduced in this manner in many cases.

Another approach, which permits introduction of anionic ligands of strong acids, is to react a complex containing coordinated chloride ion directly with an anhydrous strong acid, such as trifluoromethanesulfonic acid, in the total absence of any other solvent. One example, where HCl is released as a covalent gas and leaves the anhydrous reaction mixture and the $CF_3SO_3^-$ anion enters the coordination sphere as an O-bound monodentate ligand, is reaction with chloropentaamminecobalt(III) chloride (6.19).

$$[CoCl(NH_3)_5]Cl_2 + 3(CF_3SO_3H) \xrightarrow[\text{100\% } CF_3SO_3H]{\text{heat in excess}} [Co(NH_3)_5(OSO_2CF_3)](CF_3SO_3)_2 + 3HCl \uparrow \quad \text{(6.19)}$$

The complex product can be isolated as a solid by very slow and careful addition of cold diethyl ether to the stirring cooled reaction mixture; a great deal of heat is generated in this process, so great caution is required. However, the product in this example is very useful as a synthetic reagent. The coordinated trifluoromethanesulfonate anion is an extremely labile group when bound to almost any metal ion and thus readily replaced. Consequently, these complexes find use as reagents for the introduction of other ligands through simple substitution reactions in a poorly coordinating solvent, written in general as in (6.20).

$$[M(L_a)_x(CF_3SO_3)_y](CF_3SO_3)_z + y(L_b) \longrightarrow [M(L_a)_x(L_b)_y](CF_3SO_3)_{y+z}$$

poorly-coordinating
non-aqueous solvent

EXAMPLE:

(6.20)

Another approach is to employ the effect of heat on a solid complex. It has been known for a long time that metal complexes, if heated strongly, undergo decomposition reactions that eventually take them through to usually simple salts or oxides. Generally, ligands are lost in a series of steps, related in part to their volatility, and this can be probed using a technique called thermal analysis, which effectively amounts to following weight change with a sensitive balance during heating of a small sample. Simple neutral ligands often depart the coordination sphere as molecular species over a reasonably small and well-defined characteristic temperature range, so that heating to a controlled temperature can allow controlled conversion to occur. The simplest examples are the hydrated salts of metal ions, such as $[M(OH_2)_n](SO_4)$, which on heating lose water to form anhydrous $M(SO_4)$, usually with a distinctive colour change (such as from blue to nearly colourless, as seen for copper ion). For complexes containing water as one of several ligands, its loss tends to occur ahead of other ligands such as amines, allowing partial change of the coordination sphere. The metal centre involved still seeks to retain its original coordination geometry, so that loss of water is usually associated with replacement in the coordination sphere by an involatile anion of the original salt, such as in (6.21).

$$[Co(NH_3)_5(OH_2)]Cl_3 \longrightarrow [CoCl(NH_3)_5]Cl_2 + H_2O$$

heat the solid;
110°C, hrs

red-pink   *purple*

(6.21)

This approach permits the insertion of a wide range of stable anions apart from chloride into the coordination sphere, simply by commencing with them present as the counter-ion. At higher temperature, amine ligands can be lost in the same manner that water is lost, and may occur in a stepwise process that permits isolation of useful intermediate complexes. Eventually, at sufficiently high temperature, all ligands are lost, as in (6.22).

$$[Pt(NH_3)_4]Cl_2 \longrightarrow [PtCl_2(NH_3)_2] + 2\,NH_3\uparrow \longrightarrow PtCl_2 + 2\,NH_3\uparrow$$

white   heat the solid;   *yellow*   extend heating;
              250°C, hrs   500°C, hrs

(6.22)

This reaction of Pt(II) is general for a range of coordinated amines apart from ammonia, and appears to yield exclusively the *trans* geometric isomer (which is called, because of this exclusivity, a *stereospecific* reaction). Even chelated diamines can be substituted, as they are inevitably more volatile than any anions present; thus chelated 1,2-ethanediamine can be replaced by two chloride ions in $[Co(en)_3]Cl_3$ to form *cis*-$[CoCl_2(en)_2]Cl$, and by two thiocyanate ions in $[Cr(en)_3](NCS)_3$ to form *trans*-$[Cr(en)_2(NCS)_2](NCS)$ on heating. Anions such as thiocyanate are ambivalent ligands that can coordinate through either of two donors, in this case the S or N atoms. Because heating is involved in these syntheses, the thermodynamically stable form will always be isolated, which in the case of thiocyanate with chromium(III) is the N-bound form. This behaviour can be easily demonstrated by commencing with the less stable isomer, and observing behaviour on heating; for example pink $[Co(NH_3)_5(ONO)]Cl_2$ on mild heating changes to the thermodynamically stable yellow $[Co(NH_3)_5(NO_2)]Cl_2$, where an O- to N-donor isomerization has occurred, as also occurs in solution and discussed in Chapter 5.

6.3.4 Chiral Complexes

Many ligands are chiral as a result of asymmetric centres, which mean they exist as two different optical isomers. These ligands may be present as a racemate (a 50 : 50 mixture of the two optical forms), or else as a chiral form (that is, as one optical isomer). The presence of chirality in a ligand is 'felt' by the complex and seen in its chiroptical properties, but there are rarely any particular special requirements for synthesis involving introduction of a chiral ligand as opposed to its racemic form. However, in some circumstances the complex itself can be chiral as a result of the disposition of ligands. As a general rule, the presence of at least two chelate rings is sufficient for some octahedral complexes to be resolvable into their enantiomeric forms. The classical chiral metal complex is the octahedral $[Co(en)_3]^{3+}$ ion, which can be separated into Δ and Λ optical forms through successive recrystallization of its salt formed with an optically active anion such as *d*-tartrate (6.23). This approach is based on the (Δ,*d*) salt forming a more stable crystalline lattice than the (Λ,*d*) salt, meaning their solubility differs and one is preferentially precipitated.

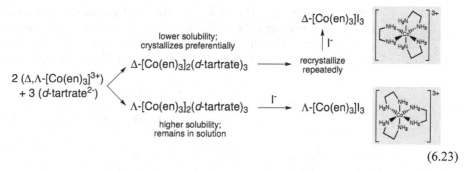

$$\tag{6.23}$$

It is also possible to separate the optical isomers through chromatography on a chiral cation-exchange resin (such as SP-Sephadex® C-25 resin) by using a chiral eluate such as a *d*-tartrate solution. Differential binding to the resin, which is itself chiral, means the complex separates on a sufficiently long column into two bands comprising the two optical forms of the complex.

Chirality is often met with polydentate ligands at a number of different centres in the complex, and separation of all optical isomers is either impractical or, in effect, impossible. In many cases, working with a racemate has no significant influence on the chemistry, and optical resolution of complexes is attempted on only very limited occasions, such as where researchers wish to record the chiroptical properties, or where a chiral complex is required to assist in achieving a chiral reaction, such as use of a chiral complex as a catalyst in synthesis of organic molecules where a particular optical isomer is sought (exemplified in Chapter 9).

6.3.5   Catalysed Reactions

Catalysis is an approach to dealing with extremely slow reactions that otherwise can be driven conveniently only by increasing the temperature, pressure and/or reaction time. The concept of catalysis is that the catalyst lowers the activation barrier for a reaction, thus allowing it to proceed at a faster rate; it is not usually considered to influence the position of an equilibrium. The catalyst may be dissolved in the reaction solution (a *homogenous* catalyst) or present as a solid (a *heterogenous* catalyst). The former has the advantage of being in the same phase as the reagents; for a heterogenous system, reaction must occur only at the surface of the solid catalysts, so that surface area is an important consideration, as is stirring to allow mass transport in the system.

Many catalysts are metals, metal oxides or other simple salts, or metal complexes. For example, formation of platinum(IV) complexes involving ligand substitution is an extremely slow process, due to the kinetic inertness of this oxidation state. However, the addition of small amounts of a platinum(II) complex to the reaction mixture leads to excellent catalysis of the reaction, assigned to mixed oxidation state bridged intermediates that promote ligand transfer.

There are also nonmetallic catalysts, of which the best known are the various forms of activated carbon. For example, isomerization of inert, chiral cobalt(III) complexes is accelerated significantly by the presence of carbon; this is assigned to reduction of cobalt(III) on the surface of the carbon to labile cobalt(II), allowing rapid ligand rearrangement, with air subsequently rapidly re-oxidizing the cobalt(II) to cobalt(III) before ligand dissociative processes can become involved.

Complexes of some metal ions show a capacity towards photo-activity, which means that they undergo different chemistry in the presence of light. This is because light can cause electron transfer from metal orbitals to ligand orbitals, leading to a reactive excited state that can undergo different chemistry; ruthenium(II) is one metal that exhibits this capacity. Usually, simple complexes exhibit only light (hv) catalysed aquation reactions, but redox chemistry can result with other reagents. Both outcomes are illustrated in the following simple example, where both aquation (6.24) and oxidation (6.25) can occur.

$$[Ru(L)(NH_3)_5]^{2+} + H_2O \underset{\text{aquation}}{\overset{hv}{\longleftarrow}} \begin{array}{l} [Ru(NH_3)_5(OH_2)]^{2+} + L \\[4pt] [Ru(L)(NH_3)_4(OH_2)]^{2+} + NH_3 \end{array} \tag{6.24}$$

$$[Ru^{II}(L)(NH_3)_5]^{2+} + H^+ \underset{\text{oxidation}}{\overset{hv}{\longrightarrow}} [Ru^{III}(L)(NH_3)_5]^{3+} + \tfrac{1}{2}H_2 \tag{6.25}$$

Examples of reactions that undergo change in oxidation state are covered in more detail in the following section.

6.4   Reactions Involving the Metal Oxidation State

Because many complexed metals ions have a range of oxidation states accessible in usual solvents, it is not surprising to find that syntheses may involve oxidation–reduction reactions. The classical example is cobalt, which is usually supplied commercially as Co(II) salts, but whose complexes are best known as Co(III) compounds. This is because Co(III) compounds are inert and usually robust, readily isolable compounds, whereas Co(II) compounds are labile and prone to rapid reactions. This lability is put to good use in synthesis, since it is convenient to use the Co(II) form to rapidly coordinate ligands initially, and then oxidize the mixture to the Co(III) form. This oxidation can often be achieved by oxygen in the air alone, depending on the ligand environment and the redox potential ($E°$) of the complex. For example in an aqueous ammonia/ammonium chloride buffered solution, Co(II) reacts in a sequential manner essentially as in (6.26).

$$\text{Co}^{2+}_{aq} + \text{NH}_3/\text{NH}_4\text{Cl} \xrightarrow{\text{excess}} [\text{Co}(\text{NH}_3)_6]^{2+} \xrightarrow{\text{O}_2} [\text{Co}(\text{NH}_3)_6]^{3+}$$

(6.26)

Substitution reactions without redox chemistry being involved are available for Co(III), but not commonly met. One example is the use of $[\text{Co}^{III}(\text{CO}_3)_3]^{3-}$ as a synthon, since the chelated carbonate ion is readily displaced by other better chelating ligands such as polyamines. An example of this type of reaction, where the entering polyamine (cyclam) is a saturated and flexible macrocyclic tetraamine, is (6.27).

(6.27)

$$[\text{Co}(\text{CO}_3)_3]^{3-} + \text{cyclam} \longrightarrow [\text{Co}(\text{CO}_3)(\text{cyclam})]^+ + 2\,\text{CO}_3^{2-}$$

Where stronger oxidizing agents than air are required to take the Co(II) form to Co(III), which usually applies where fewer N-donor and more O-donor groups are bound, hydrogen peroxide is a particularly useful oxidizing agent, because it leaves no problematical products to separate from the desired complex product. One of the problems with oxidation of Co(II) to Co(III) compounds is that in some cases a bridged peroxo complex, featuring a $\text{Co}^{III}-\text{O}_2^{2-}-\text{Co}^{III}$ linkage, forms as a stable intermediate. One can envisage its formation through a redox reaction whereby an oxygen molecule is reduced to peroxide ion, and two Co(II) ions are oxidized to Co(III). The O_2^{2-} ion in the bridge can be readily displaced by reaction with strong acid and heating, with the use of hydrochloric acid leading to monomers with $\text{Co}^{III}-\text{Cl}^-$ components replacing the bridging group (6.28).

$$\text{Co}^{2+}_{aq} + 5\,\text{L} \rightleftharpoons [\text{L}_5\text{Co}(\text{OH}_2)]^{2+} \xrightarrow{\text{O}_2} \left[\text{L}_5\overset{III}{\text{Co}}-\text{O}^--\text{O}^--\overset{III}{\text{Co}}\text{L}_5\right]^{4+}$$

(6.28)

Many other metal complexes can be chemically oxidized to higher oxidation states with an appropriate oxidizing agent. This can occur with complete preservation of the coordination sphere (6.29), which means the oxidation–reduction reaction is reversible if a suitable reducing agent is then employed for reduction of the oxidized form.

$$[IrCl_6]^{3-} \underset{reduction}{\overset{oxidation}{\underset{+e}{\overset{-e}{\rightleftharpoons}}}} [IrCl_6]^{2-} \tag{6.29}$$

Alternatively, the coordination number and/or the ligand set can change substantially in an irreversible oxidation reaction. Any reduction reaction of the product will then not regenerate the original complex.

Not only are oxidation reactions fairly common, but also one may employ reduction reactions in simple synthetic paths, provided the reduced form is also in a stable oxidation state. Two examples of reduction reactions, with the metal oxidation states included, are given in (6.30) and (6.31).

$$[Pt^{IV}(en)_3]Cl_4 \xrightarrow{+2e} [Pt^{II}Cl_2(en)] \quad (+\ 2\ (en)\ +\ 2\ Cl^-) \tag{6.30}$$

$$[Mo^{V}Cl_5O]^{2-} \xrightarrow{Zn/HCl} [Mo^{III}Cl_6]^{3-} \quad (+\ OH^-\ +\ Zn^{2+}) \tag{6.31}$$

Reduction reactions occurring along with substitution chemistry are also well known, and the examples above are such cases. Another simple example involves the $[IrCl_6]^{2-}$ ion, which undergoes both reduction (with hypophosphorous acid) and substitution in (6.32).

$$[Ir^{IV}Cl_6]^{2-} \xrightarrow[\substack{pyridine \\ boil,\ 30\ min}]{H_2PO_2^{2-}} fac\text{-}[Ir^{III}Cl_3py_3] \xrightarrow{boil,\ 6\ hrs} cis\text{-}[Ir^{III}Cl_2py_4]^- \tag{6.32}$$

Because of the slow substitution chemistry of iridium, sequential reaction steps are well separated in terms of reaction time, so that the initial product of the redox-substitution reaction undergoes further simple substitution only with prolonged heating.

Where a reactive lower oxidation state results, a key concern is the necessary protection of the reduced complex from air or other potential oxidants, as they are often readily re-oxidized. Usually, this requires their handling in special apparatus such as inert-atmosphere boxes or sealed glassware in the absence of oxygen. Where active metal reducing agents (such as potassium) are employed, special care with choice of solvent is also necessary. The nickel reduction reaction (6.33) can be performed in liquid ammonia as solvent, since the strongly-bound cyanide ions are not substituted by this potential ligand.

$$[Ni^{II}(CN)_4]^{2-} + 2\ K \xrightarrow{\substack{liquid \\ NH_3}} [Ni^{0}(CN)_4]^{4-} + 2\ K^+ \tag{6.33}$$

While the Ni(II) is reduced to Ni(0), the potassium metal is oxidized to K(I); it is a standard redox reaction. The Ni(0) complex formed exemplifies the necessity for careful handling of many low-valent complexes; it is readily oxidized by air, and also reacts with water in a redox reaction that liberates hydrogen gas.

Lower-valent metal complexes may be prepared in reduction reactions with full substitution of the coordination sphere. One example results from reduction of the vanadium(III) complex $[VCl_3(THF)_3]$ with a suitable reactive metal in the presence of carbon monoxide under pressure (6.34). This is an example of the synthesis of an organometallic compound; more examples occur in Section 6.6.

$$[\overset{III}{V}Cl_3(THF)_3] \xrightarrow[\text{heat and pressure}]{\overset{2\ Mg\ or\ 2\ Zn}{CO}} [\overset{-I}{V}(CO)_6]^- \ (+\ 2\ M^{2+} + 3\ Cl^- + 3\ THF)$$

$$\xrightarrow{H^+} [\overset{0}{V}(CO)_6] \ (+\ \tfrac{1}{2}H_2)$$

$$(6.34)$$

On some occasions, a reducing agent may not be able to reduce the metal centre in a complex, but may be sufficient to initiate chemical change in a ligand. This is, in effect, a reaction of a coordinated ligand (see Section 6.5.2), rather than a metal-centred redox reaction. A simple example involves the reduction of coordinated N_2O to N_2 and H_2O, using Cr(II) ion as reducing agent (6.35); the metal ion retains its original oxidation state throughout. The reaction is promoted through coordination lowering the N—O bond strength.

$$[Ru(NH_3)_5(N_2O)]^{2+} + 2\ Cr^{2+}_{aq} \longrightarrow [Ru(NH_3)_5(N_2)]^{2+} + 2\ Cr^{3+}_{aq}$$

$$+ 2H^+ \qquad\qquad + H_2O$$

$$(6.35)$$

Many complexes can be prepared by electrochemical reduction or oxidation under an inert atmosphere. Exhaustive reduction (coulometry) using a large working electrode can lead to a clean formation of the reduced form, if it is sufficiently stable, since electrons alone are used in the reduction process. As an example, the octahedral vanadium(III) complex $[V(phen)_3]^{3+}$ undergoes a series of sequential and reversible reduction steps (6.36) with retention of the three chelate ligands down to oxidation state of formally vanadium $(-I)$.

$$V(III) \underset{-e}{\overset{+e}{\rightleftarrows}} V(II) \underset{-e}{\overset{+e}{\rightleftarrows}} V(I) \underset{-e}{\overset{+e}{\rightleftarrows}} V(0) \underset{-e}{\overset{+e}{\rightleftarrows}} V(-I)$$

four electron reversible reduction overall
[all reductions not necessarily metal-centred]

$$[V(phen)_3]^{3+} \underset{-e}{\overset{+e}{\rightleftarrows}} [V(phen)_3]^{2+} \underset{-e}{\overset{+e}{\rightleftarrows}} [V(phen)_3]^{1+} \underset{-e}{\overset{+e}{\rightleftarrows}} [V(phen)_3]^{0} \underset{-e}{\overset{+e}{\rightleftarrows}} [V(phen)_3]^{-1}$$

$$(6.36)$$

With unsaturated ligands particularly, the location of introduced electrons is not easily assigned, and they may reside on the ligand (leading to a ligand radical) rather than the metal ion (leading to a lower oxidation state). This does not invalidate the chemistry, but does bring into question the nature of the product. Overall, electrochemistry provides a useful way of performing not only reduction reactions but also oxidation reactions. The sole concern is that exhaustive electrolysis to convert all of one form to another oxidation state takes some minutes to perform, so the kinetic stability of the product does influence the validity of the method.

When undertaking electrochemical reactions, it is important to note that the $E°$ for a complex is dependent on the set of ligands coordinated, and can change substantially with donor set. This is represented in (6.37) for the simple substitution of a $M^{n+}{}_{aq}$ complex, with the potential required to reduce the precursor complex invariably differing from that for the product complex ($E°_1 \neq E°_2$).

$$[M(OH_2)_y]^{n+} + y\,L \;\rightleftharpoons\; [M(L)_y]^{n+} + y\,H_2O$$

$$+e \Big\downarrow E°_1 \qquad\qquad\qquad +e \Big\downarrow E°_2 \qquad\qquad (6.37)$$

$$[M(OH_2)_y]^{(n-1)+} \qquad\qquad [M(L)_y]^{(n-1)+}$$

The shift in redox potential with ligand substitution is particularly obvious for cobalt, as the $E°$ for $Co^{III/II}{}_{aq}$ is near $+1.8$ V, making the aqua Co(III) complex inaccessible from $Co^{2+}{}_{aq}$ with most oxidants, whereas introduction of other donor groups lowers the potential substantially, and in some cases sufficiently to make even air (oxygen) an effective oxidant of the Co(II) complex.

6.5   Reactions Involving Coordinated Ligands

6.5.1   Metal-directed Reactions

There exist both metal-catalysed and metal-directed reactions, which require definition to distinguish the two classes. *Metal-catalysed* reactions are those in which the metal-containing species in the reaction is regenerated in each reaction cycle, so stoichiometric amounts are not required. It is the transition state of the catalysed reaction, rather than the product, which is most strongly complexed by the metal. Thus it is the rate of establishment of equilibrium (k_f and k_b) rather than the position of the equilibrium ($K = k_f/k_b$) that is altered (6.38).

$$\text{Reactants} + M_{catalyst} \;\underset{k_b}{\overset{k_f}{\rightleftharpoons}}\; \text{Products} + M_{catalyst} \qquad\qquad (6.38)$$

Homogeneous transition metal catalysts usually employ their coordination sphere as the site of the chemistry they promote. One example is the rhodium catalyst used for promoting ethene hydrogenation. The keys to the process are an addition reaction of dihydrogen to the rhodium centre and a substitution reaction that also introduces ethene to the coordination sphere in place of a solvent molecule. It is in the intermediate produced that the former adds to the latter to produce ethane which then, as an exceedingly poor ligand, departs the coordination sphere and leaves vacancies for the process to occur again (Figure 6.4). Here, the catalyst is reused, many times.

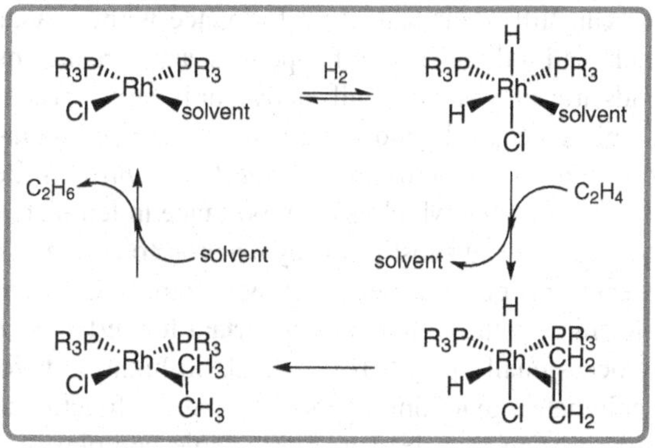

Figure 6.4
A simplified catalytic cycle for the hydrogenation of ethene, which employs a rhodium(I) catalyst.

For a *metal-directed* reaction, the product is formed as a metal complex, and stoichiometric amounts of metal are consumed in the process (6.39).

$$\text{Reactants} + \text{M} \longrightarrow [\text{Products—M}] \qquad (6.39)$$

In virtually all cases, the driving force for metal-directed reactions is the stabilization associated with the formation of a chelating ligand from monodentate ligands, or the conversion of a weakly chelating ligand to a stronger chelator. Often, the major product in the presence of the metal ion is not even detected from the same reaction in the absence of the metal ion. The metal has either caused an extreme displacement of an equilibrium, or promoted a new and rapid reaction pathway by complexation and stabilization of an otherwise inaccessible transition state.

Normally, metal-directed reactions refer to those reactions that are involved in synthesis of a larger organic molecule from smaller components, with the product being an effective ligand. In fact, these reactions result in the organic product being bound to the metal, which is therefore consumed in stoichiometric amounts. There are several general principles considered to be of importance in governing metal-directed reactions, the most important of which are:

Chelation – This is probably the most important factor. In nearly all cases, it is the formation of a (more) stable metal chelate as the primary reaction product that drives the equilibrium to favour that product.

Ligand Polarization – Nucleophilic and electrophilic reactions of organic molecules (such as condensation, hydrolysis, alkylation and solvolysis, amongst others) can be greatly enhanced by their coordination to metal ions as ligands. The metal ion can act variously as a Lewis acid, π-acid or π-donor to alter electron density or distribution on the bound organic molecule (or ligand), thus altering the character of the ligand as a nucleophile or electrophile, and hence its reactivity.

Template Effects – The metal ion can act as an 'organizer' or 'collector' of ligands into arrangements around it that are most suitable for the desired reactions.

There are, in addition, some other contributing effects. *Enantiomer discrimination* relates to the fact that ligand binding and reactivity can be affected by other ligands not actually participating in the reaction (the so-called 'spectator' ligands – like spectators at a football game can 'lift' their teams' performance without actually playing themselves, spectator ligands will influence what happens at reaction sites by their presence). If these spectator ligands are bulky and optically active, or if the metal centre is made chiral by a dissymmetric and rigid spectator ligand, differential binding of another chiral ligand, or stereospecificity of reactions, can be introduced. *Metal ion lability* is the ability of a metal ion to exchange its ligands rapidly, which is of importance in template reactions. Very slow exchange will effectively prevent substitution by reaction components and hence limit reactions occurring. However, this doesn't mean that inert complexes are not relevant, as inert metal ions that retain chirality throughout are important for certain stereospecific syntheses. *Redox effects* may occasionally play a role. Metals in high oxidation states may act in some cases as stoichiometric oxidizing agents of a ligand functional group. Also, because varying the oxidation state may lead to more stable complexes, electron transfer reactions may be assisted.

Metal ions can direct *spontaneous self-assembly* of larger and often cyclic molecules from smaller components through the above effects. Nature does this exceptionally well, but is not alone in being able to build cyclic molecules readily. Simple, one-step, comparable syntheses can be achieved in an open beaker, directed by a metal ion, as exemplified in Figure 6.5. This reaction occurs spontaneously in high yield when appropriate amounts of copper(II) nitrate, 1,2-ethanediamine, aqueous formaldehyde and nitroethane are mixed in methanol in a beaker, warmed for a short period, and then stood overnight to allow crystallization. The metal ion acts as a 'collector' of ligands as well as a promoter of ligand reactivity by means of chelation and polarization effects.

A reaction in the absence of a metal ion may differ completely from what occurs in the presence of a metal ion, as exemplified in Figure 6.6. The linear organic molecule formed in the presence of stoichiometric amounts of a metal ion is totally absent in the metal-free chemistry, where small heterocyclic ring formation occurs.

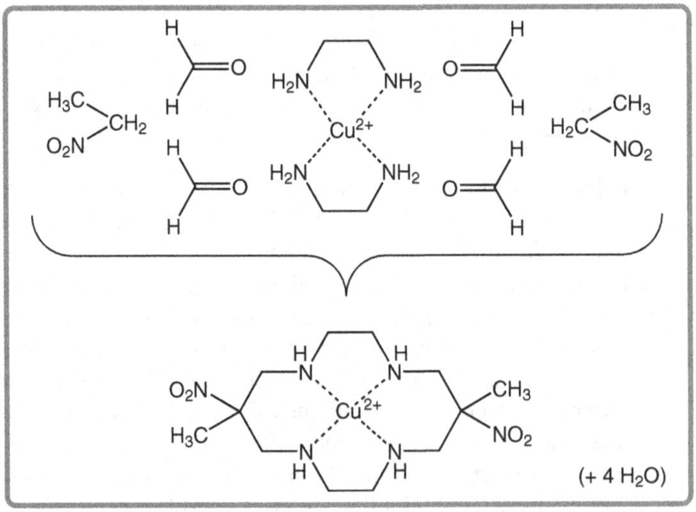

Figure 6.5
An example of a spontaneous self-assembling template reaction, in which small organic components are organized by the metal ion and undergo reaction to form a large cyclic organic product that includes the metal ion.

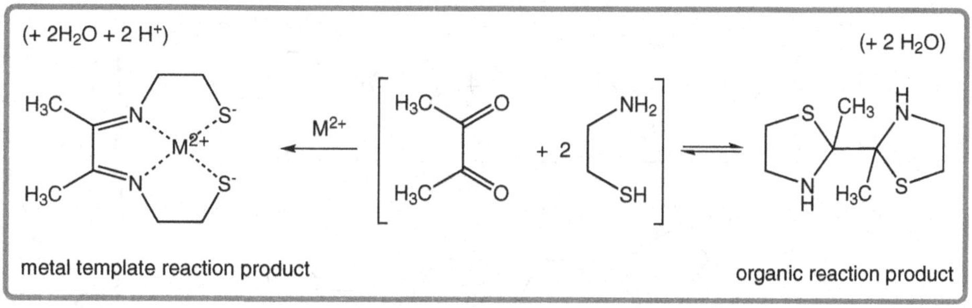

Figure 6.6
An example of the normal organic reaction route compared with the template reaction in the presence of a metal ion; a distinctly different path is followed.

Metal-directed reactions have been used to prepare a wide range of cyclic and acyclic ligand systems. Often, they involve reaction of a coordinated amine with an aldehyde or ketone. Reaction of a carbon acid anion with an electron-deficient site is also commonly featured. Zinc(II)-directed condensation between an aldehyde (R—CHO) and an aromatic nitrogen heterocycle (pyrrole) has been used to prepare substituted porphyrin rings (aromatic tetraaza macrocycles, analogous to hemes found as iron complexes in blood) since the 1960s. Once formed, the new ligands can be removed from the templating metal ion and different metal ions bound to it. This usually involves one of the following: treatment with acid to protonate the ligand and cause it to dissociate; reduction or oxidation of the metal to an oxidation state which will not bind the ligand effectively, allowing it to be removed; treatment with a strongly-binding anion (such as CN^-) that removes the metal ion competitively, leaving the free ligand; addition of another competing metal ion to which the ligand binds preferentially to a solution of the templated product, causing metal exchange (or ligand transfer to the added metal ion).

Polynuclear complexes form through self-assembly also, where both the ligand and precursor metal complex geometry play a role in the outcome – the pieces tend to fit together in a particular way like children's building blocks. An early example involves the self-assembly of 4,4′-bipyridyl and [Pd(en)(ONO$_2$)$_2$]. The two *cis*-disposed O-bound nitrate ions are readily substituted by the preferred pyridine nitrogen donors, but they impose an L-shape when including the palladium, while the ligand imposes a rod-like shape; thus an assembly of four Pd 'corner' L-shapes and four ligand 'rod' shapes creates a square 'picture frame' shape, which is now a large cyclic molecule (Figure 6.7).

The product in the above reaction is of low solubility, and its precipitation from the reaction solution drives further formation, leading to a good yield of the product. The reaction is, in effect, a sequence of substitution steps, with coordinated nitrate ions each replaced in turn by the pyridine nitrogen donors.

6.5.2 Reactions of Coordinated Ligands

It has been known for decades that a range of reactions occurs that involve chemistry of the ligands and in which metal–ligand bond cleavage is not involved. We can regard these as reactions of coordinated ligands. These early and deceptively simple studies provide fine examples of chemical detective work. One of the earliest studies probed the preparation of a H$_2$O—Co(III) species from a O$_2$CO—Co(III) precursor, a reaction which was seen to

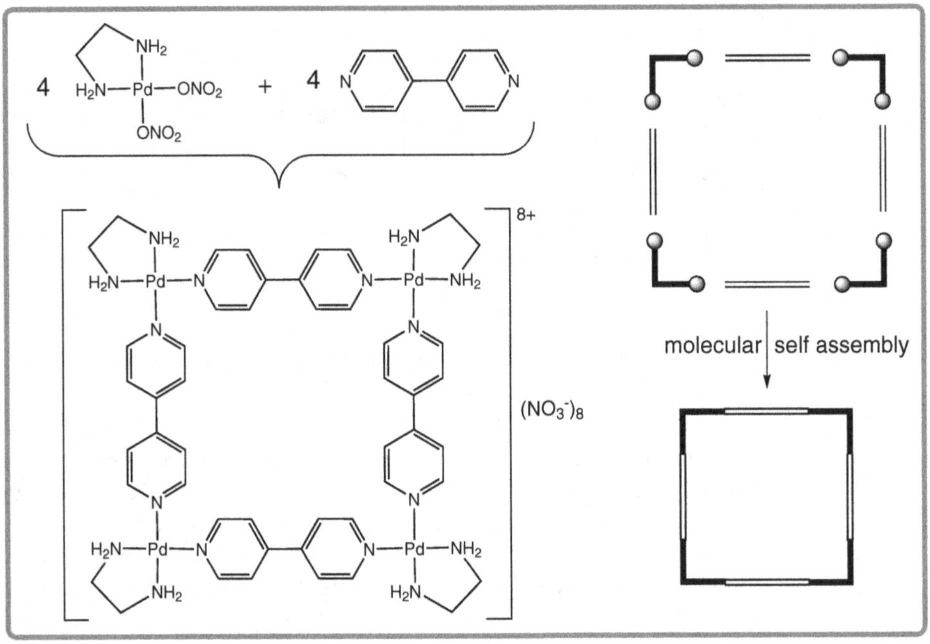

Figure 6.7
An example of a self-assembly reaction of a monomeric complex and ligand, forming a macrocyclic tetranuclear complex.

release CO_2 gas. Two quite different options for this reaction are: dissociation of carbonate ion from the complex followed by release of CO_2 from decomposition of the released anion, with a solvent water molecule entering the coordination sphere in its place; or else cleavage of a C—O bond on the coordinated ligand to release CO_2 while leaving the residual O atom bound to cobalt, with diprotonation of the residual bound oxygen dianion to form coordinated water. In the former case a completely different ligand is inserted, making it a traditional substitution reaction; in the latter case part of the original ligand remains behind with the metal–donor atom bond staying intact, making it a different class of reaction which we now define as a reaction of a coordinated ligand. The key to distinguishing these reactions was to use isotopically-labelled $^{18}OH_2$ as solvent, which showed that the product contained just the normal dominantly $^{16}OH_2$ coordinated with no entry of $^{18}OH_2$ into the coordination sphere – the CO_2 must depart from the coordinated carbonate, as in (6.40).

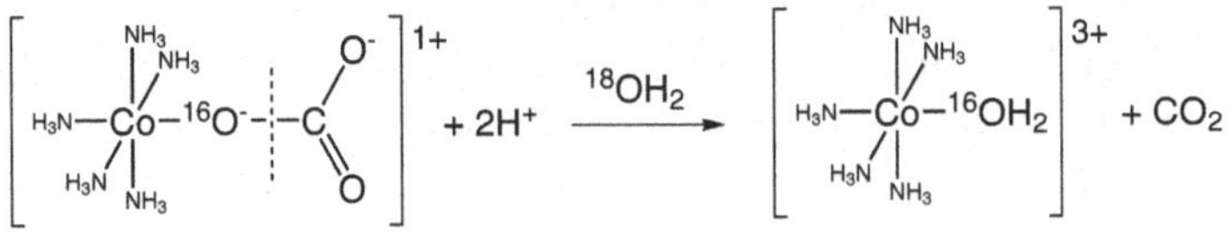

$$\tag{6.40}$$

Not only can this type of reaction occur for this and a range of other coordinated oxyanions, but also what is effectively the reverse reaction can occur, where the reactive species is a HO^-—Co(III) complex, which as the ^{18}O-labelled form retains the label essentially exclusively in the product, proving that it is a reaction between the nucleophilic coordinated hydroxide and an electrophile, as exemplified for the following well-known reaction that produces coordinated nitrite ion (6.41). In this case, the formation of the thermodynamically unstable O-bound nitrite rather than the thermodynamically stable N-bound nitrite is supporting evidence. The formed O-bound form does spontaneously isomerize to the N-bound form in a relatively slow isomerization reaction, but the rate is sufficiently slow that the O-bound form can be isolated readily.

$$\left[\begin{array}{c} H_3N-\underset{\underset{NH_3}{|}}{\overset{\overset{NH_3}{|}}{Co}}-{}^{18}OH \end{array}\right]^{2+} + NO^+ \xrightarrow{{}^{16}OH_2} \left[\begin{array}{c} H_3N-\underset{\underset{NH_3}{|}}{\overset{\overset{NH_3}{|}}{Co}}-{}^{18}O-N{\overset{O}{}} \end{array}\right]^{2+} + H^+ \qquad (6.41)$$

Other reactions of simple anions that may occur in the absence of coordination can also be observed to occur for the complexed form, with the rate of reaction usually changed significantly as a result of complexation. This is anticipated, since a coordinated ion is bonded directly to a highly-charged metal ion, which must influence the electron distribution in the bound molecule and hence its reactivity. Two well-known examples where the product is ammonia occur through either reduction of nitrite with zinc/acid or oxidation of thiocyanate with peroxide. The former example is exemplified in (6.42) below.

$$\left[\begin{array}{c} H_3N-\underset{\underset{NH_3}{|}}{\overset{\overset{NH_3}{|}}{Pt}}-N{\overset{O^-}{\underset{O}{}}} \end{array}\right]^{1+} \xrightarrow{Zn / H^+_{aq}} \left[\begin{array}{c} H_3N-\underset{\underset{NH_3}{|}}{\overset{\overset{NH_3}{|}}{Pt}}-NH_3 \end{array}\right]^{2+} (+ 2 H_2O + Zn^{2+}_{aq}) \qquad (6.42)$$

Transition metal complexes can promote reactions by organizing and binding substrates. We have already seen this in terms of metal-directed reactions. Another important function is the supply of a *coordinated nucleophile* for the reaction, which is incorporated in the product. We have already seen a coordinated nucleophile at work in the reaction discussed above of Co—$^-$OH with NO$^+$; nucleophiles, which are electron-rich entities, are best represented in coordination chemistry by coordinated hydroxide ion, formed by proton loss from a water molecule; this is a common ligand in metal complexes. Normally, water dissociates only to a very limited extent, via

$$OH_2 \leftrightarrows H^+ + {}^-OH$$

for which we define $K_w = [H^+][{}^-OH]/[H_2O]$ and for which $pK_w \approx 14$.

However, when bound to a highly-charged metal ion, its acidity is very significantly enhanced, to the extent that, at neutral pH, a coordination complex will have a significant part of its M—OH$_2$ present as M—$^-$OH, via

$$M^{n+}-OH_2 \leftrightarrows H^+ + M^{n+}-{}^-OH \qquad (pK_a \approx 5 - 9)$$

Although the coordinated hydroxide is a slightly worse nucleophile than free hydroxide, due to electronic effects of the bonded metal cation, its substantially higher concentration in the bound form at any pH more than compensates. A coordinated water molecule with a pK_a of 7 will be 50% in the hydroxide form at neutral pH, for example. Importantly, because it can often be placed adjacent to a bound substrate (thus *pre-organized* for reaction) it is very effective, and marked catalysis is commonly observed.

Although the most important, hydroxide is not the sole example of a coordinated nucleophile met in coordination chemistry. The next most important, as a result of the prevalence of ammonia as a ligand, is the amide ion. Ammonia is usually thought of simply as a base, but it has the capacity to lose protons and thus act as an acid,

$$NH_3 \leftrightarrows H^+ + {}^-NH_2 \qquad (pK_a > 20)$$

although this reaction has such a high pK_a that it is not of significance for free ammonia in water. However, as for coordinated water, acidity of ammonia is significantly enhanced through coordination,

$$M^{n+}-NH_3 \leftrightarrows H^+ + M^{n+}-{}^-NH_2 \qquad (pK_a \approx 10 - 14)$$

so that sufficient concentrations of bound amide can form to permit reaction. Overall, the coordinated amide anion is a far better nucleophile than free amide ion. Alkylamines can show the same activity as nucleophiles,

$$M^{n+}-NHR_2 \leftrightarrows H^+ + M^{n+}-{}^-NR_2$$

as long as they have at least one amine hydrogen atom to release as a proton.

Reactions of coordinated ligands with organic substrates usually occur where the organic molecule enters the coordination sphere in a position adjacent to the nucleophile, and the subsequent reaction involves attack of the coordinated nucleophile at a relatively electron-deficient site on the organic substrate. These reactions lead to a new organic molecule that is usually chelated to the metal ion. This product may depart from labile complex centres through substitution by other ligands (providing a mechanism for repeating the reaction, or catalysis), or else may occur with inert metal centres as a single stoichiometric reaction. These reactions can also induce a particular stereochemistry, and may be defined as *stereospecific* (producing exclusively one isomer) or *stereoselective* (producing an excess of one isomer). *Selectivity* can be introduced simply by preference for a particular conformation in a chelate ring equilibrium, as illustrated in Figure 6.8.

Here the λ conformation (left hand side, Figure 6.8) is preferred and not the δ conformation (right hand side, Figure 6.8) of the chelate ring, as steric clashing (of the ring methyl substituent with other axial ligands on the complex) is minimized in the former. Any subsequent reaction will 'carry forward' this selectivity into the reaction outcome, leading to selectivity in the product.

Either a coordinated ${}^-OH$ or ${}^-NH_2$ group is able to initiate chemistry with appropriate ligands present in an adjacent (*cis*) site. This reactivity was probed in detail for several

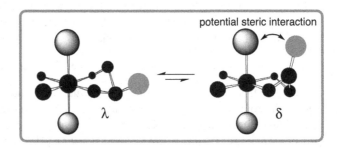

Figure 6.8
Two conformations of a substituted ethylenediamine chelate ring. The left-hand conformer has the methyl substituent on the chelate ring displaced away from the rest of the molecule (equatorial), whereas the right-hand conformer places it in an axial disposition where it may clash with another axial ligand, which is thus unfavourable.

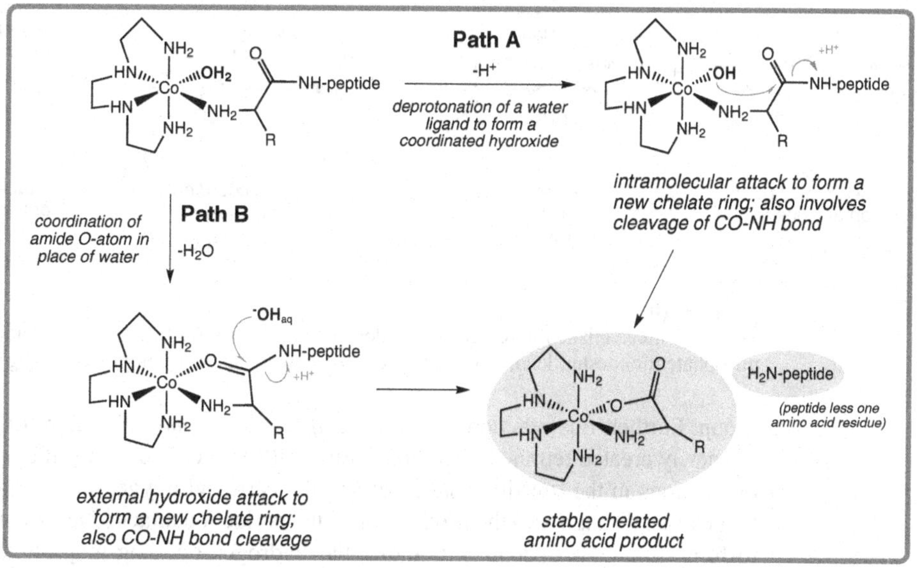

Figure 6.9
Proposed competing mechanisms for reaction of coordinated and free hydroxide with an adjacent coordinated peptide to form a new chelated amino acid ligand; the coordinated hydroxide (Path A) is more efficient.

decades commencing in the 1960s by the groups of New Zealander D.A. Buckingham and Australian A.M. Sargeson. The coordinated hydroxide ion in particular has been the subject of extensive studies. In aqueous solution, of course, free hydroxide ion is present, that can in principle compete with coordinated hydroxide ion. Buckingham studied peptide cleavage by an inert cobalt(III) complex, employing isotopic labelling to probe the origin and fate of oxygen atoms in the product. This work showed clearly that, while pathways for both internal and external hydroxide attack exist, the coordinated hydroxide is the significant player, consistent with its enhanced acidity and pre-organized location. The alternate processes proposed are illustrated in Figure 6.9.

For a coordinated $^-NH_2$ group in aqueous solution, there is no capability for competition as above that would lead to the same product, and hence the intramolecular nature of the reaction is clearly defined. The reactivity of this group is illustrated in the simple reaction with $[Co(NH_3)_5(OPO_3R)]^+$ in aqueous base (Figure 6.10). Here, the nucleophile can attack at the adjacent and relatively electron-deficient P atom, leading to a new N—P bond forming. The phosphorane intermediate formed has in effect one too many bonds around the P, relieved by a P—O bond breaking to release an alkoxide ion.

A more complicated example of a reaction featuring a coordinated $^-NR_2$ nucleophile is the reaction of N-bound aminoacetaldehyde with a coordinated polyamine (Figure 6.11). In this molecule, there are three amine groups sufficiently close (in positions adjacent or *cis* to the aldehyde) to participate in reaction, and yet the reaction occurs at only one of these three – it is therefore *regiospecific*. We discussed regiospecificity with respect to substitution reactions earlier, and it is the influence of a *trans* group that was used as the example. In this case, the regiospecificity arises in the same manner, since the *trans* chloride ion makes the secondary amine group opposite significantly more acidic than those in the other two sites, so that it is deprotonated much more readily to produce a reactive $^-NR_2$

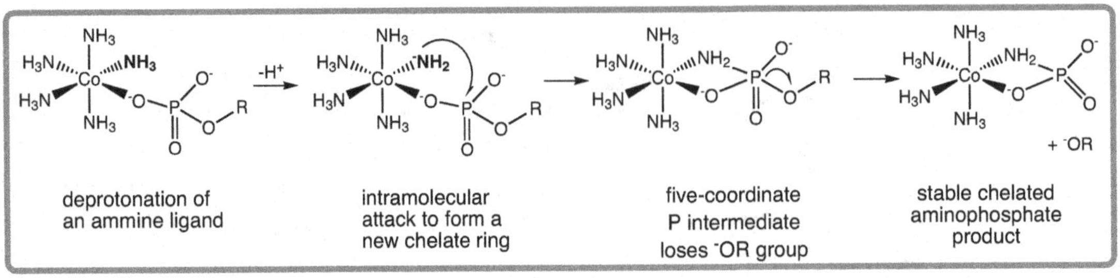

Figure 6.10
Proposed mechanism for reaction of a deprotonated ammine ligand with an adjacent coordinated phosphate monoester, leading to formation of a new chelated aminophosphate ligand.

group. Further, this reaction is *stereospecific*, since a particular chirality is introduced at the newly created tetrahedral carbon centre. This is ascribed to specific hydrogen bonding interactions in the transition state holding the carbonyl oxygen in a particular orientation that carries through into the product, once the amide ion attacks the carbon of the carbonyl to form a new N—C bond. Intramolecular hydrogen-bonding may often be important in directing stereoselectivity.

Stereospecific hydration of olefins is another reaction involving a coordinated hydroxide. This reaction with maleate monoester (Figure 6.12, top) involves essentially alkene hydration by the coordinated nucleophile, hydroxide. The reaction is *stereospecific* in terms of the *site* of reaction, as it selects five-membered ring formation exclusively over six-membered

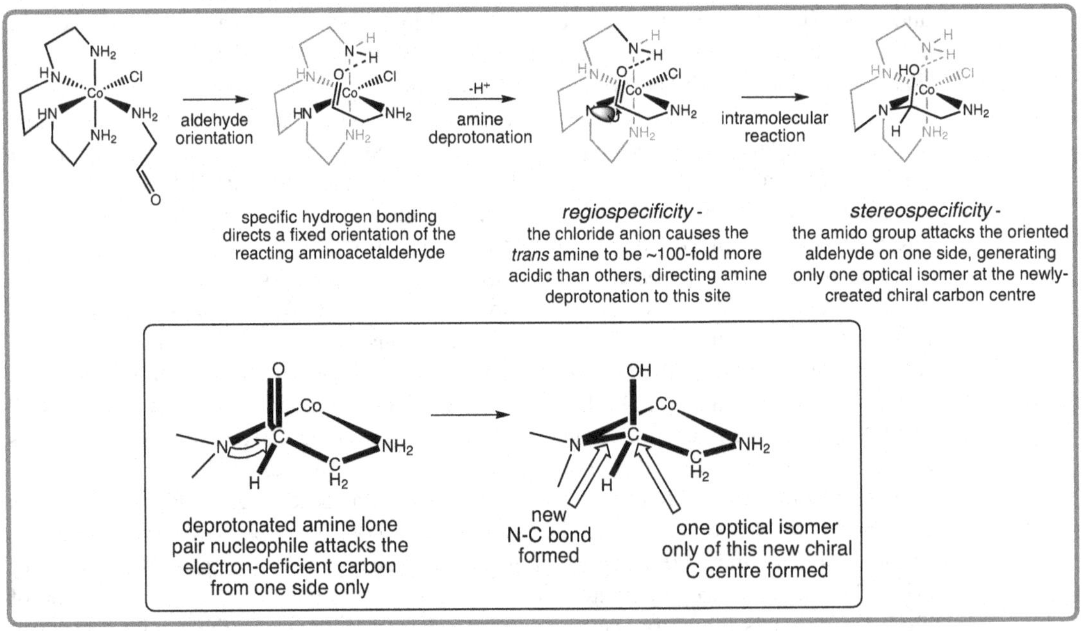

Figure 6.11
Proposed mechanism for reaction of a particular amine group (an example of *regiospecificity*) with a pendant aldehyde of a coordinated aminoaldehyde, leading to only one optical isomer of an aminol (an example of *stereospecificity*). The core reaction appears in the inside box.

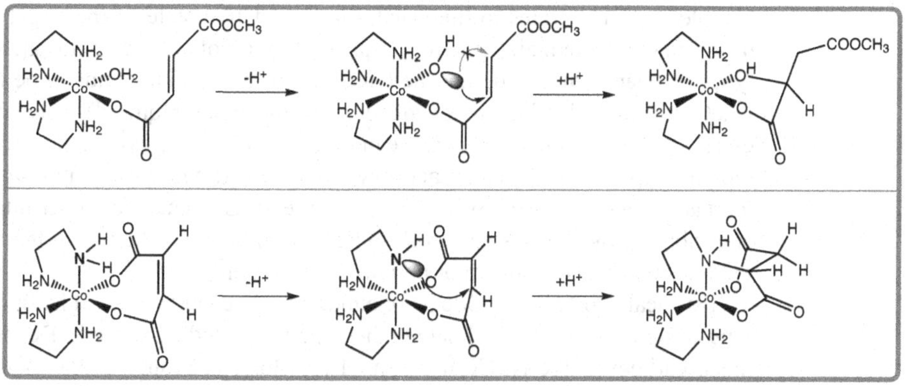

Figure 6.12
Hydration of a pendant alkene by coordinated hydroxide (top), displaying specificity in site of attack, with the five-membered chelate ring only formed, and stereospecific attack of a deprotonated amine nucleophile (bottom) at the alkene of a chelate maleate ion.

ring formation. With the transition state apparently important, it is difficult to be certain why there is exclusive formation of the five-membered ring, but which —CH= centre is more electron-deficient is likely to be involved.

Whereas the maleate monoester met above can coordinate as a monodentate ligand through the one free carboxylic acid group, the diacid can employ both acid groups to bind as a didentate chelate, forming a seven-membered chelate ring. Chelated maleate dianion bound to Co(III) can also undergo intramolecular attack, but in this case the nucleophile is a deprotonated amine group of an adjacent chelated 1,2-ethanediamine (Figure 6.12, bottom), as no coordinated hydroxide is present. The reaction has a choice of two sites for attack (at either end of the C=C), but again there is stereospecificity observed. This specificity in ring formation may arise if we require in the transition state one carboxylate to be coplanar and thus conjugated with the diene. This leads to two discrete conformations, depending on which of the two carboxylate groups is coplanar. From examining models, it appears substantially more favourable for nucleophilic addition to occur at the 'front' —CH= group, as a result of the spatial orientation of the lone pair and the preferred conformation adopted by the chelate ring leading to a closer and appropriately directed approach at that site, depicted in Figure 6.12.

It is notable that reaction does not occur between free maleate ion and free 1,2-ethanediamine, although it does occur (but only very slowly) with maleate diester and 1,2-ethanediamine. Acceleration of the reaction of the maleate resulting from coordination to a metal ion is significant, and lies in the range 10^6–10^{10}. Accelerations of this size are common for reactions of coordinated nucleophiles. The observation of stereospecificity and large accelerations suggests that these types of reactions may be relevant to the modes of action of certain metalloenzymes, where reactions are very rapid, specific and stereoselective. Such reactions are met in Chapter 8.

6.6 Organometallic Synthesis

Compounds with metal–carbon bonds usually require specialized approaches to their synthesis that differ from those discussed above for traditional Werner-type coordination complexes. This relates to the usually low oxidation state of the target compounds, that makes many of them air-sensitive (requiring the use of an inert atmosphere), the distinctive types of ligands involved, and the tendency for these compounds to be insoluble in or to react with water (leading to nonaqueous solvents being required). Solvents such as diethyl ether, tetrahydrofuran or toluene are more likely to be employed in this field. It is also common to employ sealed glass reaction vessels flushed with nitrogen or argon gas or else an inert atmosphere 'glove box', which is a large glass-fronted container fitted with portholes and rubber gloves that allow work to be carried out separated from the atmosphere in an inert gas (such as dinitrogen or helium) environment.

Historically, organometallic compounds have been known for at least as long as Werner-type complexes, with coordinated ethene first reported by Zeise in Denmark in 1827, and metal carbonyls like $Ni(CO)_4$ prepared by Mond in France in the 1890s, although it is true that the vast number of examples date from the 1950s and beyond. The two gross classes of ligands met in organometallic chemistry are: simple σ-bonded type, such as $M-CH_3$, that behave in many ways like a conventional metal–donor bond; and multi-centred π-bonded systems such as occur with an alkene that binds symmetrically side-on and involves its π-electrons in the linkage to the metal centre. This has been discussed earlier in Chapter 2.5.

As an aid to understanding the outcomes of organometallic reactions and in synthesis, there is a convenient way in which the stability of a compound can be predicted, called the 18-electron rule. In the light elements of the p block, we traditionally invoke an octet (or 8-electron) rule to probe stability. This assumes that an s and three p valence orbitals are used in bonding, and allows us to understand why CH_4 is stable whereas CH_5 is not. In d-block chemistry, it is possible to use what is in that case an 18-electron rule (using an s, three p and five d orbitals) to help predict stability of a complex. This is used mostly for organometallic complexes, and has value since it limits the number of combinations of metal and ligand that lead to the desired electron count. This concept was explained in Chapter 2 and will not be developed further here, but is invariably met in more specialized and advanced courses. It works best for low-valent metals with small neutral high-field ligands like carbonyls, but is less effective for high-valent metal ions involving weak-field ligands, and thus is not usually invoked for traditional Werner-type coordination complexes.

Metal carbonyls represent a key class of organometallic compounds, and are often a starting point for other chemistry. They tend to be monomers, dimers or small oligomers, such as $Ni(CO)_4$ and $Mn_2(CO)_{10}$, the latter involving a formal metal–metal bond. Most metal carbonyls are made by reduction of simple salts or oxides in the presence of CO, or direct reaction of CO with finely-divided metals at elevated pressure. Examples appear in (6.43) and (6.44).

$$Ni + 4\,CO \xrightarrow[\text{1 atmosphere}]{25\,°C} [Ni(CO)_4] \qquad\qquad (6.43)$$

$$Re_2O_7 + 17\,CO \xrightarrow[\text{100 atmospheres}]{200\,°C} [Re_2(CO)_{10}] + 7\,CO_2 \qquad (6.44)$$

The carbonyl ligands are able to be substituted by some other ligands, or else the coordination sphere can be expanded by addition reactions with other compounds. It is also straightforward to prepare compounds that incorporate carbonyl and other ligands such as amine ligands. This can sometimes be achieved directly, such as in (6.45).

$$Cu^{I}Cl + en + CO \longrightarrow [CuCl(CO)(en)] \qquad \qquad (6.45)$$

The product (with en = 1,2-ethanediamine chelated) can be isolated readily. The transition metal carbonyls are usually far more robust than those formed by main group elements; for example $H_3B(CO)$ decomposes below room temperature, whereas $Cr(CO)_6$ can be sublimed without decomposition.

An early example of organometallic synthesis was the reaction of $[PtCl_4]^{2-}$ with ethylene in dilute aqueous hydrochloric acid solution, which yields the classical π-bonded orange complex $[Pt(C_2H_4)Cl_3]^-$, where the Pt and two carbon atoms form an equilateral triangle (or in other words, the ethene is bound symmetrically side-on to the platinum(II) centre). A vast range of alkene complexes has been reported subsequently, including the tris(η^2-ethylene)nickel(0) which features three side-on ethylene molecules arranged in a trigonal pattern around the metal atom. Bonding in these complexes requires molecular orbital theory for a satisfactory explanation. This assigns σ-bonded character through a filled π molecular orbital of the alkene overlapping symmetrically with an empty metal orbital, and π-bonded character through overlap of an empty π^* molecular orbital on the alkene with a filled metal d orbital.

Another important class of compounds are the so-called 'sandwich' compounds, featuring a metal bound (or 'sandwiched') between two flat aromatic anions, the best known of which is the cyclopentadienyl ion ($C_5H_5^-$). These compounds can be prepared by reactions such as (6.46), performed in a nonaqueous solvent such as diethyl ether.

$$FeCl_2 + 2 Na(C_5H_5) \longrightarrow Fe(C_5H_5)_2 \qquad + 2 NaCl \qquad (6.46)$$

This compound is robust – it is air stable, able to be sublimed without decomposition, and resistant to strong acids and bases. It can undergo reversible one-electron reduction chemically or electrochemically, with a significant change in colour from yellow to blue that strongly suggests that it is a metal-centred reduction. An array of other related compounds are known, including those featuring larger aromatic ring systems, such as $Cr(C_6H_6)_2$. Examples with metals from most parts of the Periodic Table, even f-block elements, are now established.

Metal–alkyl σ-bonded compounds can be formed conveniently by reactions employing alkyl halides or alkyl magnesium bromides, such as example (6.47), performed in diethyl ether.

$$[PtBr_2(PR_3)_2] + 2 CH_3MgBr \longrightarrow [Pt(CH_3)_2(PR_3)_2] \qquad + 2 MgBr_2 \qquad (6.47)$$

These reactions are facilitated by the presence of spectator ligands such as phosphines and carbonyls. Phosphines, common co-ligands in organometallic compounds, are usually conveniently introduced through direct reaction with metal halides.

Organometallic compounds also undergo reactions of coordinated ligands readily. A simple example involves the susceptibility of coordinated carbon monoxide towards nucleophilic attack. The $[Mo(CO)_6]$ complex reacts with methyllithium (6.48), with the new ligand produced at one site also able to undergo additional reactions, not described here.

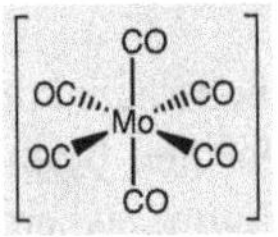

$$[Mo(CO)_6] + LiCH_3 \longrightarrow Li[Mo(CH_3CO)(CO)_5] \qquad (6.48)$$

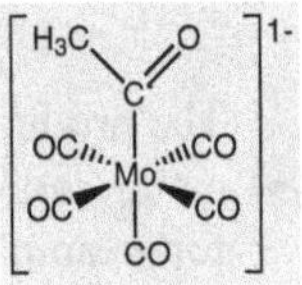

A reaction with a similar outcome involves migration of one ligand to attack another adjacent ligand (a 1,1-migratory insertion), promoted by the availability of another ligand to occupy the site vacated by the first movement. In example (6.49) below, an initial $R-M-C{\equiv}O$ component of the molecule is converted to an $-M-C(R){=}O$ component, leaving a vacant coordination site, filled by an added phosphine ligand in this case.

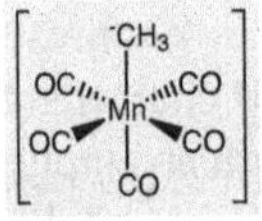

$$[Mn(CH_3)(CO)_5] + PR_3 \longrightarrow [Mn(CH_3CO)(CO)_4(PR_3)] \qquad (6.49)$$

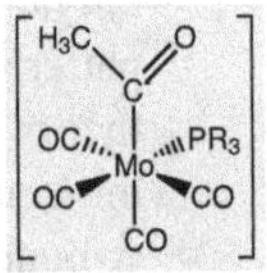

The hydride ion (H^-) is an efficient small ligand in organometallic chemistry. The first transition metal hydrides were prepared using the Hieber base reaction, exemplified in (6.50). The hydroxide adds to the carbon of one CO ligand to produce an intermediate that rapidly loses carbon dioxide, leaving the hydride ion to occupy the coordination site.

$$[Fe(CO)_5] + {}^-OH \longrightarrow [Fe(CO)_4(COOH)]^{\#} \longrightarrow [Fe(CO)_4(H)] + CO_2$$

intermediate

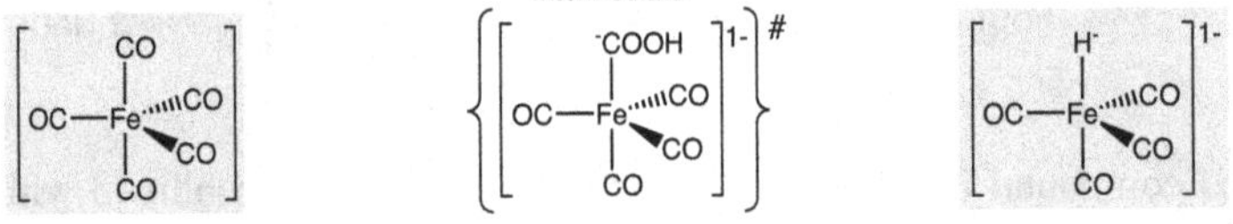

$$(6.50)$$

These are but a few examples of an array of reactions available to organometallic systems. Clearly, the level of difficulty is greater in performing many organometallic reactions compared with Werner-type coordination chemistry reactions through the special equipment that must be employed because air and/or protic solvents can lead to unwanted reactions, although the reactions themselves do not necessarily impose any greater inherent complexity. The storage and handling of products may pose problems, however, due to their reactivity, particularly in redox reactions. A detailed examination of their chemistry may be pursued through advanced and/or specialist texts.

Concept Keys

The position of a metal in the Periodic Table has an impact on the type of chemistry that element will undergo, and the complexes that can be readily synthesized.

The majority of complexes are prepared through reactions involving ligand substitution in the coordination shell.

Syntheses typically involve chemistry performed in aqueous or nonaqueous solution; however, solvent-free reactions can sometimes be employed.

Reactions may involve a deliberate change in the oxidation state of the central atom; this is particularly the case where the target complex is in a very inert oxidation state, for which prior ligand substitution in a more labile oxidation state facilitates synthesis.

Substitution reactions of low-valent complexes (such as organometallic complexes), in which the central metal is often sensitive to oxidation by air, must be performed in apparatus that ensures oxygen-free conditions.

Coordinated ligands themselves may in some cases be able to undergo chemistry while bound to a central metal.

Reactivity of a coordinated ligand can change significantly due to polarization effects that alter its acidity; this is exemplified by the acidity of coordinated water rising markedly on binding to a metal ion, the deprotonated hydroxide form being an efficient nucleophile in reactions.

Metal-catalysed reactions are accelerated processes that occur by a ligand binding and reacting to form a product that subsequently departs, leaving a regenerated complex which is able to repeat the process.

Metal-directed reactions involve the metal promoting chemistry in its coordination sphere that produces a new molecule to which the metal remains bound; thus the process occurs only once, and stoichiometric amounts of the central metal are involved.

Spontaneous self-assembly of larger molecules from smaller components can result from a metal acting as an organizer or director of components leading to reaction.

Reactions in the coordination sphere of a complex involving chelate formation may yield a sole isomer (a stereospecific reaction) or at least a preferred isomer (a stereoselective reaction).

Further Reading

Bates, R. (2000) *Organic Synthesis Using Transition Metals*, Wiley-Blackwell, Oxford, UK. Describes how transition metals can act to catalyse or direct organic reactions, with some metal-directed reactions of relevance; but more for the advanced student.

Constable, Edwin C. (1996) *Metals and Ligand Reactivity*, Wiley-VCH Verlag GmbH, Weinheim, Germany. This is a clearly written and well-illustrated text for the more advanced student, which contains some good descriptions of metal-directed reactions.

Cotton, S. (2006) *Lanthanide and Actinide Chemistry*, John Wiley & Sons, Inc., Hoboken, USA. This is one of few introductory student-focussed texts on the f-block elements, covering the field from isolation of elements through to physical properties of complexes.

Davies, J.A., Hockensmith, C.M. and Kukushin, V. Yu. (1996) *Synthetic Coordination Chemistry: Principles and Practice*, World Scientific Publishing Co. Inc., Hackensack, USA. This is a fairly comprehensive text, which an advanced student may find valuable for elucidating practical techniques and their context in experiments.

Garnovskii, A.D. and Kharisov, B.I. (2003) *Synthetic Coordination and Organometallic Chemistry*, CRC Press. A deep coverage of methods across the field, with examples; for advanced students.

Girolami, Gregory S., Rauchfuss, Thomas B. and Angelici, Robert J. (1999) *Synthesis and Technique in Inorganic Chemistry: A Laboratory Manual*, 3rd edn, University Science Books. A popular coverage of not just experiments but experimental methods, and the student may find the latter part in particular worth a read.

Hanzlik, R.P. (1976) *Inorganic Aspects of Biological and Organic Chemistry*, Academic Press, New York, USA. An ageing but still readable account of the role of metals in organic and biochemical systems; a clear account of fundamentals of metal-directed organic reactions is a useful section.

Housecroft, C.E. (1999) *The Heavier d-Block Metals: Aspects of Inorganic and Coordination Chemistry*, Oxford University Press, Oxford, UK. One of few textbooks devoted to exemplifying differences between the coordination chemistry of different rows of the d block.

Komiya, S. (ed.) (1997) *Synthesis of Organometallic Compounds: A Practical Guide*, John Wiley & Sons, Inc., New York, USA. Apart from surveying metals and their organometallic compounds and introducing important synthetic approaches in this field, detailed synthetic protocols for key reactions are given. More for the graduate student, but undergraduates may find some parts valuable.

Marusak, R.A., Doan, K. and Cummings, S.D. (2007) *Integrated Approach to Coordination Chemistry: An Inorganic Laboratory Guide*, John Wiley & Sons, Inc., New York, USA. An accessible and useful set of experiments in coordination chemistry; more for the instructor, but may assist students with understanding reactions.

Woollins, J.D. (ed.) (2003) *Inorganic Experiments*, 2nd edn, Wiley-VCH Verlag GmbH, Weinheim, Germany. Details of over 80 tested laboratory experiments in inorganic chemistry, including examples of coordination and organometallic chemistry; more for the teacher than student, but a useful resource of examples of reactions.

7 Properties

7.1 Finding Ways to Make Complexes Talk – Investigative Methods

Molecules are not inherently endowed with the capacity to communicate their characteristics. To find out about their structures and properties, we must develop ways to interrogate them efficiently and effectively – we must make them talk, or reveal their character. Working at the atomic and molecular level, this involves the application of an array of chemical and physical methods which in concert can tell us about the compound we have selected for examination. Our first task is straightforward – just what do we want to know? This may seem a simple question, but is not, since the answer will vary with the scientist. One person may simply want to know what elements a compound contains, which relates to the chemical composition; another may want to know what shape the compound takes, which is a far more complex question involving defining what components are involved and how they are assembled overall. More sophisticated questions usually require more sophisticated instrumentation to provide the answers.

It has been suggested that it is easier to make a new molecule than to discover exactly what it is you've made. In synthesis, the difficulty usually lies not in the execution, but in the separation, isolation and identification of products. This is particularly so with metal complexes, where the often large array of options for products, and their capacity to sometimes undergo further reactions in the process of separation, makes life difficult for coordination chemists. Further, species that exist as dominant components in solution may not be the same as the dominant species isolated in the solid state. All this means that defining molecular structure in both solution and the solid state requires a call on a wide range of physical methods for characterization.

The presence of a metal ion in a coordination complex provides a centre of attention for probing the properties of the assembly and for making use of special properties in diverse applications. Unfortunately, the metal also introduces complications that require the use of a range of sophisticated approaches to interrogate the complex and thus determine aspects of both the structure and properties of the complex formed. Our focus in this chapter will be restricted to a few modest techniques; in particular, ones that interact with our limited models of bonding that have been developed. Thus we will probe electronic spectroscopy and magnetic properties in part and briefly examine a few other key techniques, while leaving many other aspects untouched. However, it is inappropriate to explore these few methods in ignorance of others, so we will begin by at least identifying methods that may be of use to the coordination chemist.

Introduction to Coordination Chemistry   Geoffrey A. Lawrance
© 2010 John Wiley & Sons, Ltd

7.2 Getting Physical – Methods and Outcomes

A surprisingly large array of physical methods has been applied to metal complexes in order to provide information on molecular structure, stereochemistry, properties and reactivity. While characterization may be tackled by first isolating a complex in the solid state, this approach is complicated by differences that sometimes exist between the character of a species in solution and the solid state. In fact, it may be inappropriate to assume that the isolated solid state species even persists in solution. Thus, it may be necessary to define species in different environments. Fortunately, the vast array of chemical and instrumental methods now available makes the task of defining structure achievable, but what to choose and how a method assists definition is not always obvious.

It is not the task of an introductory text to pursue this issue to exhaustion, but it seems essential to at least acquaint you with the basic and advanced techniques and what they can achieve. This has been produced in summary form in Tables 7.1–7.4.

There are others ways of classifying the techniques that you may meet. For example, sometimes you will meet physical methods classified as sub-classes related to the core mode of examination, such as resonance techniques (NMR, ESR and Mossbauer spectroscopy), absorption spectroscopy (UV-Vis, IR and Raman), ionization-driven techniques (MS and photoelectron spectroscopy) and diffraction methods (XRD, neutron diffraction). The list in the tables is extensive, but in effect most coordination chemists employ a limited number of key techniques. While each chemist's key list will vary depending on the type of compounds worked with, it is generally true to list elemental autoanalysis, UV-Vis spectrophotometry, IR spectroscopy, NMR spectroscopy, single crystal XRD and MS as commonly employed techniques. These widely available methods are augmented by several of the vast array of others, depending on the task and target of study. Cost and availability of instruments is an issue; for example, a modern high-field NMR spectrometer may cost 100 times the price of a good quality UV-Vis spectrophotometer. The number of some types of specialist instruments world-wide may be quite small, so users may need to travel sometimes long distances to these to perform experiments.

Of course, the first task in any study is acquisition of a pure compound. For coordination chemistry, this is often a non-trivial task, as several stable compounds may form during a synthesis. Separation, by chromatography and/or crystallization, is often required (Table 7.1). Crystal growth through slow solvent evaporation, dilution with another solvent, lowering temperature or addition of other counter-ions is a common task in coordination chemistry, although it is now possible to define compounds reasonably well in solution.

Spectrometers, as a broad class, provide enhanced levels of information about coordination complexes that may not be accessible using basic techniques (Table 7.2). For example, for the complex $[Co(NH_3)_5(OOCCH_3)]Cl_2$, conductivity experiments may allow us to identify the complex as ionic, and even a 2+ cation, but little else, whereas elemental analysis my define the elemental ratio as $C_2H_{18}N_5O_2Cl_2Co$, but says essentially nothing about the actual structure of the complex or its ligands. However, a technique like IR-spectroscopy will produce vibrations consistent with the presence of ammonia and acetate ligands, UV–Vis spectroscopy will produce a spectrum consistent with Co(III) and in a CoN_5O donor environment, and NMR spectroscopy will not only confirm that we are dealing with a diamagnetic complex but also provide resonances assignable to five ammonia molecules and one acetate anion. It is often the combination of a set of experiments like the above that leads to firm definition of the complex form.

Table 7.1 Key physical methods available for separation and isolation of complexes and for determining basic information about the isolated compound.

Method	Sample and *device* requirements	Outcome expected
Separation Techniques		
Crystallization	A solution of either the pure complex or a mixture of complex species.	Selective crystallization of a pure complex; this may, but need not, follow chromatographic separation of solution species.
Ion chromatography	Solutions of soluble ionic complexes. *Ion chromatography columns packed with appropriate cationic or anionic polymer resin.*	Separation of ions mainly according to charge, and possibly assignment of charge. (Examination of separated bands of compounds directly as they exit the column by tandem instrumental methods is also possible.)
High pressure liquid chromatography (HPLC)	Liquids or solutions. *Commercial HPLC instrument with appropriate packed columns.*	Separation of neutral and/or ionic species.
Gas chromatography	Gaseous or volatile liquid samples. *Gas chromatograph instrument with capillary or packed columns.*	Separation of volatile samples, usually applicable only to some neutral low molecular weight complexes. Identification by comparison with known standards is possible.
Basic Analytical Procedures		
Elemental analysis	Pure compound, as liquid, solid or solution of known concentration. *Elemental autoanalyser (for C, H, N, S, O only usually)* or else *atomic absorption (AAS)* or *atomic emission (ICP-AES) spectrometers (for other elements).*	Percentage composition of elements determined, allowing component elements and the empirical formula to be defined.
Thermal analysis techniques [Thermogravimetric analysis, differential thermal analysis]	Solid or liquid sample. *Thermogravimetric analyser.*	Mass change with temperature; information on number of waters of crystallization, ligand loss and complex transformation with increasing temperature obtained.
Conductivity	A solution of a pure compound of known concentration. *Conductivity meter and probe.*	Ionic or neutral character of complex, and possibly the overall charge of the complex ion.
Magnetic measurements	Solid or concentrated solution. *Magnetobalance or magnetometer.* (A limited *NMR-based method* is also available.)	Defines dia- or para-magnetism; number of unpaired electrons, and some inferences about gross symmetry possible. Variable-temperature behaviour provides information on metal–metal interactions in polymetallic species.

Table 7.2 Spectroscopic and spectrometric methods available for providing information on complex formation, speciation and molecular structure.

Method	Sample and *device* requirements	Outcome expected
Spectroscopic methods		
Nuclear magnetic resonance spectroscopy (NMR) (^{1}H, ^{13}C, and most other nucleii)	Liquid or a solution in a deuterated solvent; solids can be examined. Usually for diamagnetic compounds, but paramagnetics can be examined. *Fourier-transform multinuclear NMR spectrophotometer.*	Ligand and complex environment and symmetry. Molecular structure in solution; highly detailed three-dimensional shape in solution sometimes obtainable.
Electron spin resonance (ESR; also called EPR) spectroscopy	Liquids, solids or solutions. Paramagnetic compounds. *ESR spectrometer (usually with a variable-temperature facility).*	Information on unpaired electrons, complex geometry and metal–donor environment and symmetry.
Nuclear quadrupole resonance spectroscopy (NQRS)	Solid samples. *NQR spectrometer.*	Identifies different environments for the target nucleus (particularly halides) in the crystalline lattice.
Mössbauer spectrometry	Solids or liquids; restricted to certain elements (Fe, Sn, Ru, Te). *Mössbauer instrument with γ source.*	Information on the oxidation state, and the ligand environment and symmetry around the target nucleus.
Magnetic circular dichroism (MCD)	Solid or frozen solution. *MCD instrument with cryostat for low temperature measurement.*	Symmetry and electronic structure.
Ultraviolet–visible–near infrared spectrophotometry (UV–Vis–NIR)	Usually uses solutions of known concentrations; single crystals and powders (the latter by diffuse reflectance) can be examined. *UV-Vis or UV-Vis-NIR spectrophotometer.*	Approximate complex geometry, possible ligand environment and symmetry.
Circular dichroism (CD) and optical rotary dispersion (ORD) spectroscopy	Solution of optically resolved chiral compound that absorbs in the visible and/or UV region. *CD or ORD spectrophotometer.*	Identification of the optical isomer and information on symmetry and stereochemistry.
Infrared (IR) and Raman spectroscopy	A gaseous, liquid or solid sample; solutions can be handled by subtraction of the solvent spectrum, with aqueous solution acceptable if using Raman. *Fourier-transform IR or Raman spectrophotometer.*	For simple compounds, molecular shape based on symmetry about the central atom. Functional group and ligand identification.
X-ray absorption spectroscopy	Solids usually. *X-ray diffractometer.*	Information on the molecular environment of the metal.

Table 7.2　(*Continued*)

Method	Sample and *device* requirements	Outcome expected
X-Ray absorption fine-edge spectroscopy (EXAFS) and X-ray absorption near-edge spectroscopy (XANES)	Solids or solutions; analysis of metals in biological samples possible. *X-ray diffractometer / synchrotron.*	Information on the environment around the metal; type and number of donor groups and metal bond distances.
Emission spectroscopy	Gaseous, liquid, solid or (preferably) solution of complex. *Fluorescence spectrophotometer.*	Luminescence or phosphorescence behaviour of complexes; information on electronic structure.
Microwave spectroscopy	Gaseous sample with a dipole moment necessary. *Microwave spectrometer.*	Some structural information, including bond distances for very simple molecular species.
Photoelectron spectroscopy	Solid or gaseous samples. *PES spectrophotometer.*	Information on molecular environment and donor type.
Mass spectrometry (MS)	A liquid or solid of reasonable volatility, or a gaseous sample. *Quadrupole or time-of-flight (TOF) mass spectrometer.*	Molecular mass. Compound characterization from fragmentation pattern.
Tandem techniques (general)	Where two (or even more) instruments or methodologies are joined together to allow for expansion of the analysis and collected information.	Enhanced information through the combination of methods. Detection of short-lived species can be achieved.
Tandem techniques (example): Electrospray ionization mass spectrometry (ESI-MS)	Aqueous solution of a complex (some other solvents may be used). *Tandem ESI-MS instrument.*	Molecular mass of ionic compounds. Possible identification of solution species.

Where a coordination complex is isolable as a solid, and particularly as a crystalline solid, additional methods are available to assign structure (Table 7.3), up to a detailed picture of the three-dimensional structure including accurate bond distances and angles obtained by diffraction methods. Where non-crystalline samples or solids deposited on surfaces are obtained, alternate methods can probe structure. Deliberate formation of extended solids of particular shapes defined by the way metal ions and selected ligand components self-assemble is the realm of materials science, an area of exceptional growth and promise that is taking coordination chemistry into new frontiers. An array of different physical methods is now available for investigation of such species.

The fact that metal ions in complexes often have the ability to undergo oxidation and/or reduction, with the resultant complexes having distinctly different properties as a result of the change in the metal d-electron set, means that techniques able to probe these processes have developed (Table 7.4). In coordination chemistry, the technique of cyclic voltammetry (sometimes called 'electrochemical spectroscopy' because of its capacity to rapidly probe behaviour in different oxidation states in simple solution experiments) is now commonly employed. Thermodynamic properties, such as reaction enthalpy and complex stability

Table 7.3 Spectroscopic and computational methods available for defining the three-dimensional solid state molecular structure.

Method	Sample and *device* requirements	Outcome expected
Three-dimensional Molecular and Solid State Structure		
Electron diffraction	Usually a gaseous sample; liquids and solids can be studied. *Electron diffractometer.*	Complete structural information provided the molecule is not large.
X-ray diffraction (XRD)	Crystalline sample; single crystals for full structure information. *CCD–XRD diffractometer.*	Complete three-dimensional structure, often with distances to ~0.1 pm and angle to ~0.01°.
Neutron diffraction	Crystalline sample; single crystals for full structure information. *Nuclear reactor or synchrotron with a high neutron flux beam line, and a diffractometer.*	Complete three-dimensional structure, with better definition of light atoms and better distinction between atoms of similar atomic number than achieved by XRD.
Atomic force microscopy (AFM)	Liquids or solutions, as films on a substrate. *AFM instrument.*	A view of molecular assembly on a surface, down to the individual molecule level.
Scanning electron microscopy (SEM)	Solids. *SEM instrument; samples usually gold-coated to enable imaging.*	Morphology of the sample, imaged down to near molecular level.
Positron annihilation lifetime spectroscopy (PALS)	Solids (particular nanomaterials). *Positron-emitting radioisotope and gamma ray detector system.*	Determination of electronic density and pore size, architecture and connectedness in solids.
Transmission electron microscopy (TEM)	Solids. *TEM instrument.*	Morphology of the sample. Imaging down to molecular and even atomic level possible.
Molecular modelling (MM)	*Computer software only.* Various levels and sophistication of analysis (classical MM, DFT, and semi-empirical MO) are now available.	Predictive tools for three-dimensional structures and some properties of complexes.

constants, can be routinely determined. Kinetic properties, such as rate of reactions that transform a complex from one species into another, and the mechanisms through which this occurs, can be sought employing modern equipment that measures change in a physical property such as colour or even conductivity. As distinct from the early days of coordination chemistry, what is obvious nowadays is that it is more the inventiveness and choice selection of chemists that limits our achieving clear structural assignment rather than a dearth of methods to use as probes.

7.3 Probing the Life of Complexes – Using Physical Methods

The behaviour of life on Earth that we can see with our eyes can be probed by observation and recorded using cameras. Move down from the macroscopic level through the microscopic level to the molecular level, and we eventually lose our ability to 'see' the

Table 7.4   Electrochemical and thermodynamic/kinetic analysis techniques.

Method	Sample and *device* requirements	Outcome expected
Electrochemical Methods		
Electrochemistry: voltammetry (particularly cyclic voltammetry)	Solution (aqueous or non-aqueous) of compound. *Potentiostat and electrochemistry cell.*	Metal-centred and ligand redox behaviour. Potentials and reversibility of reduction and oxidation processes. Information on kinetics of electron transfer and any decomposition processes.
Electrochemistry: coulometry	Solution (aqueous or non-aqueous) of compound. *Potentiostat/galvanostat and coulometry cell.*	Number of electrons in a particular redox process.
Kinetic and Thermodynamic Analysis Techniques		
Potentiometric and spectrophotometric titration	Solutions of a metal ion and of a pure ligand. *Auto-titration assembly and detector system (pH and/or UV-Vis spectrum).*	Metal–ligand speciation in solution, and actual complex thermodynamic formation constants.
Stopped-flow and conventional kinetics	Solutions of a known complex. *Stopped-flow instrument with UV-Vis detector (for fast reactions), or standard UV–Vis spectrophotometer; temperature-controlled cell holder.*	Information on mechanisms of reactions via measurement of reaction kinetics; rates of particular reactions and activation parameters.
Calorimetry	Solutions of a metal ion and a pure ligand. *Calorimetry cell with temperature change detection electronics.*	Heat of reaction. Information on complexation thermodynamics.

species directly under most circumstances. Thus to probe the life of complexes, or their behaviour as molecular species, we must resort to far more elaborate instrument-based techniques. Two of the most accessible spectroscopic techniques for chemists, because instrumentation is relatively cheap and readily accessible in most laboratories, are UV–Vis or electronic spectrophotometry and IR or vibrational spectroscopy. Of the more expensive instrumentation available, NMR is perhaps the most commonly employed. For both electronic and vibrational spectroscopy, the observation of a molecule happens extremely rapidly ($<10^{-11}$ s), so we get an 'instant' view, and any rearrangements are too slow to be seen or influence outcomes. For NMR, timescales are of the order of 10^{-2}–10^{-5} s, so that rearrangements where the activation energy is small (as in some fluxional molecules) may be probed, such as by varying temperature. We call these aspects the spectroscopic timeframe of the techniques; each method will have one, which influences the tasks it can perform somewhat.

Whereas IR spectroscopy is often employed in coordination chemistry to identify functional groups or ligand types, and thus differs little from its application in organic chemistry, UV-Vis spectroscopy for coordination complexes relies heavily on the special theories (crystal and ligand field theory) developed for these complexes. Thus, it will be a target for particular attention here. Another physical method closely tied up with these theories is magnetochemistry, which is again experimentally simple, and also the subject of attention below.

7.3.1 Peak Performance – Illustrating Selected Physical Methods

There are a number of techniques that rely on applying specific protocols that lead to excitation and a resonance condition, reported experimentally as the appearance of a peak(s) in a spectrum by using appropriate instrumentation. It is beyond the scope of this textbook to develop the theory behind these many methods, but it is appropriate to illustrate two of the most common techniques – IR and NMR spectroscopy.

NMR relies on change in nuclear spin through absorption of energy in the MHz range, where each atom of a particular element in a compound in a unique molecular environment will yield a signal in a unique position, reported as a chemical shift (δ). Moreover, atoms may interact with close neighbours in many cases, to produce a splitting pattern superimposed on the gross chemical shift that is characteristic of this neighbouring environment. This makes the technique powerful as a weapon in determining structure, particularly in solution. The majority of elements can produce an NMR spectrum, but the traditional element targeted is hydrogen, because ^1H is of high isotopic abundance, has simple spin characteristics, and the highest sensitivity. Another popular element used, ^{13}C, has a relative sensitivity of only $\sim 1.6\%$ compared with ^1H, and a significantly lower isotopic abundance (only $\sim 1\%$), so that its detection is consequently more demanding. Even the metals in complexes can be examined directly, but suffer from low sensitivities and/or inappropriate isotopic abundance, as well as other limitations that can lead to very broad peaks and extreme chemical shifts. Nevertheless, modern NMR instruments offer adequate to excellent determination of spectra of a vast number of elements, although ^1H, ^{13}C, ^{19}F and ^{31}P remain most commonly available in commercial instruments, and dominantly the former two are used. This means that, for coordination complexes, it is usually the organic ligands that are being probed by this technique rather than the metal or even the ligand heteroatoms. The MRI instrument used for whole-body scanning in medicine is a type of NMR instrument, but focussed on variation of the properties of water molecules in different environments as the basis of its operation.

To illustrate the NMR technique very simply, we shall draw on some simple complexes. The NMR method is most applicable to diamagnetic metal complexes, and we shall restrict examples to low-spin d^6 Co(III), which has no unpaired d electrons and is therefore diamagnetic. To remove strong signals from the solvent H atoms, spectra are routinely measured in a deuterated solvent in which D atoms replace all H atoms; further, by using an aprotic solvent like CD_3CN, any H/D exchange issues met for molecules with readily exchangeable centres like —NH— and —OH in the common NMR solvent, D_2O, are removed. If we consider highly symmetrical $[Co(NH_3)_6]^{3+}$, all six ammonia ligands are in equivalent environments, and so the ^1H NMR spectrum should yield a single peak with one specific chemical shift. If we turn to $[CoCl(NH_3)_5]^{2+}$, the four ammonia ligands around the plane of the metal are equivalent, but the one *trans* to the chloride ion is unique, so two peaks of different chemical shifts should result, in a ratio of 4:1. For $[CoCl_2(NH_3)_4]^+$, the *trans* isomer has all four ammonia ligands equivalent, and thus one peak, whereas the *cis* isomer has two different types, two opposite chloride ions and two opposite other ammonia molecules, so two peaks in a ratio of 1:1 will result (Figure 7.1). For *fac*- and *mer*-$[CoCl_3(NH_3)_3]$, the former will yield a single peak, the latter two peaks in a 2:1 ratio. Thus you may see how molecular symmetry is being defined by the NMR pattern. If we were to replace the NH_3 ligands by CH_3NH_2 ligands, we would get two sets of peaks

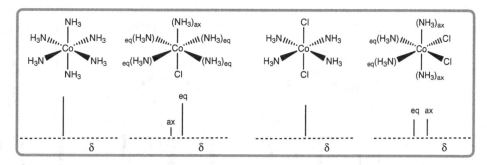

Figure 7.1
Molecular shapes and ^{1}H chemical shift patterns for chloro-ammine cobalt(III) compounds. Different environments for ammonia molecules in some complexes are defined by *eq* and *ax* subscripts, with the two types opposite different types of ligand leading to different magnetic environments and chemical shifts (δ).

in different chemical shift regions due to H_N and H_C types of protons, but the patterns for each in the absence of any coupling between centres would be identical. Were we to record the ^{13}C NMR spectra (rather than ^{1}H) of the coordination complexes, the pattern would remain the same, but in this case report the different environments of just the carbon centres, spanning a different chemical shift range ($\sim$200 ppm) than observed for protons ($\sim$10 ppm). For coordination complexes including organic molecules as ligands, the ^{1}H and ^{13}C NMR spectra in combination present a powerful technique for probing complex stereochemistry and ligand character. In a simple sense, every unique carbon will have a unique chemical shift, so that a simple count of peaks (discounting any fine structure from coupling to any neighbouring protons) provides a good start to structural assignment, drawing on symmetry considerations to assist. The NMR technique is highly advanced, with modern spectrometers offering a range of methodologies designed to assist in structural elucidation; many of these are more often met in organic chemistry.

Infrared spectroscopy is based on absorption of IR radiation associated with molecular vibrations, such as bond stretching and bond angle deformation. Again, these specific processes lead to resonances and tied absorbance peaks in specific positions characteristic of the molecule or its components. IR spectra in modern Fourier-transform instruments may be reported as absorbance spectra or the inverse transmittance spectra. For simple inorganic complexes, the molecular symmetry allows definition of expected vibrations based on the whole molecule. For more complex molecules, it is the parts of the whole, particularly organic functional groups, that are more likely to be detected and identified. For example, $[Co(NH_3)_6]^{3+}$, of O_h symmetry, and $[CoCl(NH_3)_5]^{2+}$, of C_{4v} symmetry, will produce inherently different spectra, but it is really the peaks associated with motions of the ammonia molecules themselves (asymmetric and symmetric stretching, twisting, scissoring, rotating and wagging) that dominate and are reported in the region accessible in most IR spectrometers (4000–400 cm^{-1}). When nitrite ion is introduced as a ligand, it can be O-bound or N-bound, and the Co—O—N=O isomer is of different bonding mode and symmetry to the Co—NO$_2$ isomer, and thus not surprisingly produces a different IR spectrum (Figure 7.2). Once more, as for NMR, it is the molecular components present as ligands that are commonly probed in this technique. Since the technique relies on vibrations, single atoms or ions (like Cl$^-$) cannot themselves give an IR band; however, vibrations involving the Co—Cl bond will occur (though below 500 cm^{-1}, well away from most

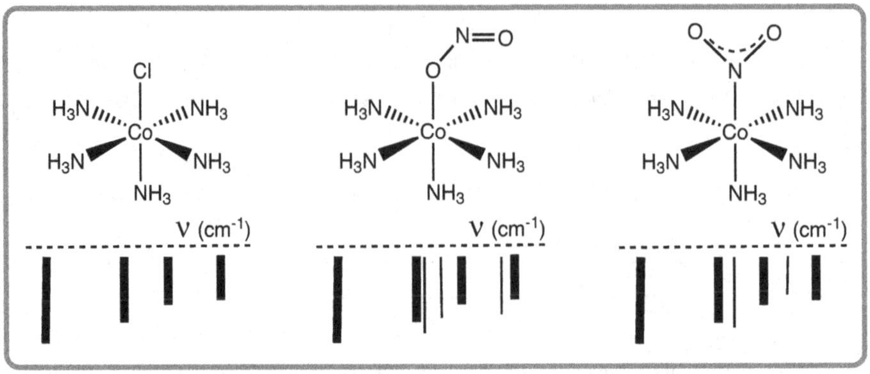

Figure 7.2
Shapes and IR spectra patterns for chloro, *O*-nitrito and *N*-nitrito cobalt(III) ammine compounds. The thick lines represent the vibrations associated with the coordinated ammonia molecule, whereas the thin lines represent the vibrations associated with the linkage isomer of nitrite (two and three bands for N- and O-bound isomers respectively).

ligand vibrations). Basically, the heavier the atomic masses of two atoms joined in a bond, the higher energy will be required to cause a stretching motion, for example. This means that the position of vibrational bands reflects the type of atom (particularly atomic mass) involved in a bond, so, for example, O—H, N—H and C—H stretching vibrations will occur at slightly different positions.

Infrared spectroscopy can also provide information about ligand binding character. This is readily illustrated with carbon monoxide as a ligand, since upon coordination to a transition metal the CO bond is weakened, and this is seen as a shift in the position of the stretching vibration (Table 7.5). Two sets of data illustrate the effect: on the left, the effect of increasing the number of metals of one type to which the CO is bound is seen, with the bond becoming progressively weaker as the number of metals bound rises; on the right, the effect of increasing the formal charge on the central metal is shown, with again a relationship between metal oxidation state and coordinated ligand bond strength apparent.

Electrospray ionization mass spectrometry (ESI—MS is a tandem technique; it employs an electrospray ionization device (ESI) to produce 'bare' ions in the gas phase from a supplied (usually aqueous) solution, and supplies these ions as a very dilute gaseous stream to the high vacuum chamber of a mass spectrometer, which is the analyser. This analyser 'sorts' and detects ions by mass/charge ratio (m/z), as cations or as anions, and will report either on request. For coordination complexes, the ESI process can be sufficiently 'soft'

Table 7.5   The variation in the IR stretching vibration of CO as a result of complexation.

Species	IR resonance (cm^{-1})	Species	IR resonance (cm^{-1})
free CO (not coordinated)	2143	free CO (not coordinated)	2143
M(CO) (monodentate-bound to a single metal)	2120–1850	$[Ni(CO)_4]$ (bound to a M(0) centre)	2060
$M_2(\mu^2$-CO) (bridged to two metal centres at once)	1850–1750	$[Co(CO)_4]^-$ (bound to a M(-I) centre)	1890
$M_3(\mu^3$-CO) (bridged to three metal centres at once)	1730–1620	$[Fe(CO)_4]^{2-}$ (bound to a M(-II) centre)	1790

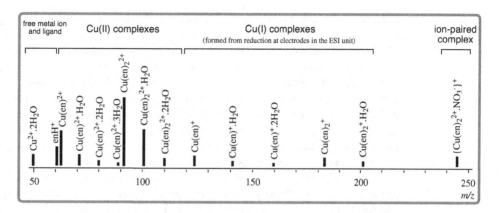

Figure 7.3
Schematic representation of the ESI-MS spectrum of compounds present in a copper(II) ion and 1,2-ethanediamine (1 : 2) mixture in water. Peaks are defined in terms of m/z (mass/charge ratio).

so as to allow cations and anions of complex species in solution to travel into the analyser in their original complex form, providing a method for detecting the mass of complex species that exist in the solution environment, which can assist in defining their structure. This capacity to define the mass of complex ions, and identify the presence of several different complexes in an originally usually dilute aqueous solution, is a key application of ESI–MS. Moreover, where ESI conditions lead to some decomposition of complex species, the pattern of products observed can assist in identifying the character of the parent species and the way it undergoes change. Where a metal exists as a mixture of several isotopes, a set of peaks for a particular complex species related in pattern to the different isotopic metal centres will result, and this characteristic pattern helps confirm the presence of the metal in a complex species, and even of several metals in an oligomer.

To illustrate the ESI–MS technique, we'll look at a simple example, involving complexes formed in solution between Cu^{2+}_{aq} and 1,2-ethanediamine (en). We know from potentiometric and spectrophotometric titrations that two complex species form dominantly during reaction, the 1:1 $Cu(en)^{2+}_{aq}$ and 1:2 $Cu(en)_2^{2+}_{aq}$ complex ions, and even know the stability constants for these species. In fact, careful and slow addition of en to Cu^{2+}_{aq} in a beaker allows one to see the colour changes as we step from dominantly Cu^{2+}_{aq} to successively the 1:1 and then 1:2 M:L species. Thus this is a well understood system, and suited to illustration and evaluation of the ESI–MS method. When a dilute solution of a 1:2 mixture of copper ion and en ligand is passed into the ESI–MS tandem instrument, a set of cations of different mass are detected (Figure 7.3). From the masses of these ions, we can assign the peaks to $(en)H^+$, $Cu^{2+}.xH_2O$, $Cu(en)^{2+}.xH_2O$ and $Cu(en)_2^{2+}.xH_2O$ (with peaks for various x values). In addition, as a nitrate salt of copper(II) was employed, an ion-paired species $\{Cu(en)_2^{2+}.(NO_3^-)\}^+$ is observed. As well, reduced complexes $Cu(en)^+.xH_2O$ and $Cu(en)_2^+.xH_2O$ formed by reduction that can occur at electrodes in the ESI unit occur as complicating peaks. What this spectrum tells us is that the solution contains various amounts of free ligand (detected as $(en)H^+$), free copper ion (detected as $Cu^{2+}.xH_2O$), 1:1 complex (detected as $Cu(en)^{2+}.xH_2O$) and 1 : 2 complex (detected as $Cu(en)_2^{2+}.xH_2O$), and an ion-paired species $(\{Cu(en)_2^{2+}.(NO_3^-)\}^+)$. This is consistent with our interpretation of the chemistry from observations and from titration experiments. Thus the ESI–MS has provided additional firm evidence for these species occurring in solution. Moreover, it does not detect any $Cu(en)_3^{2+}$, consistent with the extremely low stability of this due to Jahn–Teller distortion, nor does it find any dinuclear bridged species, even for different M : L ratios that could support such speciation. The shortcomings of the ESI–MS technique are first, that it can only detect ions and thus any neutral species remain un-sensed, and secondly, the peak height is not a direct measure of relative amount of a species, so that it is not strictly valid for defining relative concentrations of species present. However, it does provide a very simple method for probing solution species, based on molecular mass/charge. There is one additional caveat; because the method 'strips' solvent molecules from droplets to form ions, it can produce and hence detect 'bare' species such as $Cu(en)^{2+}$ which obviously would exist as $Cu(en)^{2+}_{aq}$ in solution. Solvent 'stripping' may be incomplete, so a sequence of solvated species, with peaks separated by $18/z$ (the mass of water divided by the overall charge on the ion), may be seen in some cases. Lastly, there is one additional bonus; where the metal has several significant isotopes (as occurs with Cu), the presence of the metal will, in a high-resolution instrument, lead to a characteristic pattern associated with that isotopic ratio whenever the metal is present, allowing easy assignment of metal-containing and metal-free species. Overall, it is a useful addition to the armoury of coordination chemists.

The above examples are designed to be merely illustrative of some key techniques. Details of these and other techniques may be found in advanced textbooks on physical inorganic chemistry and/or analytical chemistry. What should shine through the above, at least, is how molecular size, shape and symmetry relate to spectroscopic analysis.

7.3.2   Pretty in Red? – Colour and the Spectrochemical Series

Many, but by no means all, coordination complexes absorb light in the visible region of the spectrum, leading to their distinctive colour. The crystal field and ligand field theory that we have developed to some extent earlier in this textbook provided a reasonable interpretation of colour. The ligands lead to, for octahedral geometry, a stabilization of diagonal (t_{2g}) orbitals by $-4D_q$ ($-0.4\Delta_o$) and destabilization of axial (e_g) orbitals by $+6D_q$ ($+0.6\Delta_o$), and a separation of Δ_o; for the vast majority of complexes, Δ_o lies in the range from ~7000 to $\sim40\,000$ cm^{-1}, which places it in the near-infrared–visible–near-ultraviolet regions. Energy is required to promote an electron from a lower to a higher level, and where the energy gap between levels is equivalent to that of a region of the visible light spectrum, in achieving an excited state a select part of the coloured light spectrum is absorbed; we see the residue as colour in the complex. If we examine the octahedral splitting diagram for all of the first row transition metal ions in an octahedral field (Figure 7.4), we can appreciate the concept, and even understand why some compounds are colourless.

The simplest case is d^1, as found for Ti^{3+}_{aq}. Light of the appropriate energy is absorbed, and the electron promoted from the t_{2g} to the e_g level, to produce an excited state. If the excited state decays by transferring energy to its immediate environment in various ways (such as through molecular collisions), rather than re-emitting the light energy, then we have achieved removal of part of the coloured light spectrum, and the complex appears coloured. The Zn^{2+}_{aq} has a full complement of electrons, so that electron promotion is forbidden, and the compound is predicted and observed to be colourless. Further, Mn^{2+}_{aq} is extremely pale in colour, which we can interpret if we assign a special stability to the half-filled set of orbitals so that electron promotion is more difficult (or less 'allowed') because it would require a theoretically forbidden change in spin of the promoted electron

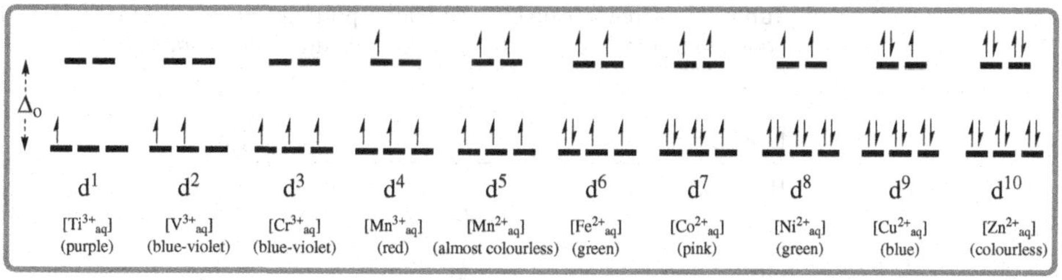

Figure 7.4
The high spin d-electron configurations for d^1–d^{10} systems, with examples of aqua metal ions of these configurations, and their colours. (Although not shown in the figure, Δ_o varies across the group of ions selected as examples.)

for it to occupy an orbital already occupied by one electron of the same spin, and hence the colour is very weak. This is a modest and not entirely satisfactory interpretation, but at least a beginning. In effect, as soon as more than one electron is present in the d orbitals, we must consider the full set of electrons rather than an individual electron. By doing this a much more satisfactory and successful interpretation of electronic spectroscopy results – but this is a task for an advanced text.

The capacity of the crystal and ligand field models to respond to and accommodate ligand-directed influences is notable. This is best exemplified by the spectrochemical series for ligands. We have discussed this earlier in Chapter 3.3. Put simply, the capacity of ligands to split the d subshell of transition metal ions is a *variable*, and thus produces different Δ_o for different ligands. The capacity of a particular ligand relative to others is not affected much by the particular metal ion involved, so the capacity of ligands to split can be ordered fairly consistently, as given below for an abbreviated list of ligands:

$$I^- < Br^- < Cl^- < F^- < OH_2 < NH_3 < NO_2^- < CN^- < CO.$$

In this spectrochemical series for ligands, those at the left split the d-orbital set least, thus favouring *high-spin* complexes, whereas those at the right split the d-orbital set most, thus favouring *low-spin* complexes. This behaviour is observable experimentally. The colour of complexes depends on the set of ligands bound, and particularly the type of donor group. Subtle variations can be detected; for example, low-spin d^6 $Co^{III}L_6$ ions vary from yellow (with $L = CN^-$) to orange (with $L = NH_3$) to blue (with $L = OH_2$), even though these ligands lie at one end of the series.

We can see the influence of ligands on the splitting by examining a straight-forward example, the d^3 system. For Cr(III), the variation in colour with ligand type is clearly identified, as complexes absorb in the visible region of the electromagnetic spectrum. The size of Δ_o for d^3 is easily determined from the position of the lowest-energy absorption maxima, and follows the trend shown in Table 7.6. It is clear that there is a variation with ligand that is rather substantial; further, as the charge on the metal ion increases, the size of the splitting increases. In effect, with complex geometry and ligand donor set held constant, the variation with metal ion for a range of ligands is relatively constant and of the order for selected metal ions

$$Pt^{4+} > Ir^{3+} > Rh^{3+} > Co^{3+} > Cr^{3+} > Fe^{3+} > Fe^{2+} > Co^{2+} > Ni^{2+} > Mn^{2+}.$$

Table 7.6 Comparison of ligand field splitting energy (Δ_o) for different ligands bound to d^3 electronic configuration octahedral metal complexes $[ML_6]^{n+}$.

	Δ_o (cm^{-1})		
Ligand	V^{2+}	Cr^{3+}	Mn^{4+}
Br^-	—	12 700	—
Cl^-	7 200	13 200	17 900
F^-	—	14 900	21 800
OH_2	12 300	17 400	~24 000
NH_3	—	21 600	—
CN^-	—	26 700	—

Whereas this latter effect can be understood in part on the basic of electrostatic arguments (ion charge and size), the variation with ligand type poses more of a problem. That there is an apparent correlation between the donor atom position in the Periodic Table and splitting energy (C > N > O > F and also F > Cl > Br) is very difficult to reconcile using a simple electrostatic view, such as met in the crystal field theory. Moreover, there is a clear trend with position in the Periodic Table for metals, with the splitting for 4d and 5d metal ions significantly larger than those for 3d. This is exemplified for the cobalt triad, where for $[M(NH_3)_6]^{3+}$ complexes, Δ_o varies from Co (22 900 cm^{-1}) to Rh (34 000 cm^{-1}) to Ir (41 200 cm^{-1}). Again, the crystal field model struggles to interpret this observation, and we are drawn into the different ligand field model to provide the better interpretation.

Apart from the position of absorption bands, it is notable that the intensity of bands varies with coordination geometry. For perhaps the two most common shapes, octahedral and tetrahedral, it is noted that the intensity of absorbance bands for tetrahedral complexes are invariable greater (up to 50-fold) than those for octahedral complexes (Figure 7.5), although both are significantly smaller than absorbances of organic chromophores. This is the result of the transitions involving d electrons moving between d orbitals (called d–d transitions) being partly forbidden under the theory that governs our understanding of their behaviour, with the 'forbiddenness' relaxed by different effects. For tetrahedral shape, there is greater relaxation of the rules. As a general guide, for a particular coordination number, the lower the symmetry the more relaxation of the rules applies and hence the larger the absorbance bands.

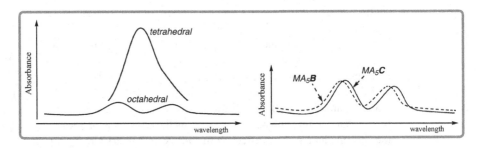

Figure 7.5
The electronic spectra of (at left) a tetrahedral versus an octahedral complex, and (at right) the shift in the spectrum of an octahedral complex on replacement of just one donor group by another with a different position in the spectrochemical series.

As a guide to how absorbance band intensity varies with selection rules that allow or forbid the transition, experimental results for some simple compounds can be compared. The high-spin octahedral d^5 $[Mn(OH_2)_6]^{2+}$ ion, which is spin forbidden and Laporte forbidden, has a very low intensity band (ε_{max} 0.1 M^{-1} cm^{-1}); the octahedral d^1 $[Ti(OH_2)_6]^{3+}$ ion, which is spin allowed and Laporte forbidden, has a moderate-sized band (ε_{max} 10 M^{-1} cm^{-1}); the d^7 tetrahedral $[CoCl_4]^{2-}$ ion, which is spin allowed and partially Laporte allowed, has a markedly greater band intensity (ε_{max} 500 M^{-1} cm^{-1}), whereas the octahedral d^0 $[TiCl_6]^{2-}$ ion, which is spin allowed and Laporte allowed (i.e. a charge-transfer spectrum), has a very large band intensity (ε_{max} 10 000 M^{-1} cm^{-1}). Above, a Laporte allowed transition is one that occurs between different orbital types, such as s→p, p→d or d→f; as a consequence, a d→d transition is Laporte forbidden, although some symmetry-based relaxation rules may operate.

It is also noted that the absorbance bands we see in the electronic spectra of d-block complexes are broad. This arises because complexes are constantly undergoing an array of molecular vibrations and rotations that, for example, are changing bond lengths slightly and thus influencing the size of Δ_o in the process. Because absorption of a photon of light is an extremely fast process compared with these minor internal structural changes in the complex, the form of the complex at the particular instant of photon capture is itself 'captured', leading to a range of energies associated with different vibrational and rotational states, so that we see an averaged outcome, and a broad peak. It is notable that f-block elements display sharp absorbance bands, as the f orbitals involved are more 'buried' and overall there is little influence of rotational and vibrational motion in that block of the Periodic Table.

7.3.3 A Magnetic Personality? – Paramagnetism and Diamagnetism

As discussed earlier in Chapter 3.3, if we examine the set of d electrons for d^4 to d^7, there is choice available as regards the arrangement of electrons in the octahedral d subshells. These configurations display options for electron arrangement, namely *high spin* and *low spin*. The differentiation depends on the size of the energy gap (Δ_o) compared with the amount of spin pairing energy (P), with a large Δ_o favouring low spin arrangements. As discussed in the last section, the size of Δ_o is ligand-dependent, especially depending on the type of donor atom, and also is influenced appreciably by the charge on the metal ion; P varies dominantly only with the metal centre and its oxidation state. In general, where $P > \Delta_o$, the complex will be high spin, whereas where $P < \Delta_o$, the complex will be low spin (Table 7.7). Note how, as the ligand changes from F^- to NH_3 for d^6 Co(III), and from OH_2 to CN^- for d^6 Fe(II) the spin state switches; if this model is correct, this should be experimentally observable. It is the magnetic properties that most clearly illustrate this behaviour.

In terms of magnetism, there are two classes of compounds, distinguished by their behaviour in a magnetic field: *diamagnetic – repelled from* the strong part of a magnetic field; and *paramagnetic – attracted into* the strong part of a magnetic field. ***Ferromagnetism*** can be considered a special case of paramagnetism.

All chemical substances are diamagnetic, since this effect is caused by the interaction of an external field with the magnetic field produced by electron movement in filled orbitals. However, diamagnetism is a much smaller effect than paramagnetism, in terms of its contribution to measurable magnetic properties. Paramagnetism arises whenever an atom, ion or molecule possesses *one or more unpaired electrons*. It is not restricted to transition metal ions, and even non-metallic compounds can be paramagnetic; dioxygen is a simple

Table 7.7 Comparison of pairing energy (P) versus ligand field splitting energy (Δ_o) for some metals ions with d^4–d^6 electronic configurations, and consequences in terms of observed spin state.

d^n	Ion	Ligands	P	Δ_o	Spin State
d^4	Cr^{4+}	$(OH_2)_6$	282	166	HIGH ($P > \Delta_o$)
	Mn^{3+}	$(OH_2)_6$	335	252	HIGH ($P > \Delta_o$)
d^5	Mn^{2+}	$(OH_2)_6$	306	94	HIGH ($P > \Delta_o$)
	Fe^{3+}	$(OH_2)_6$	360	164	HIGH ($P > \Delta_o$)
d^6	Fe^{2+}	$(OH_2)_6$	211	125	HIGH ($P > \Delta_o$)
		$(CN^-)_6$	211	395	LOW ($P < \Delta_o$)
	Co^{3+}	$(F^-)_6$	252	156	HIGH ($P > \Delta_o$)
		$(NH_3)_6$	252	272	LOW ($P < \Delta_o$)
		$(CN^-)_6$	252	404	LOW ($P < \Delta_o$)

Low spin and high spin arrangements exemplified for d^6 [Δ_o (low spin) > Δ_o (high spin)]:

low spin   high spin

example. However, this definition is significant in terms of the variable number of unpaired electrons that can be present in a transition metal complex. Since we are dealing with five d orbitals, the maximum number of unpaired electrons that can be associated with one metal ion is five – simply, there can be only from zero to five unpaired electrons, or six possibilities.

Experimentally, dia- and para-magnetism can be detected by a *weight change* in the presence of a strong inhomogeneous magnetic field. With the sample suspended in a sample tube so that the lower part lies in the centre of a strong magnetic field and the upper part outside the field, the experimental outcome is that diamagnetic materials are heavier whereas paramagnetic ones are lighter than in the absence of a magnetic field. Even the number of unpaired electrons can be determined by experiment, based on theories that we shall not develop fully here. However, we will explore the factors that contribute to the magnetic properties, usually expressed in terms of an experimentally measurable parameter called the magnetic moment (μ).

We are dealing in our model with electrons in orbitals, which are defined to have both orbital motion and spin motion; both contribute to the (para)magnetic moment. Quantum theory associates quantum numbers with both these motions. The spin and orbital motion of an electron in an orbital involve quantum numbers for both *spin momentum* (S), which is actually related to the number of unpaired electrons (n) as S = $n/2$, and the *orbital angular momentum* (L). The magnetic moment (μ) (which is expressed in units of Bohr magnetons, μ_B) is a measure of the magnetism, and is defined by an expression (7.1) involving both quantum numbers.

$$\mu = [4S(S + 1) + L(L + 1)]^{1/2} \tag{7.1}$$

The introduction of an equation involving quantum numbers may be daunting, but we are fortunately able to simplify this readily. Firstly, for *first-row* transition metal ions, the effect of L on μ is small, so a fairly valid approximation can be reached by neglecting the L component, and then our expression reduces to the so-called 'spin-only'

Table 7.8 Comparisons of predicted and actual experimental magnetic moments (in units of μ_B) for octahedral transition metal complexes, with from zero to the maximum possible five unpaired electrons.

N°. Unpaired e-	Calcd μ	Exptl μ	Typical Metal Ion
0	0	~0	low spin Co^{3+}
1	1.73	~1.7	Ti^{3+}, Cu^{2+}
2	2.83	2.75–2.85	V^{3+}, Ni^{2+}
3	3.88	3.80–3.90	V^{2+}, Cr^{3+}, Mn^{4+}
4	4.90	4.75–5.00	high-spin Cr^{2+} & Co^{3+}
5	5.92	5.65–6.10	high-spin Mn^{2+}

approximation (7.2).

$$\mu = [4S(S+1)]^{1/2} \tag{7.2}$$

where the only quantum number remaining is S. Now, we may readily replace this quantum number by using the $S = n/2$ relationship, the substitution then leading to the 'spin-only' formula for the magnetic moment (7.3).

$$\mu = [n(n+2)]^{1/2} \tag{7.3}$$

Thus we have reduced our expression to one involving simply the number of unpaired electrons. Using this 'spin-only' Equation (7.3), the value of μ can be readily calculated and predictions compared with actual experimental values (Table 7.8).

It is clear that the experimental values can be used effectively to define the number of unpaired electrons, and thus the spin state of a complex. For example, Co(II) is a d^7 system, which will have three unpaired electrons if high spin and one unpaired electron if low spin, with calculated magnetic moments of 3.88 and 1.73 μ_B respectively. If a particular complex has as experimental magnetic moment of 3.83 μ_B, then it must, by comparison with the two options, be high spin. The technique can very effectively distinguish between high and low spin states. For Co(III) (d^6), for example, $[Co(NH_3)_6]^{3+}$ (low spin, $n = 0$) has an experimental magnetic moment close to zero, whereas $[CoF_6]^{3-}$ (high spin, $n = 4$) has an experimental magnetic moment of nearly 5 μ_B, readily differentiated. The simple ligand field model that we have developed has triumphed again, and it is its applicability for dealing simply with an array of experimental data such as magnetic behaviour that has led to its longevity.

7.3.4 Ligand Field Stabilization

Yet is there any other basic evidence to support the theory that we have dwelt on so much? One of the key pillars of the theory is that there should be some inherent stabilization of compounds as a result of introducing a field of ligands around the central metal ion, expressed in terms of the so-called ligand field stabilization energy. Recall that in an octahedral field, electrons in diagonal orbitals are stabilized (by $-4D_q$) and those in axial orbitals are destabilized (by $+6D_q$) relative to the spherical field situation. By summing up the energy values for all d electrons in the set, we arrive at the overall stabilization or ligand field stabilization energy (LFSE), which will thus vary with **n** in d**n**, as shown in Table 7.9.

Table 7.9 Ligand field stabilization energies (LFSE, given in $D_q = \Delta_o/10$) for the various d^n configurations.

Configuration	d^0	d^1	d^2	d^3	d^4	d^5	d^6	d^7	d^8	d^9	d^{10}
low spin					-16	-20	-24	-18			
high spin	0	-4	-8	-12	-6	0	-4	-8	-12	-6	0

For example, for high spin d^5, we have three electrons in the t_{2g} level ($3 \times -4D_q$) and two in the e_g level ($2 \times +6D_q$), for an overall LFSE of zero; however, for low spin d^5, all five electrons lie in the t_{2g} level ($5 \times -4D_q$) for an overall LFSE of $-20D_q$. The concept is simple, and for high-spin produces a pattern as shown in the inset of Figure 7.6.

There is experimental evidence that supports this trend. For example, hydration energies for M(II) ions of the first row transition metals exhibit a W-shaped pattern, as a result of crystal field stabilization superimposed on the expected periodic increase (Figure 7.6), consistent with this model. Likewise, the measured lattice energies of transition metal fluorides (MF_2) display a very similar pattern. The predominance of low spin d^6 complexes is also consistent with the model, as there is a significant difference in LFSE between spin states in favour of low spin in the model. However, there are concerns with the model that have attracted criticism; for example, one might anticipate low spin d^5 more often than is observed. However, it is a simple and sufficient model to use at a basic level.

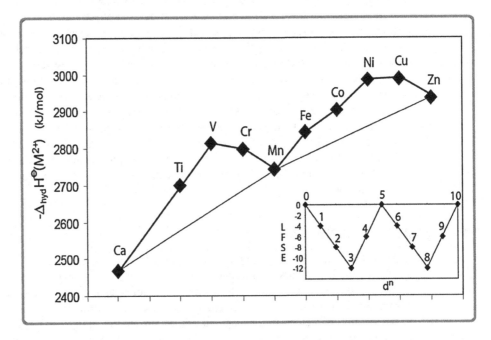

Figure 7.6
Predicted LFSE for *high-spin* d^0 to d^{10} (inset), and variation in hydration energy for first-row M^{2+} transition metal ions, which mirrors this trend, superimposed on a general increase with atomic number.

Concept Keys

Coordination chemistry relies heavily on a capacity to separate and/or isolate individual complexes from reaction mixtures, and subsequently to employ an array of physical methods to determine the nature and structure of such complexes.

Single crystal X-ray crystallography is a key physical method, as it provides an accurate three-dimensional picture of crystalline complexes in the solid state.

Solid state and solution structures need not be identical, and an array of less definitive physical methods must be employed to probe complex structure and chemistry in solution.

Key physical methods employed in coordination chemistry are NMR and IR spectroscopy, UV-Vis spectrophotometry and MS.

The colour and spectra of d- and f-block complexes is intimately tied to the number and arrangement of electrons in their valence shells and to complex stereochemistry.

The splitting of the d subshell in transition metal complexes is dependent on the type of ligands coordinated, with variation sufficient to lead, for some configurations, to high-spin and low-spin options that produce different spectroscopic and magnetic properties.

Further Reading

Ebsworth, E.A.V., Rankin, D.W.H. and Cradock, S. (1987) *Structural Methods in Inorganic Chemistry*, Blackwell, Oxford. An older, but still valid, account that may help reveal the mysteries of crystallography and three-dimensional structure determination of complexes.

Henderson, W. and McIndoe, J.S. (2005) *Mass Spectrometry of Inorganic and Organometallic Compounds: Tools, Techniques, Tips*, John Wiley & Sons, Ltd, Chichester, UK. Apart from providing students with a description of how a mass spectrometer works, this textbook provides a fine coverage of various techniques, expected outcomes and their application; coverage of both coordination and organometallic compounds appears.

Hill, H.A.O. and Day, P. (eds) (1968) *Physical Methods in Advanced Inorganic Chemistry*, Interscience Publishers, London. An old but readable coverage of an array of physical methods; the fundamentals have not changed, so it can be helpful for students.

Kettle, S.F.A. (1998) *Physical Inorganic Chemistry: A Coordination Chemistry Approach*, Oxford University Press. A readable but more advanced student textbook that covers some physical methods and their underlying theory in good depth and clarity.

Moore, E. (1994) *Spectroscopic Methods in Inorganic Chemistry: Concepts and Case Studies*, Open University Worldwide. A very brief introduction that students should find set at an appropriate level.

Nakamoto, K. (2009) *Infrared and Raman Spectra of Inorganic and Coordination Compounds. Parts A and B*, 6th edn, John Wiley & Sons, Inc., Hoboken, USA. Fully revised editions of the classic work on infrared spectroscopy of coordination and organometallic compounds, now with bioinorganic systems. Advanced in nature, but a student may find aspects valuable and informative.

Parish, R.V. (1990) *NMR, NQR, EPR and Mossbauer Spectroscopy in Inorganic Chemistry*, Ellis Horwood, Hemel Hempstead. An ageing but still valid coverage of some important resonance methods used in coordination chemistry; the title says it all.

Scott, R.A. (ed.) (2009) *Applications of Physical Methods to Inorganic and Bioinorganic Chemistry*, John Wiley & Sons, Inc., Hoboken, USA. An up-to-date introductory coverage of a wide range of techniques used in coordination chemistry; provides guidance on the selection of appropriate physical methods for tasks.

Sienko, M.J. and Plane, R.A. (1963) *Physical Inorganic Chemistry*, 2nd edn, W. A. Benjamin Inc., 166 pp. Another old but readable text covering core physical methods; again, as the fundamentals remain with us, students may find some value herein.

8 A Complex Life

8.1 Life's a Metal Ion

Because metal ions are key components of important natural inorganic species such as rocks and soils, it is perhaps not surprising that they were once dismissed as unimportant to living systems; the very name of the broader field in which they reside – inorganic chemistry – is pejorative, and implies a divide between the living world and the world of metals. Yet, as we have already noted in earlier chapters, this is not the case; metal ions play vital roles in living things of all types, even though the amount of metal ions present may be small (for example, even the most common transition metal, iron, constitutes <0.01% of a human). In fact, metal ions are ubiquitous in nature, with metals from the s, p and d block playing essential roles in living things, including at the active site of enzymes.

The three most frequently used transition metals in biological systems are iron, zinc and copper. However, most other light metals, and some heavier ones, appear in important life processes. Metal-containing biomolecules are surprisingly common and important. For example, some well known molecules which all contain a cyclic nitrogen-donor ligand binding to a metal ion are chlorophyll (a Mg^{2+} complex), hemoglobin (a Fe^{2+} complex) and vitamin B_{12} (a Co^{3+} complex). In effect, these are examples of natural coordination complexes, and all life depends on them – life, indeed, requires metal ions.

8.1.1 Biological Ligands

We shall in this chapter meet an array of different examples of metal complexes which we can term metallobiomolecules. Initially, however, it would seem appropriate to identify the basic types of donor atoms and groups that are available, since this aspect clearly governs the capacity for metal ions to form complexes with biomolecules. Overwhelmingly, the donor atoms that are offered by biomolecules are oxygen, nitrogen and sulfur. These are available from a number of sources. While some small molecules that form fairly simple complexes occur, the vast majority of molecules involved in metal ion binding are oligomeric or polymeric, with the two dominant classes being peptides and nucleotide chains (RNA and DNA).

Peptides consist of chains of amino acid residues linked through amide groups (—CO—NH—); amino acids (HOOC—CH(R)—NH_2) that form the peptide chain employ both their carboxylate and amine groups in peptide chain formation. The amide groups formed are good donor groups when deprotonated, and the substituent R-groups on the amino acid residues often carry function groups (such as —COOH, —NH_2, —OH and —SH) that are also available for coordination. Examples of coordination of simple amino acids and peptides appear in Figure 8.1.

Introduction to Coordination Chemistry   Geoffrey A. Lawrance
© 2010 John Wiley & Sons, Ltd

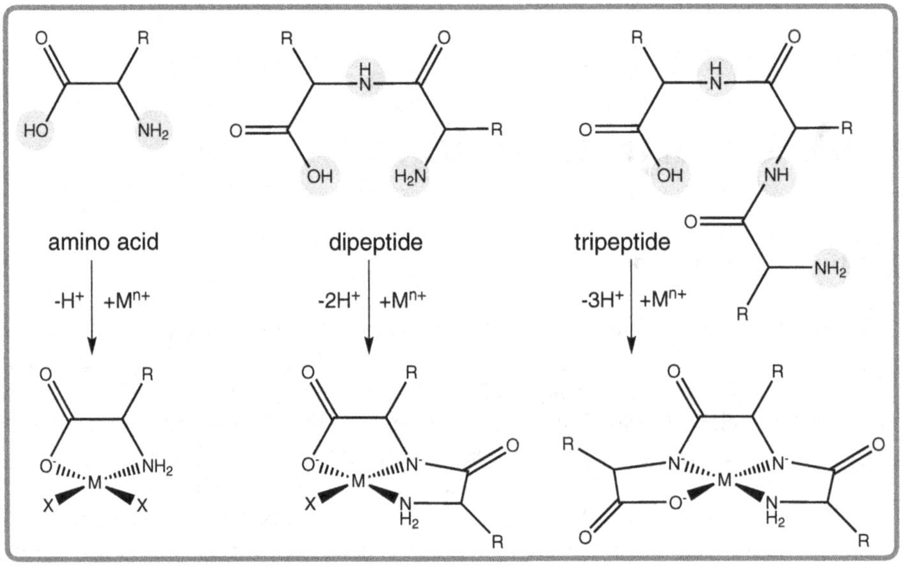

Figure 8.1
Examples of amino acid and peptide coordination to a metal ion. Deprotonation of carboxylic acid and amide groups is required for efficient coordination; potential donor groups are highlighted in the free ligands.

Nucleotide 'building blocks' of DNA/RNA polymers contain a phosphate ester, a sugar ring and an aromatic nitrogen base, and thus contain a number of potential O- and N-donor groups (Figure 8.2). RNA single chains and DNA duplex chains consist of a backbone of linked phosphate diesters $R-O-PO(O^-)-O-R$, that each offer an oxygen anion suitable for complexation or at least ion-pair binding. Moreover, the array of aromatic nitrogen bases that branch from the backbone offer nitrogen donors that have the potential to participate

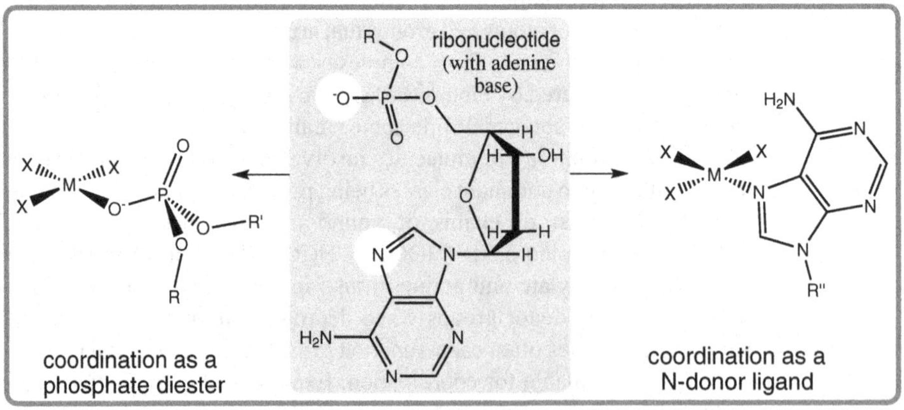

Figure 8.2
Examples of potential nucleotide coordination to a metal ion. Both oxygen and nitrogen donors (examples circled) have the ability to bind to metal ions.

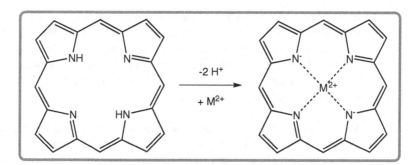

Figure 8.3
The basic heme framework, and its coordination to a M(II) ion.

in complexation, although they are usually involved in noncovalent bonding important to the polymer structure.

Another key class of molecules that are often involved in complex formation are unsaturated macrocycles, usually 15- or 16-membered rings that contain four nitrogen donors. The most common of these is the heme group, which has a framework as shown in Figure 8.3. This coordinates as a di-deprotonated ligand to metal ions, particularly iron(II), producing in that case an overall neutral complex. Additional axial coordinate bonds to groups on a peptide backbone anchor the complex unit to a biopolymer. Some metal-containing biomolecules also incorporate simple anions such as HO^-, S^{2-} or CN^- as part of their coordination spheres.

For polymeric chains, the act of coordination almost invariably requires a change in the shape of the chain as a consequence of satisfying the demands of the metal ion for a preferred stereochemistry and a set of donors with bond distances within a limited range. Thus, coordinate bond formation has consequences that clearly alter the local environment around the metal ion, but may also alter polymer chain conformation over an extended range. Since three-dimensional shape in biopolymers plays a role in function, 'natural' complexation evolved by Nature usually plays a positive role, whereas 'unnatural' complexation through the addition of 'foreign' metal ions may be deleterious to function.

8.1.2 Metal Ions in Biology

Metals occur in several forms in biology, largely in environments that we would recognize as those of coordination compounds. Functionally the most advanced class met are the metal-containing enzymes, or biological catalysts. Around 2000 enzymes are known, and a large number are metalloenzymes, with one or several metal ions present coordinated to donor groups that are part of a biopolymer. The metal ions in metalloenzymes and other metallobiomolecules are commonly the lighter, more abundant and more reactive metals of the s or d block of the Periodic Table, but not exclusively. Rarer and heavier elements (such as molybdenum) can be used. The metal ion is also not present just by accident – it has a specific role. In many cases it is found coordinated at the active site in enzymes (i.e. the site of catalysis), where the chemical reaction catalysed by the enzyme occurs.

We shall begin a brief examination of metal complexes in biology with a short overview of metals from the first row of the periodic d block found in biomolecules, and follow this with more detail from selected examples. Although the focus below is on d-block elements, it should be remembered that s-block elements, in particular, are also important; for example apart from its structural role in bones, calcium has a key role in triggering muscle action, and magnesium appears in several *transferase*, *phosphohydrase* and *polymerase* enzymes. While metals are met usually in trace amounts, balance is important in biosystems – both a deficiency and an excess of metals can cause disease. Fortunately, a balanced diet provides sufficient amounts of essential metals.

Vanadium in extremely small amounts is a nutritional requirement for many organisms, including higher animals. Some marine organisms (tunicates) accumulate vanadium, whereas some lichens and fungi contain vanadium in the active site of some enzymes. Vanadium appears in one form of the important enzyme *nitrogenase* (which converts dinitrogen to ammonium ion), and as vanadium(V) in *haloperoxidase* (which coordinates hydrogen peroxide and then oxidizes halides).

Chromium, as Cr(III), is an essential trace element in mammals, and participates in glucose and lipid metabolism. Chromium may help in maintaining normal glucose tolerance by regulating insulin action. There is no evidence of significant chromium(III) toxicity, but chromium(VI) and chromium(V) are toxic (mutagenic and carcinogenic). Cr(V) is believed to damage DNA by promoting cleavage and DNA–protein cross-linking.

Manganese is an essential element, and appears in a wide range of organisms, including humans. It occurs in important enzymes and processes, such as in photosynthesis in plants. It can have a redox role, using particularly its Mn(II) and Mn(III) oxidation states, as in *superoxide dismutase*, whose role is to destroy undesirable superoxide ion through conversion to oxygen and peroxide ion.

Iron is truly ubiquitous in living systems, being found in all life forms from bacteria through to humans. It is at the active centre of molecules responsible for oxygen transport, for electron transport and in a vast range of enzymes. In the human body, *hemoglobin* and *myoglobin* are vital components for oxygen transport, and represent 65% and 6% respectively of all iron in the body. There is a wide range of other human iron-containing proteins, however, with mainly redox roles, making use of accessible oxidation states, notably Fe(II) and Fe(III). The iron proteins tend to be classified as either hemes (featuring cyclic aromatic nitrogen ligands) or nonhemes. Iron is so significant that a special iron storage protein, *ferritin*, is used to allow rapid access to iron when required.

Cobalt is an essential element in small amounts. Cobalt complexes of macrocyclic nitrogen-donor ligands are found in many organisms, including humans, who contain about 5 mg of these cobalamins. Vitamin B_{12} is a cobalamin coenzyme required in humans for its key role in promoting several molecular transformations. Recently, a range of enzymes containing inert cobalt(III) has been discovered, distinguished by their unusual low coordination number, which provides labile reaction sites despite the inherent inertness of this ion.

Nickel is an essential element in small amounts, is a component of the important enzymes *urease*, *carbon monoxide dehydrogenase*, *hydrogenase* and *methyl-S-coenzyme M reductase*, and lies at their active sites. Nickel can exist under physiological conditions in oxidation states I, II and III, but the higher two seem most relevant.

Copper occurs in almost all life forms and it plays a role at the active site of a large number of enzymes. Copper is the third most abundant transition metal in the human body after iron and zinc. Enzymes of copper include *superoxide dismutase*, *tyrosinase*, *nitrite reductase* and *cytochrome c oxidase*. Most copper proteins and enzymes have roles as electron transfer agents and in redox reactions, as Cu(II) and Cu(I) are accessible.

Zinc is recognized as essential to all forms of life, and is the most common transition metal in the body after iron. There are 2 to 3 g of zinc in adults, compared with 4 to 6 g of iron and 0.25 g of copper. Enzymes containing zinc include *carbonic anhydrase* and *carboxypeptidase*, the first two metalloenzymes detected – now there are over 300 zinc enzymes known. Zinc serves an important structural role in DNA binding proteins, stabilizing the correct binding site. Zinc reserves are stored in the metallothionine proteins.

Heavier elements can also appear in proteins. *Molybdenum,* found geologically together with iron, has found use in a range of enzymes. Its prime role seems to be to facilitate oxygen atom transfer to a substrate, via a $Mo^{VI}=O$ species, leading to a Mo^{IV} compound and an oxygenated substrate. Enzymes with Mo present include *aldehyde oxidase, nitrate reductase, sulfite oxidase, xanthine oxidase, formate dehydrogenase* and *DMSO reductase. Tungsten*-containing enzymes are also known; in fact, Mo and W appear to find more roles in proteins than their triad parent chromium.

8.1.3 Classes of Metallobiomolecules

Biomolecules containing metals are large in number and variety. For convenience, we can divide them into three main categories: nonproteins (simple and/or small molecule species), functional proteins (transporters of electrons or specific molecules or ions, or else sites serving as storage reserves of heavily used metal ions; sometimes called transport and storage proteins), and enzymes (selective catalysts of specific reactions). Within these, there are further sub-divisions, as exemplified in Table 8.1.

8.2 Metalloproteins and Metalloenzymes

The general family of functional proteins containing metal ions is the *metalloproteins*, of which one important branch is the *metalloenzymes*. The task of a metalloenzyme, put simply,

Table 8.1 Division of metal-containing biomolecules, sub-classes and examples.

Category	Task or Role	Examples (metal ion involved)
Nonproteins	metal transport and structural	siderophores (Fe); skeletals (Ca, Si)
	photo-redox	chlorophyll (Mg)
Functional proteins	electron carriers	cytochromes (Fe); iron-sulfur (Fe); blue copper (Cu)
	metal storage and metal carriers	ferritin (Fe); transferrin (Fe); ceruloplasmin (Cu); metallothionine (Zn)
	structural	zinc fingers (Zn)
	oxygen binding	myoglobin (Fe); hemoglobin (Fe); hemerythrin (Fe); hemocyanin (Cu)
Enzymes	hydrolases	carboxypeptidases (Zn); aminopeptidases (Mg, Mn); phosphatases (Mg, Zn, Cu)
	oxido-reductases	oxidases (Fe, Cu, Mo); dehydrogenases (Fe, Cu, Mo); hydroxylases (Fe, Cu, Mo); oxygenases (Fe); nitrogenases (Fe, Mo, V); hydrogenases (Fe)
	isomerases and synthesases	vitamin B_{12} coenzymes (Co)

is usually to bind a specific organic molecule (the *substrate*) and mediate the performance of some catalysed chemical reaction on that substrate to convert it to a different product(s). The site of this reaction is called the *active site*, and features one or more metal ion(s) in a defined coordination environment that exists within a wider environment defined by the shape of the surrounding biopolymer. The role of the metal ion in metalloenzymes may involve one or more of the following:

- coordination and activation of a substrate;
- 'organization' of the environment for facilitating reactions involving the substrate;
- providing a coordinated (and highly active) nucleophile;
- causing major pK_a changes of molecules upon binding, to promote reaction;
- using the redox properties of the metal ion.

The chemistry taking place at the active site of metalloenzymes is not unusual – it is fairly standard coordination chemistry, often metal-directed organic reactions with the substrate as a coordinated ligand. The main difference over small-molecule analogues is that the three-dimensional structure of the parent protein polymer (in effect, also the 'ligand') has a big role to play in the chemistry, such as providing a *specially-shaped space* for the substrate to occupy and in which the chemistry occurs. However, in reality, even this is not unexpected, as the chemistry of even simple metal complexes is influenced by the non-participating or 'spectator' ligands around the metal. In effect, metalloenzymes are just 'supersized' coordination complexes. They are usually very large molecules, with the active site being a very small part of the overall structure. This active site can be *modelled* by the synthesis of small metal complexes. Simple models perform one of two tasks – they mimic either the shape of the biomolecule's active site, or they perform the same chemistry – sometimes even both. It is arguably not surprising that the structure at active sites is often very easily mirrored in the laboratory – after all, the enzyme must form under very mild conditions in nature.

We will examine a number of examples of metallobiomolecules, including enzymes, from a viewpoint of the structure of the metal 'core' and the chemistry that occurs there. This is peculiarly a coordination chemist's view of biosystems, but this is most appropriate in the context of this textbook. Our target here is to identify and exemplify the metal coordination chemistry met abundantly in biomolecules.

8.2.1 Iron-containing Biomolecules

Biomolecules that contain iron are the most prevalent group, which can be divided into heme and nonheme types. The dominant forms are iron porphyrin (*heme*) proteins. These appear in a large number of systems, and have diverse roles, including as oxygen transport and storage proteins (*hemoglobin* and *myoglobin*), for catalytic dehydrogenation or oxidation of organic molecules (*peroxidases*, *cytochrome P450*), and for reduction (*cytochrome c oxidase*) and electron transport (*cytochrome c*). We shall examine some selected examples.

8.2.1.1 *Myoglobin and Hemoglobin*

Myoglobin is a small mononuclear globular protein with a single heme unit attached to an α-helical protein chain (Figure 8.4). It stores oxygen in muscle tissue until it is required for oxidative phosphorylation, thus providing a ready reserve of oxygen to cope with intense respiration demands. *Hemoglobin* has the primary role of transporting oxygen (although its full function is somewhat more complicated, and also includes transporting carbon dioxide), and

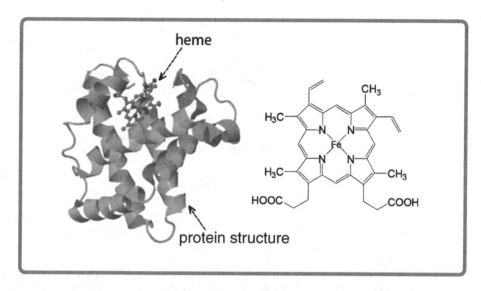

Figure 8.4
Structure of myoglobin (Protein Data Bank; DOI: 10.2210/pdb1mbn/pdb) and detail of the heme of hemoglobin.

is found in red blood cells. It is a larger and more complicated protein, referred to as an $\alpha_2\beta_2$ heterotetramer, meaning there are pairs of two classes of polypeptide chains in the proteins, each containing one heme unit that each bind dioxygen. In both myoglobin and hemoglobin, the oxygen binds to an iron(II) centre bound covalently also to the four N-donors of a porphyrin, which is an aromatic 16-membered tetraaza-macrocycle. The porphyrin is an easily-built compound in chemistry. In nature, the heme is assembled from precursor components that leave substituents on the macrocycle ring of the heme (Figure 8.4). The heme in hemoglobin is embedded in a protein crevice where it is surrounded by hydrophobic groups, and where the Fe(II) is covalently bonded at one axial site by close approach of one imidazole group nitrogen that is part of a histidine amino acid residue (called the *'proximal'* histidine) of a protein chain. Another imidazole nitrogen is located near the other 'empty' axial site, approaching but not achieving a bonding distance (this is called the *'distal'* histidine); it is at this protected site that dioxygen coordinates as a monodentate ligand, taking the iron(II) from a five-coordinate to a preferred six-coordinate octahedral complex without oxidizing the metal centre. Only the coordinate bond from the proximal histidine to the iron binds the heme to the protein; if this bond is broken, the small heme unit can be removed from the protein. Degradation is initiated by this Fe—N bond breaking, which is followed by release and destruction of the heme, accomplished by special *oxygenase* and *reductase* enzymes.

The role of the protein 'pocket' in which the heme unit sits in hemoglobin is twofold. It provides a water-resistant hydrocarbon-based environment of low dielectric constant that is unsuited to highly ionic character in the heme. This leads to the redox potential for the Fe(II)/(III) couple being altered sufficiently to make oxidation impossible by available biological oxidants. Thus dioxygen can bind, but will not oxidize the iron centre. The severe steric constraints imposed by the heme environment means that it is also impossible for another heme iron centre to approach the embedded heme, thus prohibiting unwanted formation of stable oxidized bridging oxygen-linked dimers of the form $Fe^{III}—O^{2-}—Fe^{III}$.

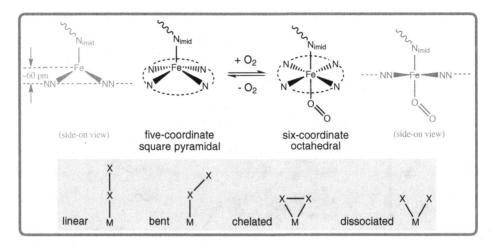

Figure 8.5
The geometry change at the iron centre of the heme unit in *myoglobin* or *hemoglobin* upon dioxygen coordination. Potential modes of coordination of a diatomic molecule are also shown, with the bent form favoured for dioxygen binding employed in the drawing of the complex.

This removes one of the common reactions of iron complexes 'in the beaker', and is an example of the role and importance of the enveloping biopolymer in the natural system.

The Fe(II) centre, in the absence of oxygen, has a five-coordinate square pyramidal geometry. When oxygen binds at the 'vacant' sixth site, opposite the histidine imidazole, the molecule changes to six-coordinate octahedral geometry. This requires the Fe to move from being displaced above the plane of the macrocycle ring (by ∼60 pm), as expected for the square pyramidal shape, to lying in the plane, as expected for an octahedral shape (Figure 8.5). In achieving this, the protein is required to adjust its conformation, to retain the required Fe—N(histidine) bond distance. Release of oxygen allows relaxation back to the five-coordinate square-based pyramidal shape around the iron. No permanent change in the protein is involved, and hence re-use of the protein can occur.

Another point to consider is the way dioxygen binds to the heme. In principle, there are four ways a diatomic molecule can bind to a metal ion: linear, bent, chelated or dissociated into two separate atoms (Figure 8.5). Examples of all four modes exist: linear is found in carbonyl complexes M—CO; bent is found in M—NO$^+$ complexes; chelated is found for CO in [IrCl(CO)(PR$_3$)], and dissociated is found for H$_2$ following its addition to some four-coordinate organometallic complexes. The very large number of atoms in a biomolecule makes it very difficult to determine the structural detail accurately by X-ray crystal structure analysis, which is not a problem encountered in the case of low molecular mass molecules. The biological structure has been supported through examining low molecular weight model compounds that bind dioxygen reversibly, and from spectroscopic methods. The 'bent' mode (Figure 8.5) is now well defined for dioxygen binding in hemoglobin and myoglobin.

8.2.1.2 *Nonheme Oxygen-binding Iron Proteins*

Whereas oxygen binding in humans and many other animals involves heme units, not all life-forms bind and carry dioxygen in this way. *Hemerythrin* is a nonheme iron protein used by *sipunculid* and *brachiopod* marine invertebrates for oxygen transfer and/or storage.

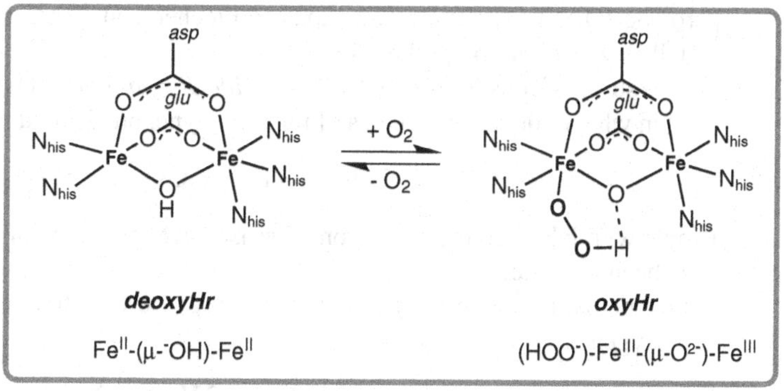

Figure 8.6
The geometry at the iron centres of the bridged dinuclear unit in *hemerythrin* in its dioxygen-free *deoxy* (left) and oxygenated *oxy* (right) forms, with changes upon dioxygen coordination shown.

The complete enzyme has a complicated polymeric structure including a large number of subunits, each of which is an active site for oxygen addition. Each subunit has a molecular weight of around 14 000 Dalton, contains two iron atoms, and binds one molecule of oxygen. This di-iron oxo protein contains octahedral iron(III) centres linked by one oxo (O^{2-}) and two carboxylato ($-COO^-$) bridges. The oxygen-free form, *deoxyhemerythrin*, is a colourless complex with two high-spin d^6 Fe(II) centres bridged by a hydroxide ion; one Fe is six-coordinate (bound to three histidine residue N-donors, two peptide carboxylate groups, and the bridging hydroxide ion), the other one is five-coordinate (bound to two histidine residue N-donors, two peptide carboxylate groups and the bridging hydroxide ion) and has the one 'vacant' site where dioxygen can bind. The dioxygen-bound form, *oxyhemerythrin*, is red-violet in colour, and now contains two six-coordinate low-spin d^5 Fe(III) centres (Figure 8.6).

The O_2 binds to the five-coordinate iron(II) centre and abstracts a H-atom from the bridging hydroxo group, forming a FeOOH group that remains strongly hydrogen-bonded to what is now a bridging oxo group. Oxygen uptake is also accompanied by one-electron oxidation of *both* Fe(II) centres to Fe(III), meaning that the dioxygen undergoes two-electron reduction to bound peroxide ion.

8.2.1.3 Oxygen Reduction

Once dioxygen arrives at a cell it is reduced to water in order to yield the energy necessary to convert adenosine diphosphate (ADP) to adenosine triphosphate (ATP), the 'energy carrier' for cell processes. This oxygen 'burning' (8.1) is mediated in turn by a series of metalloenzymes, such as *cytochrome oxidase*, which contains one heme Fe and one Cu in the active site.

$$O_2 + 4\,H_3O^+ + 4e^- \rightarrow 6\,H_2O \tag{8.1}$$

Since dioxygen reduction provides much more energy ($\sim$400 kJ mol^{-1}) than is needed to form a single ATP from ADP + phosphate ($\sim$30 kJ mol^{-1}), it is to the cell's advantage to carry out the reduction stepwise. The actual process yields six ATP (i.e. $\sim$45% efficient) per dioxygen. This involves intermediate oxygen reduction products, superoxide (O_2^-) and peroxide (O_2^{2-}), which are hazardous to biochemical systems if they 'escape'. Special metalloenzymes 'mop up' these ions.

Free superoxide is dealt with by *superoxide dismutase* (SOD) via a dismutation reaction (8.2) in which, for two molecules of the same type, one is oxidized and the other reduced.

$$2\,O_2^- + 4\,H^+ \rightarrow O_2 + 2\,H_2O_2 \tag{8.2}$$

People with inherited motor neurone disease have mutations of SOD; over 100 mutations have been identified.

Free peroxide is inherently unstable to a disproportionation reaction (8.3) to form water and oxygen.

$$2\,H_2O_2 \rightarrow 2\,H_2O + O_2 \tag{8.3}$$

This is accelerated by the iron heme protein *catalase,* a particularly efficient enzyme with one of the highest turnover numbers of all known enzymes (at $\sim 4 \times 10^7$ molecules per second). This high rate reflects the important role for the enzyme, and its capacity for detoxifying hydrogen peroxide.

8.2.1.4   Electron Carriers

There are a family of nonheme iron proteins that participate in electron transfer that all contain iron bound to sulfur of cysteine (cys, $HS-CH_2-CH(NH_2)-COOH$) amino acid residues present in a protein backbone. The simplest of these is the small protein (MW ~ 6000) *rubredoxin,* found in sulfur-containing bacteria, that consists of a protein containing about 50 amino acids and one iron bound by the S atoms of four cysteine amino acid residues. The iron is bound to four S atoms in a distorted tetrahedral arrangement. It has a $Fe^{II/III}$ redox potential of ~ 0 V, meaning it can be oxidized and reduced readily by biological redox reagents. The iron centre lies close to the surface of each protein (Figure 8.7), providing good access for interaction with and electron transfer to other compounds. This location of the metal redox centre near the exterior of a protein is common in those proteins with a redox role. This makes outer sphere electron transfer easier and faster.

In addition to this simple compound, there are a number of related compounds, the *ferredoxins,* which contain several iron centres closely linked in small Fe_mS_n clusters. In addition to cys-S, they contain 'labile S' that can be released as H_2S on addition of acid, being present in the biomolecules as coordinated and bridging S^{2-}. The 2-Fe cluster contains 2 labile S ions, and is often designated as Fe_2S_2 (although 4 other cys-S are also coordinated), whereas the 4-Fe cluster contains four labile S ions (the cluster is thus designated as Fe_4S_4) with a central Fe_4S_4 core with a distorted cubic box-like structure. A form of the latter structure with one iron 'missing' from a corner of the cube, termed a Fe_3S_4 species, is also known. In all cases, each iron centre lie in a distorted tetrahedral environment of four S-donor ligands.

8.2.1.5   Iron Storage in Higher Animals

Iron is the most important and used metal in higher animals, and thus having a ready supply of bioavailable iron is essential to their proper function. To achieve this, higher animals have developed a way of storing iron. Iron is bound and transported in the body via *transferrin* and stored in *ferritin* protein, made of carboxylate-rich peptide subunits assembled into

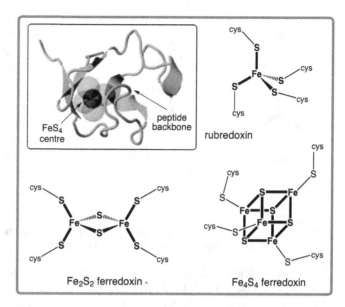

Figure 8.7

A *rubredoxin* oxygen carrier with the FeS_4 centre highlighted (Protein Data Bank; DOI: 10.2210/pdb5rxn/pdb), and drawings of the cores of rubredoxin and diiron (Fe_2S_2) and tetrairon (Fe_4S_4) ferredoxins.

a hollow spherical shell. The hollow sphere formed is $\sim$8000 pm in diameter, with walls $\sim$1000 pm thick. Channels in the sphere are formed at the intersections of peptide subunits. These channels are the key to *ferritin*'s ability to release iron in a controlled manner. The iron in the *ferritin* is stored in the core of the hollow sphere as iron(III) hydroxide in a crystalline solid, $[FeO(OH)]_8 \cdot [FeO(OPO_3H_2)]$. The best simple model for the ferritin core is the well-known mineral ferrihydrite, $FeO(OH)$.

The complete particle of *ferritin* provides a readily available source from which biological chelates can 'collect' iron and transfer it to usable sites. Iron is released by reduction to form soluble $Fe(H_2O)_6^{2+}$, that departs via the channels in the shell, after which chelates 'collect' it. Two types, fourfold and threefold channels, exist. These two types of channels have different properties, and perform different functions. The threefold channels are lined with polar amino acids (aspartate and glutamate), which allow favourable interaction with the aquated Fe^{2+} ion. This interaction allows Fe^{2+} to pass through the channel; essentially, the channel acts like an ion-exchange resin in a column. The fourfold channel is lined with nonpolar amino acid groups and is thought to be the path for the electron transfer process to form Fe(II), but this process is not fully elucidated.

8.2.1.6 *Iron Storage and Transport in Lower Organisms – Siderophores*

Iron storage and supply in bacteria, yeasts and fungi is by small molecular weight octahedral iron(III) complexes. These are categorized, because of their simplicity, as nonproteins. Most known microbial organisms biosynthesise these siderophores, which solubilize and transport environmental iron into the cells during aerobic growth. These are intensely-coloured red-brown compounds, where the chelating ligands are hydroxamates in fungi and yeasts, and either hydroxamates or substituted catechols in bacteria. These ligands

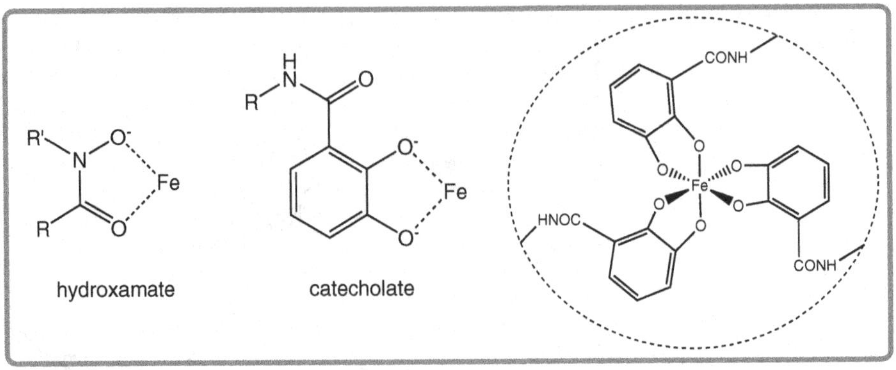

Figure 8.8
The basic hydroxamate and catecholate chelate rings, as well as an example of three catecholate chelates attached to a larger ring binding to an octahedral iron(III) centre.

are basically didentate chelates (Figure 8.8), and three are bound to form very stable six-coordinate octahedral and chiral Fe(III) complexes. In fact, the three chelate units are part of larger single molecules, which means these effectively act as hexadentate ligands, forming complexes with Fe(III) with very high stability constants. The ligands are highly selective for Fe(III), even over Fe(II). A number of common compounds are known. *Ferrichrome* is a hydroxamate complex with the hydroxamates as side chains to a peptide ring. *Ferrioxamine* complexes contain the hydroxamates as part of a simple peptide chain. *Enterobactin* is a catechol complex in which the catechols are side chains of a cyclic ester.

Hydroxamates release Fe by metal ion reduction, so the ligand can be reused once the reduced Fe(II) is dissociated. Whereas the formation constants are high for Fe(III), they are much lower for Fe(II), so reduction provides one mechanism for iron release, with other chelates then able to compete successfully for the Fe(II). However, simply complex destruction provides another mode of iron release. *Catechols*, with $E°$ near -0.75 V, are not reducible biologically (as no reducing agent with a suitable potential is available), and so iron release is by ligand destruction, meaning the ligand cannot be reused. This seems curiously inefficient, but then these species originate from very early in the evolutionary process.

There are many other examples of iron-containing proteins that we could draw upon, but this is better left to a specialist course in bioinorganic chemistry.

8.2.2   Copper-containing Biomolecules

Copper proteins display an array of functions, including: in electron transfer, involving the Cu(I)/Cu(II) couple through either an outer sphere mechanism or acting as an inner-sphere reductase; as mono-terminal oxidases, forming either water or hydrogen peroxide from dioxygen; as oxygenases, whose task is to incorporate an oxygen atom into a substrate; in superoxide degradation to form dioxygen and peroxide; and in oxygen transport.

From both a structural and spectroscopic point of view, three main types of biologically active copper centres have been found in the copper proteins (Figure 8.9). These may be distinguished according to their spectroscopic behaviour.

Type 1 has what is called a 'blue' copper centre, with the copper normally coordinated to two nitrogen and two sulfur atoms in a distorted tetrahedral shape. Blue copper proteins form perhaps the best-known examples of copper proteins. They have a far more intense blue

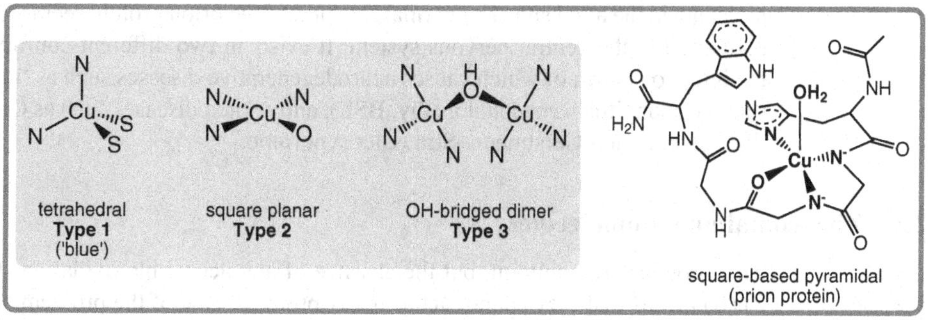

Figure 8.9
The three types of four-coordinate copper(II) centres found in most copper proteins. An example of a rarer five-coordinate square-based pyramidal geometry of copper(II) found in the prion protein appears at right.

colour than Cu^{2+}_{aq} ion, consistent with the usual observation that tetrahedral complexes exhibit more intense absorbance bands than octahedral or square planar complexes, but are similar in hue.

Type 2 has a 'non-blue' copper centre, with the copper coordinated to two or three nitrogen donors in addition to oxygen donor(s) in a square planar geometry, with the different donor set and shape together responsible for a markedly different colour compared with Type 1.

Type 3 has two copper centres closely adjacent, forming a hydroxide-bridged dimer with each copper ion in an approximately square planar geometry.

In this family of compounds, the N-donors come from unsaturated nitrogens in histidine amino acid residues, the S-donors come from methionine and cysteine residues, and the O-donors come from a carboxylic acid in the protein chain. Water, hydroxide and alkoxide oxygens are also employed as O-donor ligands.

Electron transfer copper proteins usually belong to the blue copper proteins (*Type 1*); *azurin* is a simple example. This family of proteins are also called *cupredoxins*, and they participate in many redox reactions involved in processes fundamental to biology, such as respiration or photosynthesis. The striking electron transfer capabilities of blue copper proteins have been studied extensively. *Plastocyanin*, with a tetrahedral CuN_2S_2 core, acts as the electron donor to *Photo System I* in photosynthesis in higher plants and some algae.

Non-blue copper centres are actually most common in copper proteins, and the copper centre in these adopts, as mentioned above, essentially square-planar geometry. The copper ion is bound to the imidazole nitrogen of two or three histidine residues and to O-donor ligands; there is weak additional O-donor coordination in axial sites, with the typical Jahn–Teller distortion expected of d^9 Cu(II) complexes. The *Type 2* sites are more ionic than *Type 1* sites, having mainly neutral donor groups rather than the thiolate anions of the latter. Working cooperatively with organic coenzymes, *Type 2* copper centres direct a wide range of biological oxidation reactions, which include alcohol oxidation and amine degradation. In *caeruloplasmin*, a *Type 2* monomer joins with a *Type 3* dimer to form a larger copper trimer; similar clusters appear in *laccase* and *ascorbate oxidase*.

Compounds with another stereochemistry are also observed. The copper binding site in the prion protein, an approximately 200-amino-acid residue glycosylated protein that carries one copper ion, is of a square-based pyramidal form (Figure 8.9), having one imidazole nitrogen, two amide nitrogens and an amide oxygen bound around the copper, with a water molecule in the axial site. The normal function of the prion protein seems to involve copper regulation in the central nervous system. It exists in two different conformational forms, the infectious form of which causes neurodegenerative diseases such as 'mad cow' disease (bovine spongiform encephalopathy, BSE), and related diseases such as Creutzfeldt–Jakob disease, kuru and Gerstmann–Straussler syndrome.

8.2.3   Zinc-containing Biomolecules

Zinc proteins are common, but the absence of any accessible oxidation states apart from Zn(II) means that they cannot act as redox proteins. One of the problems in working with with Zn(II) is that it is, as a d^{10} system, spectroscopically 'silent', which limits the type of studies that can be used to probe its chemistry in biomolecules. However, several zinc proteins are relatively low molecular weight species and these were characterized by crystal structure analysis some decades ago.

  Carbonic anhydrase is an enzyme that catalyses the carbonic acid/carbon dioxide hydration/dehydration reaction, and the hydrolysis of certain esters. Tetrahedral zinc(II) is present at the active site. The mechanism has been fully elucidated. *Carboxypeptidase* is a fairly small enzyme (MW $\sim$34 600) that hydrolyses (cleaves) the terminal peptide (amide) bond of a peptide chain specifically. It is selective for terminal peptides where the terminal amino acid has an aromatic or branched aliphatic substituent of L absolute configuration. Zinc(II) sits in a protein pocket, bound in a distorted square pyramidal geometry to two histidine imidazole groups, a chelated glutamic acid carboxylate and a water molecule in its resting state. Didentate coordination of the carboxylate in the resting state of the latter shown in Figure 8.10 reverts to monodentate coordination (and distorted tetrahedral geometry for the zinc ion) in the active state. The ligand environment in *carbonic anhydrase* differs somewhat, involving just histidine imidazole groups and one water group in a tetrahedral environment (Figure 8.10).

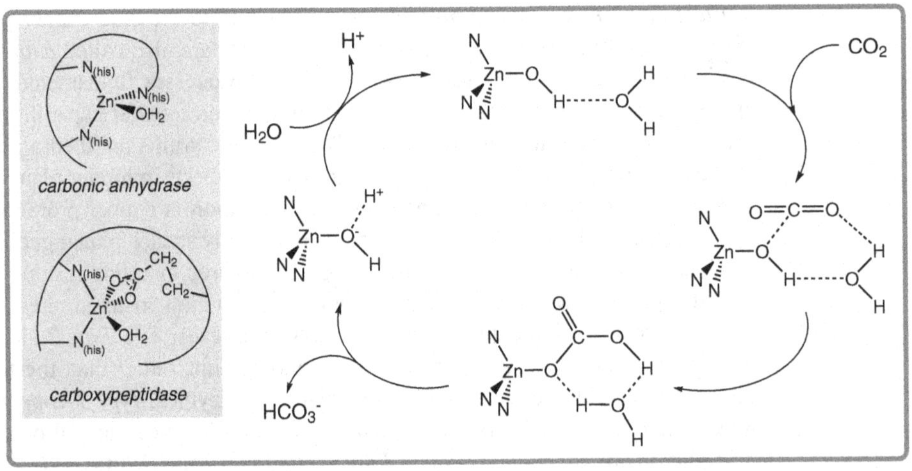

Figure 8.10
A simple representation of the tetrahedral and distorted square pyramidal ligand environments for zinc(II) in *carbonic anhydrase II* and *carboxypeptidase A* respectively; both feature a coordinated water group. A simplified mechanism for the CO_2/HCO_3^- process catalysed by *carbonic anhydrase* is also shown, at right.

The mechanism depicted for *carbonic anhydrase* (Figure 8.10, right) is an example of a biological reaction involving a coordinated hydroxide nucleophile, reminiscent of examples discussed in Chapter 6. The cyclic nature of the mechanism indicates the enzyme is able to be reused, as required in a catalytic process.

8.2.4 Other Metal-containing Biomolecules

Although iron, zinc and copper proteins are most common, there are a range of well known examples that involve other metal ions. One of the best known is *cobalamin*, a cobalt-containing molecule that is not an enzyme, but is a tightly bound prosthetic group for various enzymes which plays a central role in the catalytic reaction. It is known as a *co-enzyme*. A co-enzyme's role is to support an enzyme by operating in conjunction with it; in fact, the enzyme is inactive without it. In the case of *cobalamin*, the coenzyme is actually the site where the substrate binds and is changed to the product. The enzyme/coenzyme assembly has a primary task of promoting one-carbon transfer reactions, exemplified in Figure 8.11.

Cobalamin (or *Vitamin B$_{12}$*) is unusual – not only is it is one of Nature's rare organometal-lic compounds (forming Co–C bonds), but also it is the only vitamin known to contain a metal ion. It is essential for all higher animals, but is not found in plants. It appears to be synthesized exclusively by bacteria. *Vitamin B$_{12}$* has a similar (but not identical) ring to porphyrins. Formally, it is a cobalt(III) compound; reversible reduction to cobalt(II) [called B$_{12r}$] and cobalt(I) [called B$_{12s}$] is possible. In its active form, it contains an adenosine group as the sixth ligand of the cobalt octahedron. This can be replaced by a substrate (R), involving Co–C bonding. The chemistry in the enzyme/coenzyme system occurs on the cobalt centre, where the substrate binds and reacts in a reaction as a coordinated ligand.

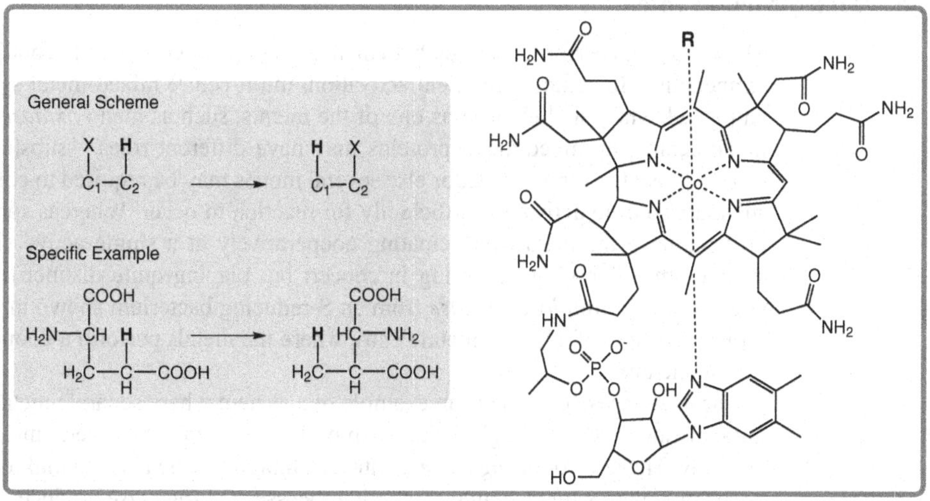

Figure 8.11
The one-carbon transfer reaction (left) facilitated by the octahedral cobalt(III) complex Vitamin B$_{12}$ (right) as coenzyme. Carbon–cobalt bonding of the substrate is involved in the mechanism, at the site designated by the R-group in the complex.

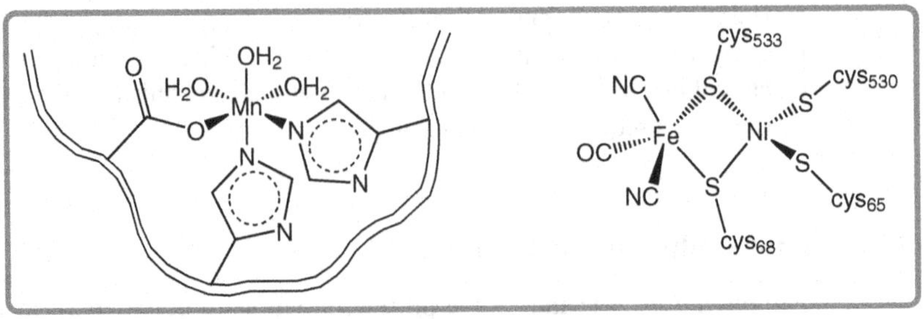

Figure 8.12
The octahedral environment of manganese in *catechol dioxygenase*, featuring three *fac*-arranged exchangeable water molecules (left), and the structure at the active site of a Ni-Fe *hydrogenase* (right), another example of a biomolecule with a metal-carbon bond. The positions of S-donor cysteine amino acid residues on the biopolymer are identified by their peptide location numbers; these can be well apart, indicating that folding of the polymer is important to bring them into appropriate positions.

The human form of *superoxide dismutase* containing manganese is a tetrameric enzyme, with four identical subunits each containing a Mn^{3+} atom. Groups adjacent to the manganese(III) centre that act as ligands are four imidazole and two carboxylate groups. Thus the Mn(III) centre is in a distorted octahedral environment of four N- and two O-donors. The catalytic mechanism for this enzyme involves cycling between the Mn(III) and Mn(II) forms. *Catechol dioxygenase* is a manganese-dependent extradiol-cleaving catechol dioxygenase. The manganese at the active site is in an octahedral MnN_2O_4 environment bound to the protein backbone by two imidazoles and a carboxylate group, with three 'free' sites carrying exchangeable water molecules (Figure 8.12). Coordinated water groups at the active site of an enzyme are often the sites for substrate substitution and subsequent activation and reaction.

8.2.5   Mixed-Metal Proteins

Many enzymes perform chemically demanding tasks under very mild conditions of pH and temperature. To achieve sufficient activation, multi-centre mixed-metal systems are often employed. Many include iron as one of the metals, such as many *oxidases*. The different metal centres in mixed-metal proteins may have different roles – substrate binding and electron transfer, for example, or else several metals may be required to collectively bind a substrate so as to activate it sufficiently for reaction to occur. Whereas some mixed metal proteins have the metals participating cooperatively at a single active site, others have separate metalloproteins working in concert but playing quite distinct roles. The active site of a nickel-iron *hydrogenase* from an S-reducing bacterium shown in Figure 8.12 is a typical example of a mixed-metal centre where the metals perform a cooperative function as a single unit.

The *nitrogenase* enzyme is an example of a system where separate metalloproteins play quite distinct roles, although within each protein there may be several metals acting cooperatively. The task for *nitrogenase* is nitrogen fixation, which is essential for plant life, so it is one of Nature's fundamental synthetic processes. *Nitrogenase* produces two ammonium ions from dinitrogen and water in the presence of oxygen at soil temperatures and normal atmospheric pressure – a chemically demanding task. It occurs only in prokaryotic cells (those without nuclear membrane or mitochondria) such as bacteria and blue-green algae. An important nitrogen fixer is *rhizobium*, which invades legume roots. Nitrogenase is a

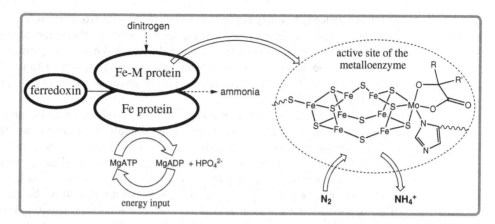

Figure 8.13
A cartoon representation of nitrogenase in operation, showing the component proteins. The active site of *nitrogenase* that resides in the Fe—Mo protein determined from a crystal structure analysis, and which contains two linked distorted M_4S_4 cubes, is also shown.

complex association of three proteins: a Fe—M protein (Fe—Mo, Fe—V, or Fe—Fe), a different Fe protein and a ferredoxin (Figure 8.13). The first two are required in combination to fix nitrogen. Ferredoxin is present as simply a reducing agent, and it can be replaced by other chemical reductants, even simple inorganic reducing agents. The overall reaction is given in (8.4).

$$N_2 + 16\,MgATP + 16\,H_2O + 8e^- \rightarrow 2\,NH_4^+ + H_2 + 16\,MgADP + 16\,HPO_4^{2-} + 6\,H^+$$
$$(8.4)$$

This reaction is thermodynamically inefficient, reflected in the complexity of the enzymic process.

The essential metalloproteins in this complex system contain Fe—S clusters somewhat like those we have met earlier in the ferredoxins. The Fe protein (MW $\sim$60 000) mediates the energy transfer mechanism for the reaction as a specific one-electron donor. A dimer of two identical subunits bridged by one Fe_4S_4 is involved, and it is thought to bind two ATP in order to deliver one electron to the adjacent Fe—M protein. The Fe—M protein is a large protein (MW $\sim$220 000) which exists as a $\alpha_2\beta_2$ tetramer, containing two Mo (or V, or extra Fe), $\sim$32 Fe centres, including four Fe_4S_4 units, and $\sim$32 S^{2-} ions. In recent years, after much effort, the crystal structure of the active site of a nitrogenase was revealed (Figure 8.13); note the linked cubic M_4S_4 clusters. It is believed that the dinitrogen binds near the centre of this linked cluster, which is the active site of the system. Nitrogenase is not a specific enzyme. It will reduce other small molecules of similar size and/or shape, as they can also bind to the active site. It promotes six-electron reductions, such as its primary target N_2 to NH_4^+ and HCN to $CH_4 + NH_4^+$, as well as two-electron reductions such as N_3^- to $N_2 + NH_4^+$ and C_2H_2 to C_2H_4.

8.3 Doing What Comes Unnaturally – Synthetic Biomolecules

To mimic nature is a demanding but appealing task. Many metalloenzymes are difficult to isolate and purify, and thus to obtain in sufficient quantities to employ in a commercially useful way. Indeed, they may not operate successfully away from their native environments. One approach is to develop synthetic compounds that mimic their chemistry. A mimic can be of two types – it may mimic the shape of the active site, or it may mimic the reaction that occurs at the active site. Arguably, one might expect the former to lead to the latter, but this is not usually the case, because the protein environment plays an important role in practically all enzymes.

To illustrate the art, we'll examine a small number of shape-related synthetic examples, commencing with our heme oxygen carrier. In principle, if successful, an artificial blood could be developed from this approach. Synthetic complexes that coordinate dioxygen are well established, as the first was described by Werner and Mylius in 1898 at the very beginning of coordination chemistry. In the laboratory, to mimic iron heme chemistry, two things must be achieved: oxygen coordination without iron oxidation; and removal of the opportunity for the formation of very stable dimers with μ-oxo linkages. This can be achieved by producing heme molecules with a scaffold on one side that prevents close approach of another heme, just what the protein achieves in a different but equally elegant manner. With close approach of two iron centres prohibited in one direction and the other axial site blocked by an imidazole ligand, dimers cannot form. Initially this was achieved with what was called a 'picket fence' porphyrin, where four bulky *tert*-butyl groups were disposed on one side of a synthetic porphyrin ring. A more sophisticated molecule, with a full 'cap' under which only a small molecule like oxygen can bind, enhanced the concept further (Figure 8.14). These are demanding organic syntheses, but their iron complexes do bind oxygen reversibly under controlled conditions, indicating that the chemistry we meet in the natural heme can be reproduced to some extent with small molecule analogues.

Model compounds that mimic the structure of ferredoxins can be prepared readily in the laboratory. This is achieved by choosing ligands that offer only S-donor groups, such as an R—S$^-$ or chelating $^-$S—R—S$^-$ species, and forming complexes with iron, a task readily achieved in a beaker under mild conditions. Some examples appear in Figure 8.15.

Mimics of *carbonic anhydrase* exemplify the genre further. Here, a simple saturated triaza-macrocycle replaces the histidine imidazole groups of the protein, yet the shape

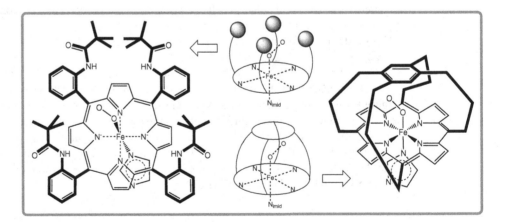

Figure 8.14
Examples of substituted synthetic hemes as their iron(II) complexes where close face-to-face approach of another heme is forbidden by the organic scaffolds introduced (and highlighted), allowing the small dioxygen molecule to bind reversibly at the protected axial site.

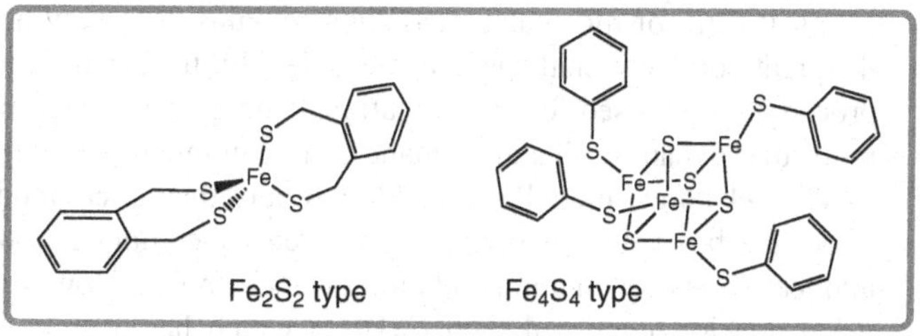

Figure 8.15
Examples of Fe$_n$S$_n$ clusters synthesized in the laboratory from an iron(II) salt reacted with simple thiol ligands and sulfide ion.

around the zinc(II) is similar (Figure 8.16). Such models lack the complexity and secondary structure of 'real' systems, but allow us to gain some understanding of the way reactions and shape adaption proceed in Nature.

More extensive examples can be found in specialist texts dedicated to bioinorganic chemistry. It is not the intent to do more here than present some highlights of this increasingly important field of coordination chemistry, sufficient to represent the field and link it to the more traditional aspects of the subject.

8.4 A Laboratory-free Approach – *In Silico* Prediction

The initial expansion of interest in model compounds just predated the rapid rise in computing power. This later development meant that three-dimensional computer modelling of compounds could at last be achieved at a rate that made it practicable. The development in computer speed gave rise to a parallel development of sophisticated computer software for modelling.

The simplest approach to modelling treats molecules as solid spheres of characteristic radii joined by springs of characteristic length, with the classical Hooke's Law dominant in the development of predictive equations for what is called *molecular modelling*. This very classical, contrived approach has proved both resilient and surprisingly useful. Perhaps our capacity, using electron microscopy, to at last 'see' individual atoms, which appear for all the world as spheres, provides some justification of the simplicity of this approach. This

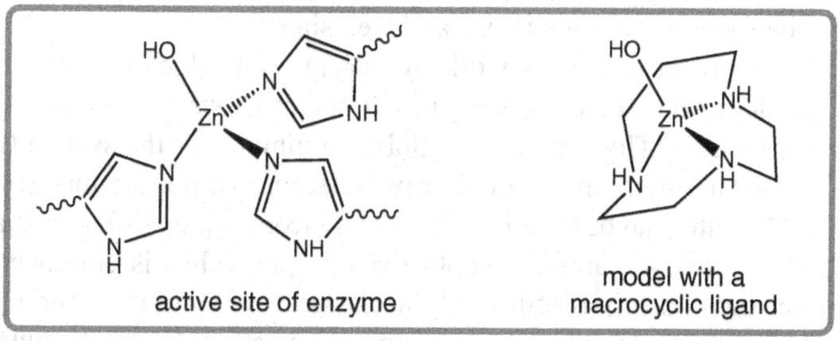

Figure 8.16
Comparison of the zinc(II) ion environment of the enzyme *carbonic anhydrase* and of a small molecule model that employs a triaza-macrocycle to supply the N-donor ligand set.

classical form of molecular modelling remains both very accessible and of value, since the great computational speed achievable with the simple equations employed allows the procedure to be used for conformational energy searching, molecule-protein docking and molecular dynamics that all demand a large number of energy calculations.

More elaborate modelling based on a pure theoretical model for the atom, an *ab initio* (from the beginning) approach, is limited by computer power, and as yet finds limited application in coordination and biological chemistry. However, a middle ground approach, using density functional theory (DFT) which has a more limited theoretical base, has grown in popularity and applicability. These so-called '*in silico*' (or silicon chip-based) approaches may present the future for an important aspect of chemistry – prediction of function. A screen-based model to probe a large family of related compounds certainly is more attractive than many hours of tedious repetitive laboratory work involving synthesis and testing to discover the most efficient compound for a particular task.

Let's explore the simplest type of 'in silico' technology, the so-called *molecular mechanics*, which treats atoms and their electron sets as hard spheres and bonds as springs. Because molecules have preferred bond distances and angles, all deformations away from ideal cost energy, and it is the sum of all deformations and other unfavourable nonbonded interactions that add up to an overall energy associated with stability. Comparison of calculated energies for different isomers of the same molecule, for example, allows prediction of the most stable isomer. This is a valid application, as is the calculation of energy for different conformations of a molecule. However, comparisons of energy between different molecules is inappropriate, since they will contain different numbers of atoms and bonds and thus will differ inherently, so that their relative energies offer little real meaning.

The total strain energy of a system can be considered to arise from bond length distortion (stretching or compressing a bond away from its ideal, represented here as E_{str}), bond angle distortion (bending bonds to open out or close up an ideal angle; E_{bend}), torsional effects (twisting of groups around a particular bond relative to each other; E_{tor}) and nonbonded contributions (van der Waals attraction, steric repulsion and electrostatic attraction or repulsion; E_{nb}). This can be expressed by (8.5).

$$E_{total} = E_{str} + E_{bend} + E_{tor} + E_{nb} \qquad (8.5)$$

The origins of these effects are represented simplistically in Figure 8.17. Each of the contributions may be represented by simple classical equations (e.g. $E_{bend} = \Sigma \{k_\theta(\theta - \theta_0)^2\}$, where k_θ is the force field parameter and θ_0 the ideal angle, summed for all angles in the system under examination). The full sets of equations are not pursued here, but can be found in specialist texts and web sites.

Software packages work by varying atom locations by small amounts, performing a calculation, and comparing it with the previous calculation. A stepwise process leads to a minimum. This may be the global minimum for the system (that is, the best solution), or a local minimum (a low-energy situation, but not the lowest-energy minimum available). The latter can be tested for by starting from a quite different shape and allowing the process to repeat. An array of sophisticated approaches is employed in software packages, but in the end this is still really a classical ball-and-stick modelling method, with inherent limitations. Despite this, the approach is efficient and the outcomes, with 'training' of the system through the use of appropriate force field parameters, are surprisingly good if using packages developed with suitable parameters for metal ions as well as nonmetallic elements. 'Docking' (directed noncovalent bonding) between even large biomolecules and complexes

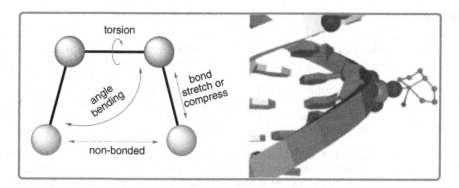

Figure 8.17
Representation of effects operating in molecular mechanics (left). An example of binding of a copper complex to a phosphate group on the backbone of a DNA strand optimised using molecular mechanics methodology appears at right.

can be probed, for example (Figure 8.17). Modelling is set to expand its reach from a dominant use with pure organic systems to include metal complexes, now that software packages appropriate for dealing successfully with metal ions have been established.

Concept Keys

Biomolecules offer a range of potential donors for metal ions, including carboxylate, amine, thiolate, phosphate and aromatic nitrogen groups.

A range of particularly the lighter metal ions from the *s* and *d* blocks find biological roles in nonproteins, functional proteins and enzymes.

Iron, zinc and copper are the most common of the transition metal ions in the human body, and play key roles at the active sites of electron carrier, structural and oxygen binding proteins, as well as in cleavage and redox enzymes.

Overall, a limited number of key structural motifs are met in the molecular components acting as ligands in metallobiomolecules. The oxygen binding proteins *myoglobin* and *hemoglobin* are examples of iron complexes of one class, unsaturated macrocyclic tetraamines called hemes.

Metal complexes lie at the heart of proteins involved in the vital life processes of oxygen storage, transport and reduction in all life forms.

Polynuclear metal units are common in biomolecules, as exemplified by the ferredoxins, which contain up to four iron centres in an iron-sulfur cluster.

Mixed-metal polynuclear units are also met in enzymes, with each metal ion usually playing a different role in the chemistry performed by the enzyme.

The chemistry in metalloenzymes usually occurs at the metal centre(s), and involves effectively reactions of coordinated ligands.

Laboratory preparation of synthetic molecules that mimic the shape and/or function of metallobiomolecules may be practicable.

Computer-based molecular modelling is finding application in the study of biomolecules and their interactions.

Further Reading

Bertini, I., Gray, H.B., Stiefel, E.I. and Valentine, J.S. (2006) *Biological Inorganic Chemistry: Structure and Reactivity*, University Science Books, Mill Valley, CA, USA. A recent and broad account of bioinorganic systems, from a perspective that provides good exposure of students to metal environments and active sites in metalloproteins.

Cramer, C.J. (2004) *Essentials of Computational Chemistry: Theories and Models*, John Wiley & Sons, Ltd, Chichester, UK. For the more advanced student, a reasonably pitched introduction to *in silico* chemistry; doesn't avoid equations, but does explain their significance and the context of their use.

Kraatz, H.-B. and Metzler-Nolte, N. (eds) (2006) *Concepts and Models in Bioinorganic Chemistry*, John Wiley & Sons, Ltd, Chichester, UK. A broad student-focussed coverage of the field of bioinorganic chemistry, which includes metalloenzyme model coordination chemistry; individual chapters are authored by experts in their field to a standard pattern that gives the book cohesion and good readability.

Lippard, S.J. and Berg, J.M. (1994) *Principles of Bioinorganic Chemistry*, University Science Books, Mill Valley, CA, USA. An ageing but accessible introduction for students; since fundamentals don't change markedly, it is still relevant.

McDowell, L.R. (2002) *Minerals in Animal and Human Nutrition*, 2nd edn, Elsevier, New York, NY, USA. For those wanting to stray more deeply into the role of metals in nutrition, this is a useful resource, with good coverage of core aspects.

Trautwein, A. (1997) *Bioinorganic Chemistry: Transition Metals in Biology and Their Coordination Chemistry*, Wiley-VCH Verlag GmbH, Weinheim, Germany. An overview of biological systems, with a coordination chemistry perspective consistent with the approach in this chapter.

9 Complexes and Commerce

It would be inappropriate, before concluding, to leave the subject without some mention of the roles that coordination complexes play in the working world. Otherwise, one could be left with the impression that the field is little more than an academic plaything. Of course, we have seen in Chapter 8 how Nature has made fine use of coordination complexes, but this goes on essentially without human intervention. What is important to appreciate is that coordination chemistry and coordination complexes lie at the core of an array of important applications in industry, medicine and other fields. Given the inventiveness of humans, it should come as no real surprise to find that development of new compounds often leads to an analysis of their potential applications, and these can be surprisingly diverse. Here, we shall touch on a few areas and examples, to give but a flavour for the field; a full serving can be pursued in specialist texts and reviews, if desired.

9.1 Kill or Cure? – Complexes as Drugs

Medicine is an area of high activity and interest, partly because health issues are of strong interest to us, as we seek to achieve and retain the best possible quality of life. Recovery from and/or control of disease often involves the use of drugs, which may be natural products, synthetic organic compounds and, though less often met, synthetic coordination complexes. To distinguish the latter from the more common organic drugs, those containing metal complexes are sometimes defined as *metallodrugs*. The use of metals in medicine has a history thousands of years old, at least back to the ancient Egyptians (who used copper compounds in potions) and Chinese (who used gold compounds), and perhaps beyond. In the distant past, some treatments may have caused more damage than provided a cure, but to be useful in modern medicine a compound must provide a measured beneficial effect while displaying low toxicity. The Therapeutic Index defines relative benefit versus toxicity, expressed as LD_{50}/ED_{50}, or the ratio of the dose required to kill 50% of a host versus the dose needed to produce an effective therapeutic response in 50% of the host. Before any new drug can be introduced, it must go through a series of clinical trials covering, for human use, four phases; needless to say, given the expense and time involved in this process, a new drug has to be markedly better than any others already on the market to succeed commercially.

Pharmaceutical companies have appeared somewhat resistant to the development of metallodrugs in the past, perhaps concerned about toxicity, specificity and metal accumulation in patients over extended periods of use, but also reflecting relative unfamiliarity with the background chemistry. They have also benefited from serendipity as regards discovery of some current metallodrugs, allowing them to put aside rational design in approaches to specific targets. Changes are afoot, however, with growing recognition in recent decades that

Introduction to Coordination Chemistry   Geoffrey A. Lawrance
© 2010 John Wiley & Sons, Ltd

Table 9.1 Some medically-related applications of metal coordination complexes.

Function	Compounds
Bioassay	
Fluoroimmunoassay	europium(III) complexes
Diagnostic Imaging	
Gamma radiolysis contrast agents	radioactive ^{99m}Tc complexes
Magnetic resonance contrast agents	gadolinium(III) complexes
Radiopharmaceuticals	short half-life radioactive metal salts and complexes
Anticancer Drugs	platinum complexes; titanium complexes; photosensitive porphyrin complexes
Other Selected Treatments	
Wilson disease and Menkes disease	copper complexes
Arthritis	gold complexes
Blood pressure control	iron complexes
Diabetes – insulin mimics	vanadium complexes
Antimicrobials	Ag, Hg, Zn and Bi compounds
Ligands as drugs	
Metal intoxication treatment	polydentate ligands (such as EDTA)

metallodrugs can be sophisticated alternatives to conventional pure organic drugs, tied not only to reactivity and stereochemical differences but also with access to nuclear, magnetic and optical properties simply not available with organic drugs.

9.1.1 Introducing Metallodrugs

Metal-containing coordination compounds that show a capacity to cure or control a disease have grown remarkably in number and range of applications in recent decades. Their form covers a wide range of metal ions, ligands and stereochemistries. Some examples appear in Table 9.1. It is not the role of an introductory text to cover applications in depth, but it is appropriate to illustrate their applications and show how they link with our basic concepts of coordination chemistry.

9.1.2 Anticancer Drugs

Cancer, as one of our most significant diseases, is the focus of intensive international study and treatment development. It is interesting to discover that platinum-based chemotherapeutic drugs are amongst the most active and commonly used clinical agents for treating a range of advanced cancers. Diamminedichloroplatinum(II) (called *cisplatin* in medical use; Figure 9.1) was the first to be used clinically, and remains one of the largest-selling drugs on the market. This is an old coordination compound, first reported by Peyrone in 1845, with its structure defined by Werner. The potential anticancer properties of cisplatin were discovered serendipitously by Rosenberg in 1965, and the drug was introduced clinically in 1971. It is routinely widely used, including for treating ovarian, testicular, bladder, head, neck and small-cell lung carcinomas. In particular, cisplatin has made testicular cancer eminently curable, with more than 90% of sufferers now cured (a celebrated case being American cyclist Lance Armstrong, a multiple Tour de France winner).

Cisplatin acts by binding to DNA and inhibiting replication in the cancer cell. Substitution reactions of coordinated chloride ligands are the key chemistry in reactions of cisplatin in a human cell. Most cisplatin circulates in the blood unchanged over a short timeframe, as

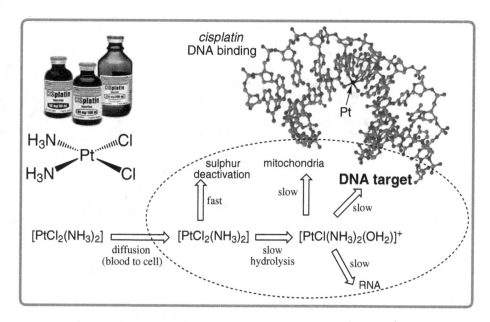

Figure 9.1
The drug cisplatin, its reaction pathways in the cell and an example of coordination to DNA, leading to distortion of the double helix (cisplatin-DNA binding figure reproduced from the Protein Data Bank; DOI:10.2210/pdb1aio/pdb).

the high chloride ion concentration in blood suppresses coordinated chloride hydrolysis and hence further reaction. Once across the cytoplasmic membrane of a cell, however, it meets a much reduced chloride concentration (as low as 4 nM) and undergoes a series of hydrolysis reactions, forming species including $[PtCl(NH_3)_2(OH_2)]^+$, where one chloride ligand has been replaced in a substitution reaction by a water ligand. The biological target of the complex is DNA in the cancer cell; however, it is not particularly discriminating in its chemistry, and can participate in coordination chemistry with sulfur-containing groups and other functionalities that lead to its 'capture' and inactivation; it binds so strongly to these that it is no longer available for further chemistry. The interaction sought (Figure 9.1) is with the heterocyclic nitrogen bases that form part of DNA – cytosine, guanine, adenine and thymidine. These types of N-donors are excellent ligands for platinum(II); guanine is believed to be a particularly favourable donor. Binding is covalent – coordinated chloride ion hydrolysis is necessary first, to provide a reasonably labile site (coordinated water) for substitution chemistry that leads to the introduction of the N-donor DNA base into the inner coordination sphere. The cationic aqua complexes formed following initial chloride hydrolysis (particularly $[Pt(NH_3)_2(OH_2)_2]^{2+}$) enhance the activity of the drug as a cytostatic agent due to an ionic attraction to the negatively-charged DNA helix, which has an anionic phosphate backbone. However, it is the structural changes arising from coordination of N-donor DNA bases in place of water ligands that are the key to their activity.

In addition, the cisplatin may bind to other proteins and biomolecules, sometimes together with binding to DNA. In achieving binding to two bio-sites, it is using both *cis* coordination sites originally occupied by chloride ions, so that hydrolysis of both chloride ion ligands is necessary. Possible binding modes are shown in Figure 9.2. Which of these is most important is under continuing debate, but intrastrand processes are favoured. Cisplatin

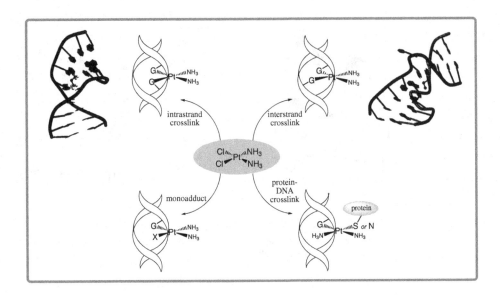

Figure 9.2
Various modes of coordination of the platinum(II) drug cisplatin to bases in DNA strands, following initial chloride hydrolysis; a guanine N-donor (G) is favoured. Actual models determined from structural studies of intra- (at left) and inter-strand (at right) coordination to DNA oligomers (drawn from the Protein Data Bank; DOI:10.2210/pdb1ddp/pdb and 10.2210/pdb1ksb/pdb) are also shown.

can form intrastrand crosslinks between adjacent bases, preferring two guanine bases (a cisplatin–guanine–guanine intrastrand DNA adduct is depicted in Figure 9.1). This coordination mode cannot be adopted by inactive transplatin, suggesting that it is indeed of importance. This crosslinking has been shown to cause unwinding and duplex bending, perhaps sufficient to attract high-mobility group damage-recognition proteins, which bind and further inhibit replication.

The core problem with cisplatin is that dose-related toxicity occurs, leading to an array of problems for patients, such as joint pain, ringing in the ears and hearing difficulties, and general weakness. The maximum tolerated dose of cisplatin is around 100 mg per day for up to only five consecutive days. While the drug displays high activity, these unfortunate side-effects led to a search for a better analogue (Figure 9.3). *Carboplatin* [*cis*-diammine(1,1-cyclobutanedicarboxylato)platinum(II)] was introduced as a second-generation drug in 1990. It shows about a fourfold reduction in side effects to the kidneys and nervous system. Because the drugs must circulate in the bloodstream before finding their target, slower degradation *in vivo* is believed to be one factor of importance. Both cisplatin and carboplatin suffer from the emergence of acquired resistance with time. Subsequently, *oxaliplatin* [(1,2-cyclohexanediamine)(oxalato)platinum(II)] was introduced as a third-generation drug. It is active in all phases of the cell cycle, and binds covalently to DNA guanine and adenine bases dominantly via intrastrand crosslinking. It is thought that DNA mismatch repair enzymes are unable to recognize oxaliplatin–DNA adducts in contrast with some other platinum–DNA adducts, as a result of the bulkier nature of the complex. The complex is inactivated by thiol protein binding, alteration in cellular transport and increased DNA repair enzyme activity. Development of a new-generation drug to address inactivation has led to *satraplatin*, the

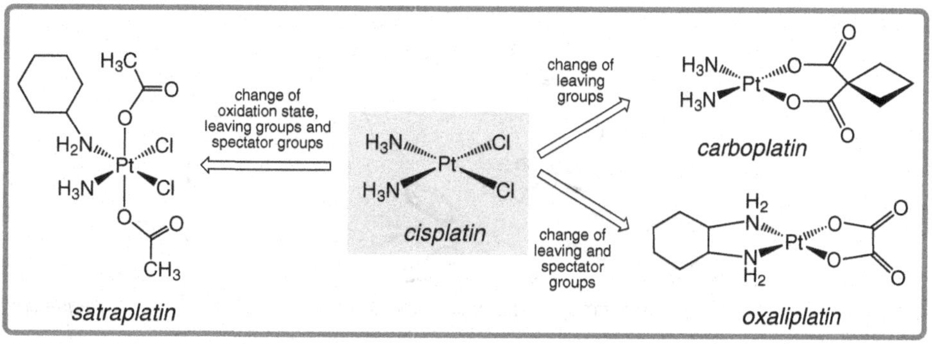

Figure 9.3
Evolution of the first generation platinum drug cisplatin into later generation drugs.

first orally-administrable platinum drug – in this case a platinum(IV) complex. A mode of action that involves reduction to Pt(II) *in vivo* has been proposed. This drug shows no neurotoxicity or nephrotoxicity, with bone marrow supression the dose-limiting adverse effect; it is under consideration for approval as a drug in prostrate cancer treatment.

Platinum complexes are currently the best-selling anticancer drugs in the world, with billion-dollar annual sales. Nevertheless, the search for better platinum drugs goes on, including development of some polynuclear complexes. Chemotherapy with platinum drugs can be effective, but may have severe side effects because the drugs are toxic and cannot discriminate sufficiently between normal and cancer cells. Development of new drugs centred on the introduction of components that deliver the drugs specifically to cancer cells will assist in specificity and reduction of side effects.

9.1.3 Other Metallodrugs

Although platinum drugs have a well-established place in the array of treatments used for cancer sufferers, they are not the only coordination complexes that exhibit activity. Another class of complexes under examination as anticancer drugs are those of titanium. One titanium(IV) anticancer drug, *budotitane* ([Ti(bzac)$_2$(OEt)$_2$], Figure 9.4) has qualified for clinical trials, and a range of organometallic compounds such as titanocene dichloride (Figure 9.4) have shown promise. Ruthenium complexes are also under investigation as potential anticancer drugs.

A gold(I) complex *auranofin* has been developed for the treatment of rheumatoid arthritis. This is a typical linear two-coordinate complex of gold(I), with 'soft' S and P donors to match the 'soft' metal ion (Figure 9.4). The bulky ligands here promote compatibility in the bio-environment, and activity is influenced by the substituents on the RS$^-$ and R$_3$P ligands. Gold complexes have also been examined for antimicrobial, antimalarial and anti HIV (human immunodeficiency virus) activity. A copper(II) complex (Figure 9.4) is effective as a competitive inhibitor of HIV-1.

Vanadium compounds, mainly square pyramidal complexes of the type [O=V(O—R—O)$_2$], act as insulin mimetics with potential in diabetes control if toxicity issues can be overcome; one is in use as an orally-administered agent in animals. A pentagonal bipyramidal complex of manganese(II), [MnCl$_2$(N_5)], where N_5 is a pentaaza macrocycle with four amine nitrogen donors and one pyridine nitrogen donor, acts as a

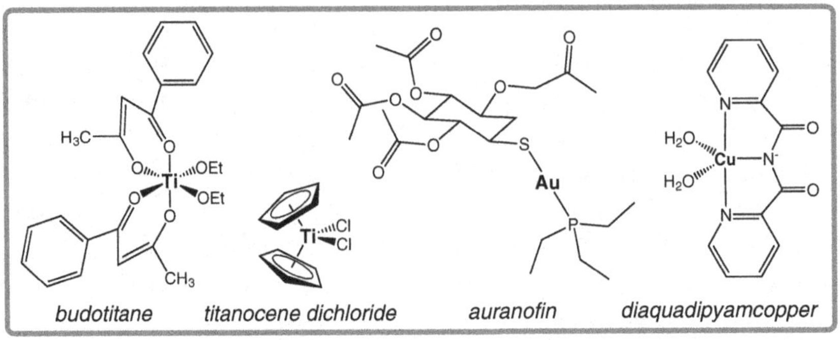

Figure 9.4
The structure of two titanium(IV) potential anticancer drugs, the gold(I) arthritis drug auranofin, and a copper(II) HIV-1 protease inhibitor.

superoxide dismutase mimetic. The variety of complexes under study as potential drugs is astounding, and suggests that more advances will be reported in the short-term future.

9.2   How Much? – Analysing with Complexes

Complexation, as we have seen already, causes change in the properties of a metal ion. Where such a change is reflected in a property that varies linearly (or at least in a known manner) with respect to complex concentration, there is the makings of an analytical method. This may be applied to detection of the metal ion, or alternatively to the detection of a ligand species that binds the metal ion selectively. It may even be possible to have complexation as a sensor of another process or species that is 'switched on' by a changed noncovalent interaction tied to the complexation event (such as an ion-pairing interaction). The two examples given here relate to changing luminescence intensity associated with complexation, but in principle any measurable process can be employed. Luminescence (fluorescence or phosphorescence) is a useful sensor, where it occurs in compounds, as it provides usually high sensitivity due to low background 'noise' levels. This therefore permits detection of the luminescent species down to very low concentrations, with instrumentation able to detect it widely available at reasonable cost.

9.2.1   Fluoroimmunoassay

The discovery by Weissman in 1942 that europium(III) complexes of β-diketone O-donor chelates absorb light in the UV region but emit light in the visible region set in train research on the analytical applications of fluorescent f-block complexes that continues to this day. These complexes work by light being absorbed by the ligand (at ~360 nm), which undergoes promotion to an excited state from which there is a transfer to metal orbitals that also leaves the metal in an excited state; energy is released at a different wavelength (as visible light at ~615 nm) when the complex reverts to its normal or ground state. The lanthanide ions that can fluoresce strongly are Sm^{3+}, Eu^{3+}, Tb^{3+} and Dy^{3+}, because the excitation energy level of these four metals lies slightly lower than the excited levels of the ligands and thus they are able to readily accept the energy transfer. With water ligands in particular, a mechanism for rapid decay of the excited state other than by fluorescence operates, and so it is only in the presence of appropriate ligands that the lifetime of the excited state is sufficient to lead to strong fluorescence, with applicability assisted by a sharp emission peak. The emission process for the complexes is relatively slow, and the complex is said to have a long fluorescent lifetime (albeit still in the sub-second range), attributed in part to the 4f orbital being shielded by outer orbitals like the 5s. This long lifetime is a key to analytical applications, since a process is employed that waits a short time following an excitation pulse to let other unwanted but short-lived fluorescence die away, allowing the lanthanide-based fluorescence to be measured essentially free of interference; these are called time-resolved fluorescence analysis methods. Typically, a 'wait time' of ~200 μs followed by a measurement time of ~400 μs is employed, permitted because the fluorescent life span of the lanthanide complex is hundreds of microseconds, making it possible to detect the target fluorescence with good sensitivity after fading away of the short-lifespan background fluorescence of impurities and organic species.

Fluoroimmunoassay makes use of the above behaviour. One of the common commercial methods is dissociation-enhanced fluoroimmunoassay (DELFIA). In this, a nonfluorescent Eu(III) EDTA-like complex is attached by a simple chemical reaction to an antibody or antigen, in a process called labelling. An immunoreaction is next initiated to bind the target, and then a β-diketone and trioctylphosphine oxide (TOPO) mixture are added to the immunocomplex formed, at pH ~3, to promote release of the Eu(III) from the antibody and its complexation as the strongly fluorescent complex [Eu(β-diketonate)$_3$(TOPO)$_2$], which is then measured by time-resolved fluorescence methods. The signal size relates to the amount of europium complexed, which in turns relates directly to the amount of the specifically formed target immunocomplex. This process is represented schematically in Figure 9.5.

The ease of handling tied to the high detection sensitivity of fluorescent lanthanide probes has led to their being used in base sequence analysis of genes, DNA hybridization assay and fluorescent bioimaging, apart from in immunoassay. They are rapidly replacing organic

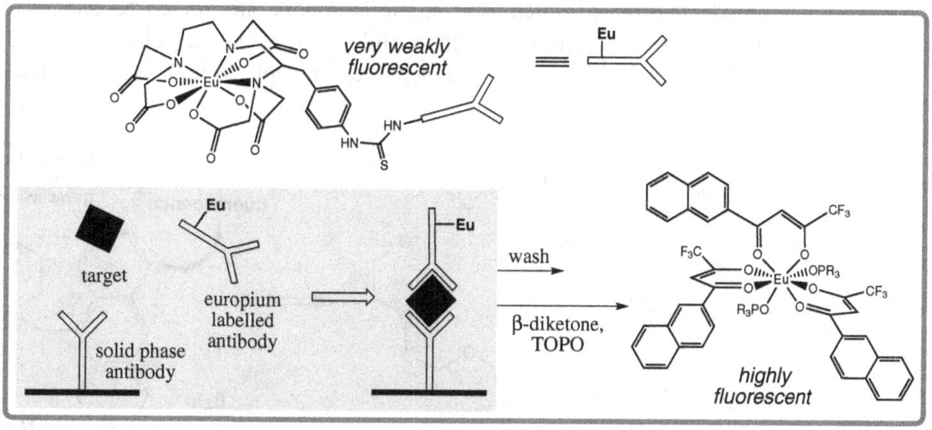

Figure 9.5
Chemistry involved in the fluoroimmunoassay process. Europium ion is bound to an EDTA-like pendant on an antibody, which traps the target pathogen at a second surface-anchored antibody with high specificity. Following washing to remove the test solution, the europium in the trapped adduct is released on chelate addition to form a highly fluorescent complex in solution that is sensed by fluorescence spectroscopy. Signal intensity is directly related to target pathogen concentration.

fluorescent dyes, which have the disadvantage of short fluorescent lifetimes, which means they are compromised by interference signals. This is a fine example of simple coordination compounds at work.

Understanding the photochemistry and photophysics of coordination complexes resulting from the interaction of complexes with light provides a fundamental base for development of other applications. Apart from biomedical imaging and therapy applications of the type exemplified above, applications under active development or the subject of preliminary research include the use of coordination complexes in photocatalysis, optical information transfer and storage, optical computing, analytical sensing and harvesting of solar energy. The penultimate of these fields is exemplified below.

9.2.2   Fluoroionophores

Fluoroionophores are large ligands that are described as supramolecular systems; they consist of two components joined together covalently – an *ionophore* ('ion-lover') molecule linked to a light-sensitive *fluorophore* ('light-lover'). The role of the first unit is to capture metal ions, whereas the role of the second unit is to capture light and emit it at a different wavelength as fluorescence, which can then be detected and measured. It is the impact of a captured metal on the fluorescence process that is at the core of the analytical application. These fluoroionophores are designed to sense metal ions selectively; ion recognition involves the ionophore binding the metal ion through coordinate covalent bonds, whereas the signalling process of the fluorophore depends on photophysics. Typically, the process involves metal ion control of photo-induced electron transfer (Figure 9.6). Selective metal-ion detection using a fluoroionophore arises because of the variation of colour (or fluorescence wavelength) with metal ion. Even the various alkali and alkaline earth ions can be distinguished.

Phosphorescence or fluorescence of complexes in the visible region has also permitted other developments of note. For example, Pt(II), Cu(II), Rh(III) and Ir(III) complexes of polydentate ligands such as aromatic N-donors, and mixed N- and P- or C-donor compounds produce emissions in the visible region, and are being developed as optical light emitting diodes (OLEDs) or phosphorescent optical light emitting diodes (PHOLEDs).

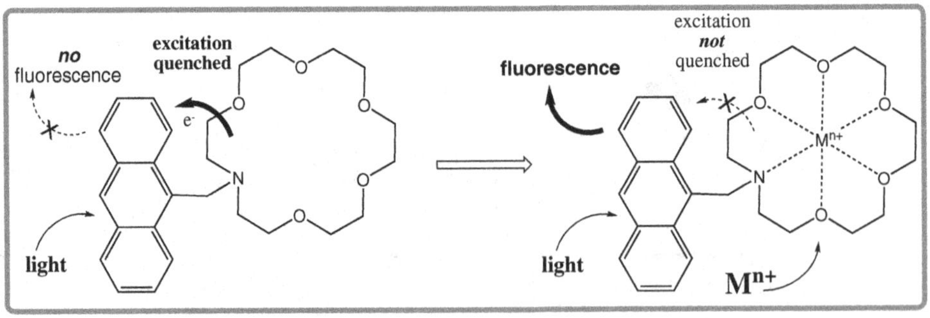

Figure 9.6
A cartoon of a fluorescent 'switch', turned on or off (quenched) depending on the absence or presence of a metal ion. The ionophore (the cyclic polyether) is the metal-binding component, the fluorophore (the fused-ring aromatic unit) is the component activated by light. Complexation stops electron transfer that otherwise quenches fluorescence.

Ruthenium complexes used to lead research in photochemistry of metal compounds, but rhodium complexes have recently overtaken them as the key target compounds due to their applications in OLEDs. This is a lively and ever-changing field; for example, over 90% of luminescent iridum(III) complexes have been reported only in the six years to the beginning of 2009. With their luminescence 'tuneable' through ligand choice, iridium complexes are firm candidates for optical display applications.

9.3 Profiting from Complexation

9.3.1 Metal Extraction

An ore body is a commercially viable concentration of one or usually several metals. This concentration reflects the amount of an element in the Earth's crust; for a common element like iron, an ore body could be ~50% iron, whereas for a rare element like gold it could be much less than 1% (Table 9.2). However, digging an ore out of the ground is but the first step in the release of the metal and its conversion into commercially useful forms. Whereas some common metals are recovered by very simple redox reactions that do not involve coordination chemistry to any marked extent, some metals do rely heavily on coordination chemistry for their extraction and recovery.

Pure titanium metal is made by chlorination of the oxide TiO_2 to form the tetrahedral complex $TiCl_4$. This is then reduced in a redox reaction with magnesium metal to yield free titanium metal as a powder. To provide a continuous loop of reagents, the $MgCl_2$ also formed in this reduction step is electrolyzed to produce chlorine and magnesium metal. With the chlorine and magnesium re-used fully, this is a good example of an industrial process with negligible waste products; of course, energy is consumed, so it still carries an environmental impact.

Gold recovery is a simple example where complexation plays a key role. Gold ore, where the gold is present as finely-divided metal that cannot be recovered mechanically, is slurried in the presence of air and added cyanide ion. The elemental gold is oxidized by dioxygen and complexed by cyanide to form a gold(I) cyanide complex (9.1).

$$Au + 2\ CN^- + {}^1\!/_2 O_2 + H_2O \rightarrow [N{\equiv}C{-}Au^I{-}C{\equiv}N]^- + 2\ {}^-OH \qquad (9.1)$$

Table 9.2 Abundance and commercial ore concentrations of selected economic metals.

Metal	Abundance (%)[a]	Ore grade (av. %)[b]	Conc. factor[c]	Value ($US/kg)[d]
Al	8	28	3.5	1.4
Fe	5	25	5	<1
Ni	0.007	0.5	70	5.1
Cu	0.005	0.4	80	4.4
Au	0.000 000 4	0.000 1	250	29 500
Mn	0.09	35	390	2.4
Zn	0.007	4	570	1.4
Sn	0.000 2	0.5	2 500	12.7
Cr	0.01	30	3 000	1.5
Pb	0.001	4	4 000	1.4

[a] Average abundance in the Earth's crust.
[b] Approximate minimum exploitable grade of deposits, as a percentage of the ore.
[c] Concentration of an ore grade deposit relative to the crustal average.
[d] Value in 2009 – varies regularly; the value also affects the concentration factor acceptable for a commercial mining operation.

This water-soluble complex is then captured following filtration by adsorption onto carbon, and subsequently recovered by further processing. The linear two-coordinate coordination complex of Au(I) is at the core of the process. Other ligands are being developed to replace the environmentally dangerous cyanide, such as thiourea ($S=C(NH_2)_2$), which also acts as a good monodentate ligand and forms a linear two-coordinate complex.

This latter chemistry is an example of what is called *hydrometallurgy*, or the treatment of ores in aqueous solution. It is in this medium that most of the coordination chemistry of metallurgical processes can be found. A decrease in the grade and increase in the complexity of available ores of some metals has increased the need to develop new hydrometallurgical processes. The two key steps for ore treatment are leaching to dissolve the target metal and recovery (often involving a precipitation reaction) of the purified dissolved target metal. Although simple acid or base leaching can be successful, some ores require the addition of complexing agents (ligands) to assist, such as the dissolution of copper oxide in ammonia/ammonium chloride (9.2).

$$CuO_{(s)} + 4\ NH_4^+ + 2\ {}^-OH \rightarrow [Cu(NH_3)_4]^{2+}{}_{aq} + 3\ H_2O \tag{9.2}$$

This is only a complexation reaction, as the copper is already in its desired oxidation state. The gold recovery process described above involves both oxidation and complexation. Another example of such a reaction is nickel sulfide pressure leaching, where it is the sulfur anion that is oxidized, the nickel remaining in its Ni(II) form throughout (9.3).

$$NiS_{(s)} + 2\ O_2 + 6\ NH_3 \rightarrow [Ni(NH_3)_6]^{2+}{}_{aq} + (SO_4{}^{2-})_{aq} \tag{9.3}$$

Complexation also plays a role in the commercial process for the separation of the platinum group metals, involving formation of chloride and ammonia complexes; this separation of so many related elements is a demanding procedure. Platinum metal concentrates are usually dissolved in the strong mixed acid *aqua regia* as the first step, to separate the soluble fraction (Au, Pd, Pt species) from the insoluble fraction (Rh, Ir, Ru, Os and Ag species). Solvent extraction of the soluble fraction with dibutyl carbitol separates gold from palladium/platinum species. The latter soluble complexes are the square planar $H_2[PdCl_4]$ and the octahedral $H_2[PtCl_6]$. Addition of ammonium chloride precipitates sparingly soluble $(NH_4)_2[PtCl_6]$, leaving the palladium complex in solution. From the separated complexes, the free metals are recovered through different redox reactions.

Solvent extraction of a metal ion in aqueous solution with another immiscible solvent containing ligands allows the target metal ion to be separated from other unwanted metal ions in the original leach solution; it is a process sometimes met in hydrometallurgy. One example is the recovery of uranyl ion ($UO_2{}^{2+}$), which uses a phosphate diester dissolved in kerosene to complex and extract the uranyl ion into the organic phase, from where it can be recovered. The extraction chemistry involves reaction (9.4), where two $R_2PO_2{}^-$ chelates bind to the uranyl cation to form a neutral complex that is highly soluble in the organic phase, into which it shifts.

$$UO_2{}^{2+}{}_{(aq)} + 2\ R_2PO_2H_{(org)} \rightarrow [UO_2(O_2PR_2)_2]_{(org)} + 2\ H^+{}_{(aq)} \tag{9.4}$$

Similar chemistry, but with a different chelate ligand, is used to purify copper(II). The copper complex is loaded into the organic phase at a pH between 2 and 4, then back-extracted into water as the $Cu^{2+}{}_{aq}$ ion using dilute sulfuric acid at pH 0. It is recovered as either copper sulfate or, with following electrochemical reduction, copper metal. Solvent extraction involving complex formation is also employed in the separation of cobalt–nickel mixtures.

9.3.2 Industrial Roles for Ligands and Coordination Complexes

Coordination chemistry offers many examples of applications in industry beyond those addressed above. Recent developments in chemistry promise greater applications in the future. Here, a limited number of examples from two fields are presented to give a sense of opportunities for coordination complexes in commercial roles.

9.3.2.1 Complexes as Catalysts and Asymmetric Catalysts

The chemical industry as we currently know it would be markedly different without transition metal catalysts, as these play roles in a wide range of processes. The key task of a catalyst is to accelerate a reaction by effectively lowering the activation barrier for the reaction. Apart from acceleration, a catalyst may also be able to induce optical activity in an organic product if it includes a chiral ligand. The success of an asymmetric catalyst is defined by the enantiomeric excess, which is the difference in percentage yields of the major and minor enantiomers of the product. If 90% of one optical isomer forms and 10% of the other, the enantiomer excess is 80%; obviously, the closer that this value is to 100% (which means stereospecificity is achieved) the better. Asymmetric synthesis in industry depends fully upon transition metals as the active site of the catalysis.

One of the best known coordination complexes that acts as a catalyst of chiral epoxidations is the Jacobsen catalyst. This is a manganese(III) complex of a sterically demanding chiral ligand (Figure 9.7), which is used in conjunction with the oxidant hypochlorite (ClO$^-$). The oxidant initially oxidizes the complex to form an oxomanganese(V) compound, which then is able to deliver the oxygen to an alkene to form an epoxide, returning to the Mn(III) state. The epoxidation occurs highly steroselectively, with enantiomer excesses of usually greater than 95% achieved routinely. It is believed that the substrate is bound to the metal ion in the activated state, with oxygen transfer occurring in the chiral environment of the tetradentate N$_2$O$_2$ ligand leading to chirality in the epoxide product. Since the Mn(III) complex is regenerated, it is able to act as a catalyst, undergoing many 'turnovers' before degradation through ligand dissociation reactions.

The above example is of a homogeneous catalyst, which is one that is dissolved in solution for the reaction. One of the best understood and long established coordination complexes

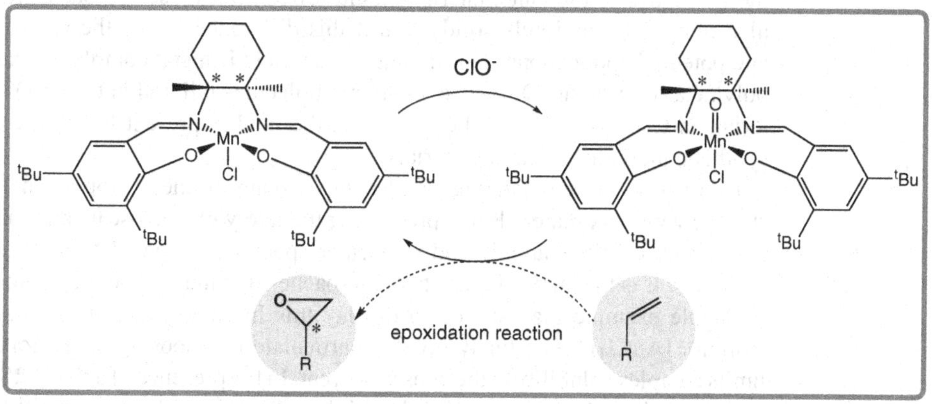

Figure 9.7
Chiral epoxidation with Jacobsen's catalyst. Chiral centres in the ligand and product are marked (∗).

acting as a catalyst is the simple square planar Rh(I) complex [RhCl{P(C$_6$H$_5$)$_3$}$_3$], Wilkinson's catalyst, used for the hydrogenation of terminal alkenes (see Figure 6.1). It involves an intermediate where both hydrogen and the alkene are coordinated, allowing an intramolecular reaction between these to form an alkane product – effectively a reaction of coordinated ligands, but one where the product can depart, so that the original complex is reformed, allowing the process to occur again, a key requirement for catalysis.

There are a number of others in common industrial use, such as those used for polymerization of alkenes. One example of an organometallic homogeneous catalyst is the Zr(IV) complex [Zr(CH$_3$)(η^5-Cp)$_2$X], which operates by binding alkene monomers to the metal prior to addition to a growing carbon chain. A similar coordination of substrate is involved in the use of [Co(CO)$_4$H] as catalyst in the hydroformylation of alkenes to aldehydes. Likewise, the [Rh(CO)$_2$I$_2$]$^-$ ion, formed *in situ*, catalyses the carbonylation of methanol to acetic acid.

Another catalyst type is the heterogeneous catalyst, which remains as a solid and promotes chemistry at the surface. To function well, they require high surface areas per unit mass. Metal oxides and hydroxides are common examples. A vanadium(V) oxide is employed in the formation of ammonia from nitrogen and hydrogen under elevated temperature and pressure, for example. Polyoxometallate metal clusters, which are oxo-ligand coordination complexes employing dominantly O^{2-} and HO$^-$ as ligands, have some catalytic roles.

The two types of catalysts merge with the development of what are called 'tethered' catalysts, where a homogeneous catalyst is amended so that it is able to be attached covalently to an inert surface, such as silica. This is also called a supported catalyst. Having the active component available as part of a solid can assist processes where carrying the catalyst forward in solution to another stage of the process may lead to contamination or catalyst destruction. Further, surface attachment also can alter catalytic activity favourably in certain cases.

9.3.2.2   Complexes as Nanomaterials

Nanotechnology is one of the new frontiers of chemistry – the development of new materials with well-defined structure within the size range of from 1 to ∼100 nm. In particular, the properties of the nanomaterial should differ from those of both conventional molecular compounds and bulk solids, and it this difference that is the key to their attraction and potential applications. This definition can include metal complexes, particularly larger polymetallic systems. Of course, many biomolecules fall within the definition of nanomaterials in terms of size, but the interest in the field at present lies in the development of synthetic more than natural materials.

Fabrication of nanomaterials divides into two approaches – 'top down', which relies on starting with bulk materials and processing them to yield nanoscale materials, and 'bottom up', which employs atomic and molecular species aggregated to form larger nanoscale materials. It is the latter of these two approaches that may involve coordination chemistry. A simple example may suffice to display this in application. It is possible to produce (cation)$^+$[AuCl$_4$]$^-$ as a finely-divided particulate by choosing an appropriate cation that limits complex solubility in the chosen solvent. In the presence of a thiol (RSH), substitution of chloride ligands on the gold(III) by thiolate (RS$^-$) ligands can occur. When the resultant gold(III) thiolate is treated with a suitable reducing agent, the Au(III) is reduced to form nanoscale metallic gold particles that remain coated with thiol, of type {Au$_x$(RSH)$_y$}. This

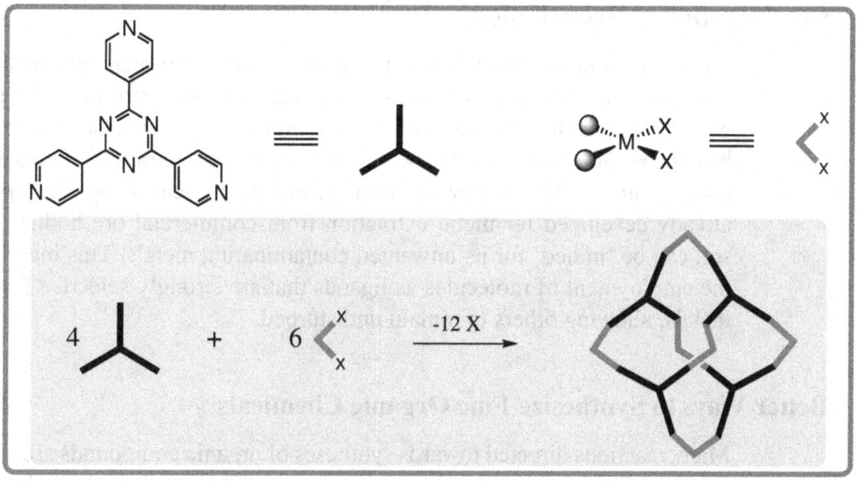

Figure 9.8
Assembly of rigid precursor units into a larger nano-assembly is directed by the shape of components.

represents a new form of gold particles with special properties, whose size and surface character can be controlled through manipulation of reaction conditions and reagents.

Building polymetallic clusters into new nanomaterials through combination of particularly shaped ligands and monomer complex components is an area of growing development. This is an extension of concepts we have already developed in Chapter 6 (see Figure 6.4). This is really molecular Lego – there is only a limited number of ways ligand and complex precursors of particular and usually rigid shapes can combine as larger units, just like Lego blocks can only fit together in certain ways. An example of the concept is illustrated in Figure 9.8.

Loss of the two *cis* X-groups from the complex allows coordination of terminal pyridine groups from the Y-shaped ligands in their place. However, as a result of the rigid ligand shape, any ligand can only attach as a monodentate to a single metal, and so a network of attachments is built up to satisfy the demands of the ligand for tridentate coordination and the metal for filling two adjacent coordination sites. A large number of complex, shape-directed nanosized clusters have been developed using this type of approach.

9.4 Being Green

If you are familiar with the Muppet philosopher Kermit the Frog, you'll probably be aware of his view that it's not easy being green; however, it seems that chemists are developing some solutions to this problem. Key new driving forces in applied chemistry are environmental sensitivity, social responsibility and energy saving processes. Some may say that greenness has been thrust upon industry by government, but for some time there has been recognition in industry that a waste product is also wasted profit, leading to process development that attempts to find a use for everything produced in a plant. It isn't our purpose here to interrogate the environmental credentials of industry, but rather to look for situations where coordination chemistry plays a role.

9.4.1   Complexation in Remediation

Soils contaminated with heavy metal waste are a problem in the urban industrialized environment, often making significant parcels of land unsuitable for re-use. An obvious way forward is the extraction of the contaminating metals from the soil, but this needs to be done without removing the benign and dominant natural metal ions in the soil. Selective complexation offers a way forward; in effect, one can adopt the approaches used and already developed for metal extraction from commercial ore bodies – the contaminated soil can be 'mined' for its unwanted contaminating metals. This may be best achieved by the employment of molecules as ligands that are strongly selective for the target metal or metals, allowing others to remain undisturbed.

9.4.2   Better Ways to Synthesize Fine Organic Chemicals

Many reactions directed towards syntheses of organic compounds are conducted in organic solvents. There are several approaches to removal of the inherently dangerous and harmful organic solvents under development: water-based reactions, use of nonvolatile and more benign ionic liquids as solvents, and solid state reactions using microwave energy are three popular areas of development. Although these need not draw on coordination chemistry, they represent the likely future of synthesis of so-called fine organic chemicals (relatively limited amounts of high purity compounds). Where coordination chemistry can play a role lies in the development of water-based chemistry, since ionic coordination complexes are inherently soluble in water. Thus, employing metal-directed reactions where metal ions template the syntheses of larger organic molecules from small components offers a way forward. Where metal complexes are currently employed as catalysts of organic reactions, most operate in organic solvents because they are simply most stable in such solvents, rather than any inherent insolubility of organic substrates in water; the development of compounds that are stable and active in an aqueous environment is a clear target. At this stage, examples are limited and commercialization not yet attempted, so opportunities exist for the entrepreneurial coordination chemist.

9.5   Complex Futures

If there is one certainty – apart from death and taxes – it is that the human race cannot predict the future well. Society generally has a very poor record of prediction, partly because not enough effort is put into developing a reliable model on which to base prediction. Sometimes advances simply come out of left field; a now classical example is the person who, extrapolating from the growth in horse-drawn traffic in New York in the late nineteenth century, calculated that its streets would eventually become waist-deep in manure. However, his calculation failed to predict the development of the automobile (so that now the streets are merely shoulder-deep in cars). Scientists have had greater success at prediction because they are prepared to develop a model, test and refine the model, and then apply it. Where we also get it wrong is in trying to predict beyond the limitations of our models and available information; for example, 50 years ago one could feel comfortable on the basis of known chemistry in saying 'all cobalt(III) complexes are six-coordinate' because no other coordination number had been reported, whereas today we know that this is not true. So it is with some temerity that one can even tackle something called 'complex futures'; but of course, consistent with human frailty, that's not going to stop us trying – and setting some challenges for future coordination chemists at the same time.

9.5.1 Taking Stock

The basic models of coordination chemistry that we use today are growing old, but ageing gracefully. What is remarkable about the crystal field theory, for example, is not that it has been in use for so many decades but that it works so well at all, given its simplistic basis. Metals, at least the accessible ones, are limited in number, but the chemistry of some is at yet not well explored. For example, it isn't too many years ago that examples of organometallic complexes of f-block elements were very few in number; while some people could apparently explain this from a theoretical standpoint, laboratory chemists simply beavered away and found their way well into the field. Ligands, the very body of coordination chemistry, have evolved through advances in synthesis to the level where it is practicable to routinely produce the 'designer' ligand – made for a particular purpose. While molecular synthesis lies at the heart of the field, there is growing interest in the '*in silico*' complex – modelled on a computer chip, but not necessarily made in the laboratory. Classical molecular modelling, which essentially is based on Hooke's Law and treats atoms as solid spheres joined by springs, gives a surprisingly valid view of shape and isomer preference, even as it is being overtaken by more sophisticated approaches such as density functional theory. Complexes are finding their way into a wide field of endeavour, including medicine, solid state chemistry and nanotechnology. For a field that some thought was 'all done' by the 1970s, coordination chemistry continues to expand and surprise. One good reason for this continued growth is that, despite all our sophistication, we really know very little still; there is plenty remaining out there to be discovered.

9.5.2 Crystal Ball Gazing

While it would be pleasant to think that we have a developed, sophisticated theory and knowledge of coordination compounds, the reality is that we have no more than scratched the surface. The wealth of research papers that continue to use words such as 'remarkable', 'unusual' and even 'unprecedented' suggest that our voyage of discovery has hardly begun.
So where could this voyage take us? Here are a few possibilities:

- ligands that bind with exceptionally high selectivity and rapidity to a particular metal, allowing new ways of recovering and analysing the metals to be developed;
- synthetic complexes that offer specificity as catalysts of selected organic reactions and that can operate in aqueous solution, eliminating the need for organic solvents;
- in a similar vein, low molecular weight synthetic analogues of natural complexes that function as effectively as metalloenzymes in promoting certain specific reactions;
- metallodrugs that have ligand environments that allow delivery to the site of action *in vivo* with very high selectivity, thus limiting side effects and lowering required doses;
- ligands that, through their purpose-designed strong and selective complexation abilities but low production cost, allow *in situ* mining of ore deposits in the ground with circulated aqueous solutions without the need for removing the ore;
- robust phosphorescent complexes tuneable through ligand selection to produce any desired colour of light, for use in optical displays;

- new magnetic materials for higher density data storage media – magnetic materials that are 'molecular magnets', being composed of low molecular weight polymetallic complexes;
- solid phases with unusual properties derived through the construction of new nanomaterials from small component complexes.

The list could go on, but perhaps you get the idea well enough, and you may be able to think of others to add. A complex to catalyse the conversion of carbon dioxide to a useful product, perhaps?

All seems a little unlikely? Well, one word that should be used in science with great caution is 'impossible'. In fact, most of the above are under some stage of current development. Coordination chemistry has thrown up some wondrous outcomes in the past century of effort, so who can know where it is going to take us. Just book your ticket and get on board.

Concept Keys

Coordination complexes find wide commercial applications, from being used in medicine to serving as the mode by which metals are extracted from ores.

Medically-related applications of coordination complexes include use in bioassay, diagnostic imaging and as drugs.

Metal complexes find important roles in the fight against cancer. In particular, cisplatin and its homologues are used a great deal and to good effect, and initiate their activity through binding to the aromatic amine bases of DNA in cancer cells.

Fluoroimmunoassay makes use of the fluorescent properties of europium complexes, and is one example of how a strong spectroscopic response has been adopted to provide an analytical method, in this case in bioanalysis.

Hydrometallurgy makes use of redox and complexation reactions for the recovery of valuable metals from ores. This is exemplified by the dissolution of gold metal in ores through oxidation by air and complexation by cyanide.

The synthetic chemical industry is heavily reliant on the use of metal complexes as catalysts, including for the preparation of optically pure chiral compounds.

Metal complexes can act as a starting point for what is termed 'bottom up' synthesis of nanomaterials, through commencing with simple complexes that aggregate into large nanoscale products.

Environmentally-sensitive 'green' chemistry is a developing commercially-relevant field, with roles for coordination chemistry included in current research.

Further Reading

Büchel, K.H., Moretto, H.-H. and Woditsch, P. (2000) *Industrial Inorganic Chemistry*, 2nd edn, Wiley-VCH Verlag GmbH, Weinheim, Germany. A lengthy but authoritative overview of industrial inorganic chemistry in one volume. Apart from production aspects, materials, energy use and economic considerations are addressed.

Elvers, B., Hawkins, S., Arpe, H.-J. *et al.* (1997) *Ullmann's Encyclopaedia of Industrial Chemistry*, Parts A (28 volumes) and B (8 volumes), 5th edn, Wiley-VCH Verlag GmbH, Weinheim, Germany.

This is the premier source book for all facets of industrial chemistry, with detailed but accessible chapters giving coverage of many processes and key chemical reagents.

Evans, A.M. (1993) *Ore Geology and Industrial Minerals: An Introduction*, 3rd edn, Blackwell Science Publications, Oxford, UK. For those who wish to read beyond this text and deeper into geochemistry, this is an accessible and student-friendly book.

Gielen, M. and Tiekink, E.R.T. (2009) *Metallotherapeutic Drugs and Metal-based Diagnostic Agents: The Use of Metals in Medicine*, John Wiley & Sons, Inc., Hoboken, USA. Provides a comprehensive and very recent advanced account of medicinal uses of metals, for those who yearn for a more detailed and current coverage.

Hadjiliadis, N. and Sletten, E. (eds) (2009) *Metal Complex – DNA Interactions*, Wiley-Blackwell, Oxford, UK. Overviews metal–DNA interactions and mechanisms, metallodrugs and toxicity; includes a deep coverage of *cisplatin*.

Jones, C.J. and Thornback, J.R. (2007) *Medicinal Applications of Coordination Chemistry*, RSC Publishing, London. A recent, readable account of the state of use of coordination compounds for pharmaceutical and medical use.

Steed, J.W., Turner, D.R. and Wallace, K. (2007) *Core Concepts in Supramolecular Chemistry and Nanochemistry*, John Wiley & Sons, Ltd, Chichester, UK. Fundamentals of the frontier research fields of supramolecular chemistry and nanochemistry are covered in a fairly concise manner, explaining the evolution of the fields from more basic foundations; suitable for students who want a clear and accessible introduction to these areas.

Appendix A   Nomenclature

From very early times, alchemists gave names to substances, although these names gave little if any indication of the actual composition and or structure, which is the aim of a true nomenclature. This was eventually addressed in the early days of 'modern' chemistry in the late eighteenth century, and modern nomenclature evolved from that early work. Since nomenclature evolved along with chemistry, it was far from systematic even up to the beginning of the twentieth century. In large part, our current approach in coordination chemistry derived from nomenclature concepts introduced by Werner to represent the range of new complexes that he and contemporaries were developing, providing both composition and structural information. His system of leading with the names of ligands followed by the metal name, as well as also employing structural 'locators', is still with us today. Although an international 'language' for organic molecules commenced from a meeting in 1892, it was some time later that a systematic international inorganic nomenclature developed, and it was as late as 1940 that a full systematic nomenclature was assembled. Constant development in the field has demanded evolution of nomenclature, and the international rules were revised or supplemented in 1959, 1970, 1977, 1990 and again early in the twentyfirst century; like all languages, chemical language continues to evolve.

In describing chemical substances, we are dealing with a need for effective communication using an appropriate language. In a sense, chemical nomenclature is as much a language as is Greek or Mandarin, albeit a restricted one with a very specific purpose; it has an organized structure, 'rules of grammar', conventions, and undergoes continuous evolution. One advantage is that it is a universal language, governed by rules set in place by the International Union of Pure and Applied Chemists (IUPAC). This body produces 'dictionaries' for chemical nomenclature that serve in much the same way as a conventional dictionary, thesaurus or grammar rule book, and which are updated regularly. The object of the nomenclature adopted is to provide information on the full stoichiometric formula and shape of a compound in a systematic manner. For coordination chemistry, we need to deal with a number of aspects – the ligands (of which there may be more than one type), the central metal (or metals, in some cases), metal oxidation state(s), ligand distributions around the metal(s) and counter-ions (if the compound is ionic). Collectively, these place a great deal of demand on the nomenclature system, to the point where it has become both sophisticated and difficult to use. We shall try and provide just a basic and introductory nomenclature here, even neglecting some more advanced aspects of naming for the sake of brevity and clarity.

Molecules can be described in terms of a structural drawing, a written name or a formula. These are in a fashion all representations of the same thing – a desire to express the character of a chemical compound in a manner that will be understandable to others. Let's examine these options for two very simple examples (Figure A.1):

Introduction to Coordination Chemistry   Geoffrey A. Lawrance
© 2010 John Wiley & Sons, Ltd

molecular drawing	*molecular formula*	*molecular name*
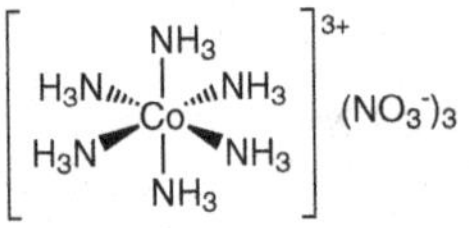 $\left[\begin{array}{c} \text{NH}_3 \\ \text{H}_3\text{N}_{\prime\prime\prime\prime}\text{Co}^{\prime\prime\prime\prime}\text{NH}_3 \\ \text{H}_3\text{N} \quad \text{NH}_3 \\ \text{NH}_3 \end{array}\right]^{3+}$ (NO$_3^-$)$_3$	[Co(NH$_3$)$_6$](NO$_3$)$_3$	hexaamminecobalt(III) nitrate
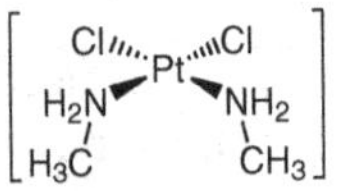 $\left[\begin{array}{c} \text{Cl}_{\prime\prime\prime\prime}\text{Pt}^{\prime\prime\prime\prime}\text{Cl} \\ \text{H}_2\text{N} \quad \text{NH}_2 \\ \text{H}_3\text{C} \quad \text{CH}_3 \end{array}\right]$	*cis*-[PtCl$_2$(NH$_2$CH$_3$)$_2$]	*cis*-dichlorobis(methanamine)platinum(II)

Figure A.1

These three representations of the same thing serve in different circumstances. The structural drawing of the molecule perhaps gives the clearest view of the complex under discussion, since it can be fairly easily 'read' by people with modest chemical training. The molecular formula is the most compact representation, and the molecular name provides a word-like form of representation for use in running text; however both carry some higher demands in terms of rules for full interpretation.

Let's look at the first example above, in terms of information that can be 'read' from the representations.

The molecular drawing:

- the set of square brackets around the cobalt-centred species separates it from the remainder of the drawing, defining it as the coordination complex unit – square brackets are the classical delineator of a complex unit, though these are not always included, particularly for isolated complex entities;
- the metal ion is obviously six-coordinate, and from the shape of the drawing apparently octahedral – it is also a 3+ cation;
- it has six NH$_3$ (ammonia) molecules bound, apparently all uniformly via the N atom;
- there are additional molecules, not bound to the metal, present – their presence in this representation suggests the complex may carry these as counter-ions, as it is an ionic complex.

The charges on the cation and anion have been inserted above, but do not invariably appear. In such circumstances as a result, some additional interpretation is required, namely:

- in the absence of ion charges, it is necessary to deduce that '(NO$_3$)$_3$' means three NO$_3^-$ anions, which must be inferred from chemical knowledge sufficient to assign 'NO$_3$' as nitrate monoanion;
- a second less obvious outcome relying on the above is then that, if three NO$_3^-$ anions are present, the complex unit must have an overall charge of 3+ to balance these anion charges;
- further, if one draws on chemical knowledge that ammonia is neutral, this means the 3+ charge lies on the cobalt centre, and, by further extension, that a 3+ charge on a metal equates with an oxidation state of three, or we are dealing with a cobalt(III) ion.

The molecular formula:

- like the molecular drawing, we can eventually establish that we are dealing with a $[Co(NH_3)_6]^{3+}$ cation and three NO_3^- anions, with the metal being cobalt(III);
- importantly, the set of square brackets here also defines the presence of a complex unit, and everything within the square brackets 'belongs' to the one complex unit, linked by covalent bonds – the central atom (metal) which is the focus of bonding invariably comes first in the formula.

The molecular name:

- in this case, we notice there are two 'words', one for the cation and the other for the anions;
- the metal and its oxidation state are defined clearly;
- the ligand name is defined clearly, along with a prefix that relates to the number of ligands;
- the anion name is defined clearly, although the number must be worked out from the charge on the cation, demanding a knowledge of the charge (or lack of charge) of the ammonia ligands.

What we hope is clear from all of the above is that all three approaches require some basic knowledge of aspects of coordination chemistry to extract the full information from the names; this is, of course, an inevitability of any language – some concepts of context and 'unspoken' rules are necessary.

However, let's move on to see how we can interpret molecular formula and names, and then step forward to some simple rules of nomenclature and examples. How we can 'read' inorganic nomenclature is exemplified below. Note how the metal is placed first in the formula representation of the complex unit and last in the written name; some other aspects are obviously common to both representations (Figure A.2).

Some basic rules for naming compounds are given below. These may not make you an expert, but will allow some understanding of how names are generated; only extensive use (as with any language) brings expertise. Fortunately, structural formulae are now frequently met and allow an escape from the more demanding naming methodology.

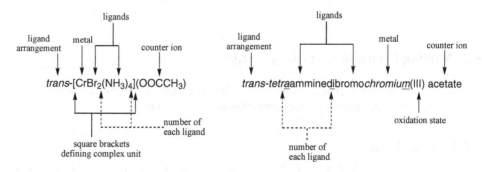

Figure A.2

Table A.1

Number of donor groups coordinated	Name	Number of donor groups coordinated	Name
0	*free ligand*	6	hexadentate
1	monodentate	7	heptadentate
2	didentate	8	octadentate
3	tridentate	9	enneadentate
4	tetradentate	10	decadentate
5	pentadentate	*many*	polydentate

A.1.1 Polydenticity

When dealing with polydentate ligands, as is common in coordination chemistry, the concept of *denticity* is important in assisting development of the broader nomenclature. Denticity simply refers to the number of donor groups of any one ligand molecule that are bound to the metal ion; they can be (but are not always) separately identified in nomenclature, but knowledge of denticity assists in name development. Where coordination is by just one donor group, we are dealing with monodentate ligands. The terminology is outlined above, and in effect represents the form of nomenclature for discussing the way ligands are bound to metal ions; it is simply easier to say 'a tridentate ligand' rather than 'a ligand bound through three donor groups' (Table A.1).

To assist in recognition of polydenticity in a written name, the number of any polydentate ligands (where >2 are present) ≥2 bound to a metal ion is represented itself by a sequence of different prefixes (inserted immediately in front of the relevant ligand name) to those used for simple monodentate ligands, as given below (Table A.2). The need to use these only arises for written names.

Table A.2

Number of monodentates	Prefix to ligand name	Number of polydentates[a]	Prefix to ligand name
1	*no prefix*	1	*no prefix*
2	di	2	bis
3	tri	3	tris
4	tetra	4	tetrakis
5	penta	5	pentakis
6	hexa	6	hexakis

[a]Also used for large and or complicated monodentate ligands, or where ambiguity may arise.

A.2 Naming Coordination Compounds

An abbreviated, simple set of rules for naming coordination complexes follows. The complete set of rules for nomenclature is very extensive.

A.2.1 Simple Ligands

The names of *neutral* ligands are usually unchanged in naming a coordination compound containing them. For example, the molecule urea is still called urea as a ligand.

However, there are some important exceptions that affect some very common ligands, the two most common being:

- water (H_2O) becomes **aqua**;
- ammonia (NH_3) becomes **ammine**.

The names of coordinated *anions* are changed to always end in **-o**. Examples of common anionic ligands are given below (Table A.3).

Table A.3

Free anion	Coordinated anion	Free anion	Coordinated anion
Cl^-, chloride	chlorido	HO^-, hydroxide	hydroxido
O_2^-, superoxide	superoxido	CN^-, cyanide	cyanido
SCN^-, thiocyanate	thiocyanato	CO_3^{2-}, carbonate	carbonato
NO_3^-, nitrate	nitrato	NO_2^-, nitrite	nitrito
N_3^-, azide	azido	NH_2^-, amide	amido
ClO_4^-, perchlorate	perchlorato	SO_4^{2-}, sulfate	sulfato

A.2.2 Complexes

Neutral complexes have only a single 'word' name.

Ionic complexes are written as two 'words', with one word name for the cation and one word name for the anion. For *all* ionic complexes, invariably, the *cation* is named *first*, and the anion last. (This follows the convention for simple salts like sodium chloride.)

These basic rules means all component parts of a cation, anion or neutral name are run together without any breaks. These components will now be defined.

Ligands are arranged *first*, in *alphabetical order*, followed by the name of the metal atom. The *oxidation number* of the central metal atom is included at the end of the name, following the metal name, as a Roman numeral in parentheses.

Alternatively, it is permitted to enter the overall charge on the complex in parentheses at the end (e.g. metal(+2)) in place of the oxidation state (e.g. metal(II)).

There is no ending modification to the metal name in neutral or cationic complexes. However, in anionic complexes the name of the metal is modified to an *-ate* ending (e.g. molybdenum to molybdate, or zinc to zincate), as a signal that the complex unit is anionic.

Note that in writing the formula representation of a complex, the metal is written first at left after the opening square bracket, and then the *identical* approach of *alphabetical order* to that reported above for the 'word' name is used. Ligands are listed in alphabetical order of the first letter of the formula or standard abbreviation that is used for each ligand (thus, for example Cl before NH_3 before OH_2, and (en) before NH_3). Any bridging ligands in polynuclear complexes are listed after terminal ligands, and if more than one in increasing order of bridging multiplicity (i.e. number of bonds to metals). The formula is completed with a closing square bracket. If there are any counter-ions, their standard formulae are written before (cations) or after (anions) the complex formula.

A.2.3 Multipliers

Commonly, complexes contain more than one of the single or several ligand types attached to a central metal ion/atom. The *number* of a given ligand needs to be defined in the name, and is indicated by the appropriate Greek prefix, as tabulated above (Table A.2). The rule, in summary, is di- (2), tri- (3), tetra- (4), penta- (5), hexa- (6) *if the ligand is monoatomic or one of the simple ligands above. Polyatomic* (and multidentate) ligands are placed in parentheses, and their number is indicated by a (different) prefix outside the parentheses: bis- (2), tris- (3), tetrakis- (4), pentakis- (5), hexakis- (6). Apart from multidentate organic ligands, simpler ligands may also use this prefix format if it helps avoid ambiguities.

In formula representations, the number of each ligand is included as a subscript after the symbol (except for the case where there is only one of a ligand).

Examples:

$[Co(NH_3)_6]Br_3$	hexaamminecobalt(III) bromide (*or* hexaamminecobalt(3+) bromide *using charge instead*)
$Li_2[PtCl_6]$	lithium hexachloridoplatinate(IV)
$[RhCl_2(NH_3)_4]NO_3$	tetraamminedichloridorhodium(III) nitrate
$K_3[Cr(C_2O_4)_3]$	potassium tris(oxalato)chromate(III)
$[IrBr(H_2NCH_2COO)_2(OH_2)]$	aquabromobis(glycinato)iridium(III)
$Na[PtBrCl(NH_3)(NO_2)]$	sodium amminebromidochloridonitritoplatinate(II)
$[Pt(PPh_3)_4]$	tetrakis(triphenylphosphane)platinum(0)

A.2.4 Locators

Structural features of molecules associated with stereochemistry and isomerism are indicated by prefixes attached to the name, and *italicized*, and separated from the rest of the name by a hyphen.

Typical of this type are the geometry indicators *cis-*, *trans-*, and *fac-*, *mer-*. Note that a more elaborate but highly functional form of stereochemical descriptors exist in the nomenclature system; we shall avoid its use here.

Examples:

cis-$[PtCl_2(py)_2]$	*cis*-dichloridobis(pyridine)platinum(II)
mer-$[IrH_3\{P(C_6H_5)_3\}_3]$	*mer*-trihydridotris(triphenylphosphane)iridium(III)

A.2.5 Organic Molecules as Ligands

These take *exactly* the names they carry as an organic molecule in general (except anionic ones are usually altered slightly to carry the usual **-o** ending). Thus an understanding of organic nomenclature is vital as well. Some organic molecules have well accepted and approved 'trivial' (non-IUPAC) names, and these are also able to be carried forward into the complex name in this form.

Examples:

$H_2N\!-\!CH_2\!-\!CH(CH_3)\!-\!NH_2$	propane-1,2-diamine
$[Co(H_2N\!-\!CH_2\!-\!CH(CH_3)\!-\!NH_2)_3](ClO_4)_3$	tris(propane-1,2-diamine)cobalt(III) perchlorate
$H_2N\!-\!CH_2\!-\!CH(CH_3)\!-\!COOH$	alanine (or alaninato as the anion)
$[Cu(H_2N\!-\!CH_2\!-\!CH(CH_3)\!-\!COO^-)_2]$	bis(alaninato)copper(II)

The sole variation of nomenclature you may meet where a ligand is coordinated is the addition of κ nomenclature in the molecular formula, which is done to provide more information by defining more exactly the way a ligand is bound, that is the number of donors attached and their type. As an example, instead of just writing $[Co(H_2N-CH_2-CH(CH_3)-NH_2)_3](ClO_4)_3$ as above, you may find occasionally that it is written as $[Co(H_2N-CH_2-CH(CH_3)-NH_2-\kappa^2-N,N)_3](ClO_4)_3$ where the κ^2 identifies that two (the superscript number) unconnected groups are acting as donors, and the N,N identifies them both as nitrogen donors. In a simple case like this example, the additional information is not really needed by most people, and so tends to be left out. However, for mixed-donor polydentate ligands where there are options for coordination, such as an excess of donor groups over sites available to choose from, the additional specification is necessary. Given that you will meet only simple examples at this level, this is an extension that is best simply set aside at present.

Note that there is also a nomenclature for dealing with connected donor atoms bound to a metal, the η nomenclature. For example, the ligand $H_2C=CH_2$ bound side-on so that both connected carbons atoms are effectively equally linked to a metal would be designated as $(\eta^2-C_2H_4)$. This is elaborated for organometallic compounds later.

A.2.6 Bridging Ligands

Some ligands, with more than one lone pair of electrons, can bind two metals simultaneously, and are then termed *bridging* ligands, leading to polynuclear complexes.

These ligands are identified by the Greek letter mu (μ) to indicate the ligand is bridging together with a superscript n to indicate the number of metal atoms to which the ligand is attached (but only if $n > 2$).

Examples:

$[Fe_2(CN)_{10}(\mu\text{-}CN)]^{5-}$	μ-cyano-bis(pentacyanidoferrate(III))
$[Fe_2(CO)_6(\mu\text{-}CO)_3]$	tri-μ-carbonyl-bis(tricarbonyliron(0))
$[\{Co(NH_3)_4\}_2(\mu\text{-}Cl)(\mu\text{-}OH)]^{4+}$	μ-chlorido-μ-hydroxido-bis(tetraamminecobalt)(4+)

This simple group of rules clearly does not by any means cover the full set of naming rules that apply to deal with modern coordination chemistry. However, they go some of the way to allowing you to 'navigate' around nomenclature for the relatively simple complexes you will likely meet. It is a difficult task to name complicated compounds – which is why, even in the chemical research literature, people sometimes choose to avoid it as much as possible. For some, nothing compares with a drawing of the complex molecule and a trivial name(s) for the ligand(s) involved!

A.2.7 Organometallic Compounds

The same basic rules apply to the naming of organometallic compounds as to traditional Werner-type coordination compounds.

One commonly met feature in organometallic compounds, alluded to earlier, relates to defining the number of atoms involved in bonding of a ligand like cyclopentadienyl ($C_5H_5^-$), which can involve one, three or all five connected carbon atoms in close (bonding) contact to the metal. The distinction is defined in terms of *hapticity* (η), which reports, for example, one (η^1), three (η^3) or all five (η^5) of the carbon atoms of the ligand coordination as a prefix to the ligand representation (*e.g.* η^5-$C_5H_5^-$). This concept is analogous to the use of κ, outlined earlier, to designate the number of donors of a polydentate ligand actually bound in Werner-type complexes.

Further Reading

The IUPAC publishes a number of books and reports on nomenclature. Some are available with free access on-line through their web site. http://old.iupac.org/publications/epub/index.html

Appendix B   Molecular Symmetry: The Point Group

Molecular shape plays an important role in the spectroscopic properties of complexes. Even if we compare two simple two-coordinate linear complexes, one with two identical ligands and one with two different ligands, it should be immediately apparent that they do not look the same. For one, both sides of the molecule are equivalent, for the other they are different (Figure B.1). Turn one round by rotation 180° around an axis perpendicular to the bond direction, and it looks identical to what you had in the first place; do this for the other and it is clearly not the same. Further, consider the situation where the linear molecule with two identical donors is compared with an analogue where the X—M—X unit is bent. Now, consider a 180° rotation about an axis containing the X—M—X unit. For the linear molecule, the outcome is indistinguishable from the starting situation, whereas for the bent molecule this is not the case. In each of these examples, the linear molecule has undergone what is called a *symmetry operation*. The nonsymmetrical and bent molecules have clearly yielded a different view, and thus have behaved differently. What we are seeing is that this process we have undertaken has distinguished between what are clearly different molecules. In effect, we can use any molecule's behaviour to a series of operations like those presented below to define the molecular shape (or symmetry). Molecules can be described in terms of a symmetry *point group*, effectively a combination of a limited number of what are termed *symmetry elements* based on considering a set of operations like those introduced above. There are a limited number of both individual symmetry elements and combinations of these (called the *point group*). The symmetry elements that can operate are defined below in Table B.1. A legitimate *symmetry operation* employing these elements occurs when the molecular views before and after the operation are indistinguishable. Operations occur relative to an x, y, z coordinate system arranged around a particular point in the molecule, selected with regard to defining symmetry elements and maximizing operations in such a way that the z axis forms the principal axis, passing through the molecular centre and being the axis around which the highest order operation occurs.

Determination of all possible symmetry elements provides the ability to then assign the *point group* for the element, which (as the name suggests) is based on the symmetry around a point in the molecule that either coincides with the central atom or the geometric centre of the molecule. A flow chart (Figure B.2) is frequently used to assist in the assignment of all symmetry operations for a molecule. Once all symmetry operations have been identified, the point group evolves. Point groups are limited in number, and are identified in Table B.2. To exemplify the concepts, we shall examine two systems: the trigonal bipyramidal MX_5 and the square planar MX_4. The shapes, axes and operations involving these are shown below in Figures B.3 and B.4.

Introduction to Coordination Chemistry   Geoffrey A. Lawrance
© 2010 John Wiley & Sons, Ltd

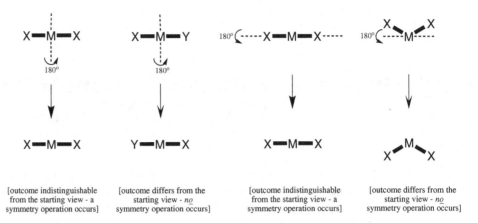

Figure B.1
Two operations with linear MX_2 and MXY molecules that distinguish their differing symmetry.

For MX_5, the principal axis passes through the M and the two axial X groups; the M is the point about which the point group is defined. The three X groups coplanar with the metal lie in the *xy* plane; one axis (designated *x*) passes through one M—X bond, which then requires the other to pass through M but not include the other X groups. Around the *z* axis, rotation by 120° leads to an indistinguishable arrangement; this is thus a C_3 axis. Rotation by 180° about each of the three M—X bonds in the *xy* plane leads to an indistinguishable arrangement; thus there are three C_2 elements. These operations alone define the molecules as belonging to a **D** point group. However, there are also three vertical planes of symmetry each containing one M—X bond in the *xy* plane and the C_3 axis (only one of which is shown as a dotted square in the figure, for clarity), as well as one horizontal plane that is the *xy* plane containing the central equatorial MX_3 part of the molecule (shown as a dotted triangle in the figure). Overall, then this means the molecule belongs to the $\mathbf{D_{3h}}$ point group.

The square planar MX_4 has some similarities to the trigonal bipyramidal MX_5 as along the *z* axis there is a C_4 operation (rotation by 90° leads to formation of an indistinguishable arrangement), and there are four C_2 elements (two about each in-plane X—M—X, and two about imaginary axes bisecting the *x* and *y* axes in two directions). There are four vertical planes containing the C_4 axis (only one of which is shown as a dotted square in the figure, for clarity) and one horizontal plane of symmetry (shown as a dotted square incorporating the four X groups in the figure). Following the flow chart, this leads to the $\mathbf{D_{4h}}$ point group.

Table B.1   Symmetry operations and elements.

Element	Description
E	*Identity.*
σ	*Plane of symmetry.* The two types are reflection in a plane containing the principle axis (σ_v and σ_d) and reflection in a plane perpendicular to the principal axis (σ_h).
C_n	*Rotation axis.* The subscript n denotes the *order* of the axis, which is the angle of rotation (360°/n) to achieve an indistinguishable image.
S_n	*Improper rotation axis.* This involves sequential steps of rotation by 360°/n, followed by a reflection in a plane perpendicular to the rotation axis.
i	*Inversion.* This occurs through a centre of symmetry.

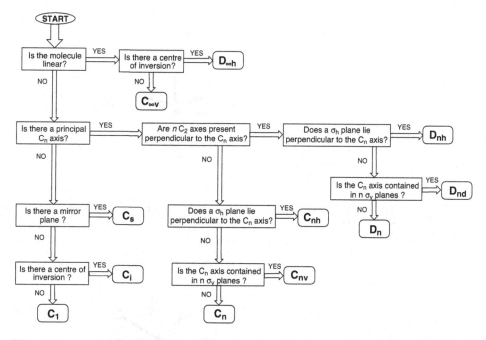

Figure B.2
Simplified flow chart for assignment of a symmetry group. Special high symmetry groups (T_d, O_h, I_h) are not included in this chart, and need to be identified separately; it applies to other and generally lower symmetry molecules, which are more often met in reality.

It is possible with the flow chart and careful examination of drawings and or three-dimensional models of a complex to assign the point group with reasonable rapidity and success after some practice. To assist further, a list of the point groups of basic higher symmetry structures are collected in Table B.3 below. Those shown assume identical ligands in all sites. You should assume that these shapes with a mixture of different ligands will be of lower symmetry and have a lower symmetry point group. This aspect is illustrated

Table B.2   Point groups.

Point group	Symmetry elements involved
C_s	One plane of symmetry.
C_i	A centre of symmetry.
C_n	One n-fold rotation axis.
D_n	One n-fold rotation axis (about the principal axis) and n horizontal twofold axes.
C_{nv}	One n-fold rotation axis (about the principal axis) and n vertical planes.
C_{nh}	One n-fold rotation axis (about the principal axis) and one horizontal plane.
D_{nh}	One n-fold rotation axis (about the principal axis, as for D_n), one horizontal plane, and n vertical planes containing the horizontal axes.
D_{nd}	One n-fold rotation axis (about the principal axis, as for D_n), and vertical planes bisecting angles between the horizontal axes.
S_n	Systems with alternating axes ($n = 4, 6, 8$).
$C_{\infty v}, D_{\infty h}$	Linear systems with an infinite rotation axis.
T_d, O_h, O, I_h, I	Special groups: tetrahedral, octahedral, cubic and icosahedral

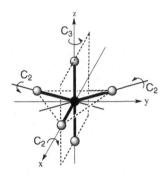

Figure B.3
The coordinate system and symmetry operations for trigonal bipyramidal MX_5, of D_{3h} symmetry point group.

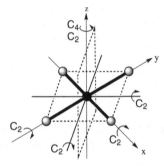

Figure B.4
The coordinate system and symmetry operations for square planar MX_4, of D_{4h} symmetry point group. Note that there is a C_2 operation colinear with a C_4 around the z axis.

in Figure B.5 for moving from square planar MX_4 (D_{4h}) to square planar *trans*-MX_2Y_2 (D_{2h}) and *cis*-MX_2Y_2 (C_{2v}).

Distortions of common geometric shapes lead necessarily to changes in the point group. For example, an octahedron (O_h) can undergo three types of distortion. The first is *tetragonal distortion*, involving elongation or contraction along a single C_4 axis direction, leading to D_{4h} symmetry. The second is *rhombic distortion*, where changes occur along two C_4 axes

Table B.3   Point groups for several molecular shapes.

Coordination number	Stereochemistry	Point group
2	linear	$\mathbf{D_{\infty h}}$
3	trigonal planar	$\mathbf{D_{3h}}$
	trigonal pyramidal	$\mathbf{C_{3v}}$
	t-shaped	$\mathbf{C_{2v}}$
4	tetrahedral	$\mathbf{T_d}$
	square planar	$\mathbf{D_{4h}}$
5	trigonal bipyramidal	$\mathbf{D_{3h}}$
	square-based pyramidal	$\mathbf{C_{4v}}$
6	octahedral	$\mathbf{O_h}$
	trigonal prismatic	$\mathbf{D_{3h}}$

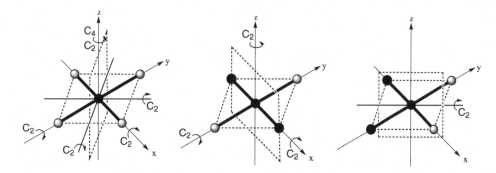

Figure B.5
The coordinate system and symmetry operations for the family of square planar complexes MX_4, *trans*-MX_2Y_2 and *cis*-MX_2Y_2.

so that no two sets of bond distances along each of the three axes are equal, leading to D_{2h} symmetry. The third is *trigonal distortion*, involving contraction or elongation along one of the C_3 axes to yield a trigonal antiprismatic shape, reducing the point group to D_{3d}. Similarly, other common stereochemistries may distort, leading to shapes with different point groups.

Just as molecules have certain symmetry, molecular orbitals likewise have symmetry. Orbital labels such as σ and π relate to the rotational symmetry of the orbital, whereas the labels a, e, e_g, t and so on met in complexes arise from considering orbital behaviour in the context of all symmetry operations of the point group of the molecule. A *character table*, which defines the symmetry types possible in a particular point group, provides a way of assigning these labels, but this will not be pursued here.

Further Reading

Kettle, S.F.A. (2007) *Symmetry and Structure: Readable Group Theory for Chemists*, 3rd ed, John Wiley & Sons, Inc., Hoboken, USA. An accessible revised and updated introduction for students, relying on a diagrammatical and non-mathematical approach.

Ogden, J.S. (2001) *Introduction to Molecular Symmetry*, Oxford University Press. A readable and well-presented book suitable for students at any level.

Vincent, A. (2001) *Molecular Symmetry and Group Theory: A Programmed Introduction to Chemical Applications*, 2nd edn, John Wiley & Sons, Inc., New York, USA. This popular textbook may serve the more advanced reader well, with a slightly deeper mathematical base but a well-staged programmed learning approach.

Index

Introduction to Coordination Chemistry   Geoffrey A. Lawrance
© 2010 John Wiley & Sons, Ltd